Advances in Industrial Control

Other titles published in this series:

Concepción A. Monje · YangQuan Chen
Blas M. Vinagre · Dingyü Xue · Vicente Feliu

Fractional-order Systems and Controls

Fundamentals and Applications

 Springer

Dr. Concepción A. Monje
Universidad Carlos III de Madrid
Departamento de Ingeniería de Sistemas
y Automática
28911 Leganés Madrid
Spain
cmonje@ing.uc3m.es

Dr. Blas M. Vinagre
Universidad de Extremadura
Escuela de Ingenierías Industriales
Departamento de Ingeniería Eléctrica,
Electrónica y Automática
06071 Badajoz
Spain
bvinagre@unex.es

Dr. YangQuan Chen
Utah State University
Department of Electrical
and Computer Engineering
84322-4160 Logan Utah
USA
yangquan.chen@usu.edu

Dr. Vicente Feliu
Universidad de Castilla-La Mancha
ETS de Ingenieros Industriales
Departamento de Ingeniería Eléctrica,
Electrónica, Automática y Comunicaciones
13071 Ciudad Real
Spain
vicente.feliu@uclm.es

Prof. Dingyü Xue
Northeastern University
Faculty of Information Sciences
and Engineering
110004 Shenyang
China
xuedingyu@mail.neu.edu.cn

ISSN 1430-9491
ISBN 978-1-84996-334-3   e-ISBN 978-1-84996-335-0
DOI 10.1007/978-1-84996-335-0
Springer London Dordrecht Heidelberg New York

British Library Cataloguing in Publication Data
A catalogue record for this book is available from the British Library

Library of Congress Control Number: 2010934759

Cover design: eStudioCalamar, Girona/Berlin

Printed on acid-free paper

Springer is part of Springer Science+Business Media (www.springer.com)

Advances in Industrial Control

Series Editors

Professor Michael J. Grimble, Professor of Industrial Systems and Director
Professor Michael A. Johnson, Professor (Emeritus) of Control Systems and Deputy Director

Industrial Control Centre
Department of Electronic and Electrical Engineering
University of Strathclyde
Graham Hills Building
50 George Street
Glasgow G1 1QE
United Kingdom

Series Advisory Board

Professor E.F. Camacho
Escuela Superior de Ingenieros
Universidad de Sevilla
Camino de los Descubrimientos s/n
41092 Sevilla
Spain

Professor S. Engell
Lehrstuhl für Anlagensteuerungstechnik
Fachbereich Chemietechnik
Universität Dortmund
44221 Dortmund
Germany

Professor G. Goodwin
Department of Electrical and Computer Engineering
The University of Newcastle
Callaghan
NSW 2308
Australia

Professor T.J. Harris
Department of Chemical Engineering
Queen's University
Kingston, Ontario
K7L 3N6
Canada

Professor I.H. Lee
Department of Electrical and Computer Engineering
National University of Singapore
4 Engineering Drive 3
Singapore 117576

Professor (Emeritus) O.P. Malik
Department of Electrical and Computer Engineering
University of Calgary
2500, University Drive, NW
Calgary, Alberta
T2N 1N4
Canada

Professor K.-F. Man
Electronic Engineering Department
City University of Hong Kong
Tat Chee Avenue
Kowloon
Hong Kong

Professor G. Olsson
Department of Industrial Electrical Engineering and Automation
Lund Institute of Technology
Box 118
S-221 00 Lund
Sweden

Professor A. Ray
Department of Mechanical Engineering
Pennsylvania State University
0329 Reber Building
University Park
PA 16802
USA

Professor D.E. Seborg
Chemical Engineering
3335 Engineering II
University of California Santa Barbara
Santa Barbara
CA 93106
USA

Doctor K.K. Tan
Department of Electrical and Computer Engineering
National University of Singapore
4 Engineering Drive 3
Singapore 117576

Professor I. Yamamoto
Department of Mechanical Systems and Environmental Engineering
The University of Kitakyushu
Faculty of Environmental Engineering
1-1, Hibikino,Wakamatsu-ku, Kitakyushu, Fukuoka, 808-0135
Japan

This work is dedicated to:

My parents, Julio Manuel and Julia, and brother, Julio Manuel, for their loving support, and Marina, for opening up such an amazing world to me.
— Concepción Alicia Monje Micharet

The memory of my father, Hanlin Chen,
and my family, Huifang Dou, Duyun, David, and Daniel.
— YangQuan Chen

My wife, Reyes, my sons, Alejandro and Carlos, and my parents, Blas and María, for giving me the time and support for working on this book.
— Blas Manuel Vinagre Jara

My wife, Jun Yang, and daughter, Yang Xue, for their understanding, and my parents, Guohuan Xue and Mengyun Jia, for their love.
— Dingyü Xue

My wife, María Isabel, for her patience and understanding during all my academic career, and my parents, María Teresa and Sebastián.
— Vicente Feliu Batlle

Series Editors' Foreword

The series *Advances in Industrial Control* aims to report and encourage technology transfer in control engineering. The rapid development of control technology has an impact on all areas of the control discipline. New theory, new controllers, actuators, sensors, new industrial processes, computer methods, new applications, new philosophies…, new challenges. Much of this development work resides in industrial reports, feasibility study papers, and the reports of advanced collaborative projects. The series offers an opportunity for researchers to present an extended exposition of such new work in all aspects of industrial control for wider and rapid dissemination.

One of the main objectives of the *Advances in Industrial Control* monograph series, as described above, is to allow authors to present a considered or reflective view of a body of work that they have recently developed. *Fractional-order Systems and Controls: Fundamentals and Applications* by Concepción A. Monje, YangQuan Chen, Blas M. Vinagre, Dingyü Xue, and Vicente Feliu perfectly exemplifies a monograph that fulfils this objective. A look at the Acknowledgements and References sections of the monograph shows that the authors have been contributing steadily to the growth of research in the systems and controls applications of fractional calculus since the late 1990s with regular contributions appearing in the journal and conference literature throughout the first decade of the millennium. From this wealth of experience and research, the authors have drawn together the various themes and research outcomes to produce a systematic presentation of their work. A recent search by the Series Editors showed that there are indeed few such monographs on fractional-order controllers designed to promote a clear understanding of the systems theory, the controller design procedures and, importantly for the *Advances in Industrial Control* series, present demonstrations of practical applications of the new fractional-order controller techniques.

The monograph is divided into six parts. In Part I, the reader is guided through the mathematical concepts of fractional calculus and the system and control implications of these concepts. This part closes with a chapter on the fundamentals of fractional-order controllers that introduces the conceptual framework to support the next three parts of the monograph. An extended presentation of different fractional controller design methods occupies Parts II, III, and IV of the book, covering some seven chapters in total. The last two parts of the monograph cover implementation aspects of the new controllers (Part V) and finally, Part VI contains five chapters of demonstrations and applications of fractional-order control.

This is a wide-ranging but top-down presentation of fractional-order systems, controllers and applications. The ordering and the partitioning of the material will assist the reader in easily finding particular topics to study. The chapters are very focussed, particularly in the fractional-order control design chapters of Parts II, III, and IV. This will aid the reader who is solely interested in a particular control design method, such as fractional-order lead–lag compensators. The range of applications in Part VI covers system identification, a flexible robot arm (flexible beam), a canal (hydraulic) control system, mechatronic controllers and finally, a power electronics application. In these chapters, experimental rigs and practical experience play an important role in demonstrating the veracity of the new controllers. There is a nice demonstration that it is not just the number of tunable parameters in a controller that is important, but that the design flexibility and performance benefits of fractional-order controllers flow from the inherent structure of the controller itself.

Naturally, such an overarching presentation of fractional-order systems and controllers will appeal to a wide range of readers. The control academic will be intrigued to see how fractional-order systems and control enhance the prevailing linear systems paradigm. The graduate researcher will find many opportunities to develop new directions for research in fractional-order control from the results presented in this monograph. The control engineer and the industrial practitioner will be able to use the monograph to investigate the potential of the new controllers through study of the design and the applications chapters. All readers will find that the book chapters give good support by providing plenty of illustrative examples, and MATLAB® code and SIMULINK® block diagrams to assist them in replicating the results given.

The Editors are very pleased to welcome this volume into the *Advances in Industrial Control* series of monographs, and expect the volume to become an essential entry to every academic and industrial control engineer's library.

Industrial Control Centre   *M.J. Grimble*
Glasgow   *M.A. Johnson*
Scotland, UK
2010

Preface

The aim of this work is to provide an introduction to the basic definitions and tools for the application of fractional calculus in automatic control. It is intended to serve the control community as a guide to understanding and using fractional calculus in order to enlarge the application domains of its disciplines, and to improve and generalize well established control methods and strategies. A major goal of this book is to present a concise and insightful view of the current knowledge on fractional-order control by emphasizing fundamental concepts, giving the basic tools to understand why fractional calculus is useful in control, to understand its terminology, and to illuminate the key points of its applicability.

Fractional calculus can be defined as the generalization of classical calculus to orders of integration and differentiation not necessarily integer. Though the concepts of non-integer-order operators are by no means new, the first meeting devoted to the topic took place in 1974, in New Haven, Connecticut, USA. Even at such an event, fractional calculus was a matter of almost exclusive interest for few mathematicians and theoretical physicists. However, circumstances have changed considerably since then. On the one hand, in the last 3 decades the general interest in such a tool has experienced a continuing growth, and at present we can find many conferences, symposia, workshops, or special sessions, as well as papers and special issues in recognized journals, devoted to the theoretical and application aspects of fractional calculus. On the other hand, as can be observed in such conferences and journals, motivation for this growing interest has been the engineering applications, especially the control engineering ones.

Control is an interdisciplinary branch of engineering and mathematics that deals with the modification of dynamic systems to obtain a desired behavior given in terms of a set of specifications or a reference model. To obtain the desired behavior, a designed controller senses the operation of the system, compares it to the desired behavior, computes corrective actions based on specifications or reference models, and actuates the system to obtain the desired change. So, in order that the dynamics of a system or process might be properly modified, we need a model of the system, tools for its analysis, ways to specify the required behavior, methods to design the controller, and techniques to implement them. Since the usual tools to model dynamic systems at a macroscopic level are integrals and derivatives, at least in the linear systems case, the algorithms that implement the controllers are mainly composed of such tools. So, it is not hard to understand that a way to extend the definitions of integrals and derivatives can provide a way to expand the frontiers for their applicability.

Fractional-order control is the use of fractional calculus in the aforementioned topics, the system being modeled in a classical way or as a fractional one. From a certain point of view, the applications of fractional calculus have experienced an evolution analogous to that of control, following two parallel paths depending on the starting point: the time domain or the frequency domain. Whilst the applications in dynamic systems modeling have used, except in some cases of electrochemistry, the time domain, the applications in control have been developed, mainly and from the very beginning, in the frequency domain.

It is our hope that this book will be read by, and of interest to, a wide audience. For this reason, it is organized following the structure of a traditional textbook in control. Therefore, in Part I, after the introduction in Chapter 1, Chapter 2 gives the fundamental definitions of fractional calculus, having in mind our goal of providing a stimulating introduction for the control community. Therefore, the mathematical prerequisites have been kept to a minimum (those used in a basic course of control: linear algebra, including matrices, vectors, and eigenvalues; classical calculus, including differential equations and concepts of homogeneous and particular solutions; complex numbers, functions, and variables; and integral transforms of Laplace and Fourier), avoiding unnecessary intricate mathematical considerations but without an essential loss of rigor. Chapter 3 is devoted to state-space representations and analysis of fractional-order systems, completing the fundamental definitions given in Chapter 2. Chapter 4 is a detailed exposition of the core concepts and tools for the useful application of fractional calculus to control, based on the generalization of the basic control actions. In Part II, there is a complete study of fractional-order PID controllers, dealing with definitions, tuning methods, and real application examples given in Chapters 5–7. Part III focuses on the generalization of the standard lead-lag compensator. Chapter 8 presents an effective tuning method for the fractional-order lead-lag compensator (FOLLC), and Chapter 9 proposes a simple and direct auto-tuning technique for this type of structure. Part IV provides an overview of other fractional-order control strategies, showing their achievements and analyzing the challenges for further work. Chapter 10 reviews some important fractional-order robust control techniques, such as CRONE and QFT. Chapter 11 presents some nonlinear fractional-order control strategies. Part V provides methods and tools for the implementation of fractional-order controllers. Chapter 12 deals with continuous- and discrete-time implementations of these types of controllers and Chapter 13 with numerical issues and MATLAB implementations. Finally, Part VI is devoted to real applications of fractional-order systems and controls. The identification problem of an electrochemical process and a flexible structure is presented in Chapter 14; the position control of a single-link flexible robot in Chapter 15; the automatic control of a hydraulic canal in Chapter 16; mechatronic applications in Chapter 17; and fractional-order control strategies for power electronic buck converters in Chapter 18. In the Appendix, additional useful information is given, such as Laplace transform tables involving fractional-order operators.

We would like to thank Professor (Emeritus) Michael A. Johnson and Professor Michael J. Grimble, Series Editors of Advances in Industrial Control Monograph Series of Springer London. Without their invitation, encouragement, and wise recommendations and comments, this book project would not have been possible. Thanks are also due to Oliver Jackson and his Editorial Assistants for Engineering, Ms. Aislinn Bunning and Ms. Charlotte Cross of Springer London, who have helped us through the review, copy-editing, and production process with care and professional support.

We would also like to acknowledge the collaborations of Prof. Igor Podlubny, Prof. Ivo Petráš, Prof. Richard Magin, Prof. Hongsheng Li, Dr. Gary Bohannan, Prof. Kevin L. Moore, Dr. Hyo-Sung Ahn, Dr. Jun-Guo Lu, Dr. Larry Ying Luo, Prof. Yan Li, Dr. Chunna Zhao, Dr. Antonio J. Calderón, Dr. Inés Tejado, Dr. Fernando J. Castillo, Dr. Francisco Ramos, Prof. Raúl Rivas, and Prof. Luis Sánchez.

Last but not least, we wish to thank the small but growing community of researchers on fractional calculus and its applications. We acknowledge the benefits we have obtained from interacting with this open-minded community in the past years.

University of Extremadura, Spain   *Concepción Alicia Monje Micharet*
Utah State University, USA   *YangQuan Chen*
University of Extremadura, Spain   *Blas Manuel Vinagre Jara*
Northeastern University, China   *Dingyü Xue*
University of Castilla-La Mancha, Spain   *Vicente Feliu Batlle*
January 2010

Acknowledgements

This first monograph on fractional-order controls is based on a series of papers and articles that we have written in the past 10 plus years. Therefore, it has been necessary at times to reuse some material that we previously reported in various papers and publications. Although in most instances such material has been modified and rewritten for this monograph, copyright permission from several publishers is acknowledged as follows.

Acknowledgement is given to the Institute of Electrical and Electronic Engineers (IEEE) to reproduce material from the following papers:

©2009 IEEE. Reprinted, with permission, from YangQuan Chen, Ivo Petras and Dingyü Xue, "Fractional Order Control – A tutorial." *Proceedings of The 2009 American Control Conference* (ACC09), June 10-12, 2009, St. Louis, USA, pp. 1397–1411, Lead Tutorial Session paper, DOI:10.1109/ACC.2009.5160719 (material found in Chapter 13).

©2002 IEEE. Reprinted, with permission, from Dingyü Xue and YangQuan Chen, "A Comparative Introduction of Four Fractional Order Controllers." *Proceedings of The 4th IEEE World Congress on Intelligent Control and Automation* (WCICA02), June 10–14, 2002, Shanghai, China, pp. 3228–3235 (material found in Chapter 4).

©1998 IEEE. Reprinted, with permission, from B. M. Vinagre, V. Feliu, and J. J. Feliu, "Frequency Domain Identification of a Flexible Structure with Piezoelectric Actuators by Using Irrational Transfer Functions." *Proceedings of the 37th IEEE Conference on Decision and Control*, Tampa, FL, USA, Dec. 1998. pp. 1278–1280 (material found in Chapter 14).

©2009 IEEE. Reprinted, with permission, from Li, H., Luo, Y. and Chen, Y. Q., "A Fractional Order Proportional and Derivative (FOPD) Motion Controller: Tuning Rule and Experiments." *IEEE Transactions on Control*

Systems Technology, DOI: 10.1109/TCST.2009.2019120 (material found in Chapter 6).

©2006 IEEE. Reprinted, with permission, from Dingyü Xue, Chunna Zhao, and YangQuan Chen, "A Modified Approximation Method of Fractional Order System." *Proceedings of the 2006 IEEE International Conference on Mechatronics and Automation (ICMA06),* 25–28 June 2006, Luoyang, China, pp. 1043–1048, DOI: 10.1109/ICMA.2006.257769 (material found in Chapter 12).

Acknowledgement is given to the American Society of Mechanical Engineers (ASME) to reproduce material from the following paper:

YangQuan Chen, Tripti Bhaskaran, and Dingyü Xue, "Practical Tuning Rule Development for Fractional Order Proportional and Integral Controller." *ASME Journal of Computational and Nonlinear Dynamics,* April 2008 – Volume 3, Issue 2, pages 021403-1 to 021403-8 DOI:10.1115/1.2833934 (material found in Chapter 5).

Acknowledgement is given to the American Society of Civil Engineers (ASCE) to reproduce material from the following paper:

V. Feliu, R. Rivas, L. Sánchez Rodríguez, and M. A. Ruiz-Torija, "Robust Fractional Order PI Controller Implemented on a Hydraulic Canal," *Journal of Hydraulic Engineering, ASCE,* vol. 135, no. 5, pp. 271–282, 2009 (material found in Chapter 16).

Acknowledgement is given to The Institution of Engineering and Technology (IET) to reproduce material from the following paper:

C. A. Monje, F. Ramos, V. Feliu, and B. M. Vinagre, "Tip Position Control of a Lightweight Flexible Manipulator Using a Fractional Order Controller." *IET Control Theory and Applications,* vol. 1, no. 5, pp. 1451–1460, 2007 (material found in Chapter 15).

Acknowledgement is given to International Federation of Automatic Control (IFAC) to reproduce material from the following paper:

C. A. Monje, B. M. Vinagre, A. J. Calderón, V. Feliu, and Y. Q. Chen, "Self-tuning of Fractional Lead-lag Compensators." *Proceedings of the 16th IFAC World Congress,* Prague, Czech, July 4 to July 8, 2005 (material found in Chapters 8 and 9).

Acknowledgement is given to Elsevier to reproduce material from the following papers:

Concepción A. Monje, Blas M. Vinagre, Vicente Feliu, and YangQuan Chen, "Tuning and Auto-tuning of Fractional Order Controllers for Industry Applications." *Control Engineering Practice*, vol. 16, pp. 798–812. doi:10.1016/j.conengprac.2007.08.006 (material found in Chapter 7).

V. Feliu, R. Rivas, and L. Sánchez Rodríguez, "Fractional Robust Control of Main Irrigation Canals with Variable Dynamic Parameters," *Control Engineering Practice*, vol. 15 no. 6 pp. 673–686, June 2007 (material found in Chapter 16).

V. Feliu, R. Rivas, F. Castillo, and L. Sánchez Rodríguez, "Smith Predictor Based Robust Fractional Order Control Application to Water Distribution in a Main Irrigation Canal Pool," *Journal of Process Control*, vol. 19, no. 3, pp. 506–519, March 2009 (material found in Chapter 16).

V. Feliu, R. Rivas, and F. Castillo, "Fractional Order Controller Robust to Time Delay for Water Distribution in an Irrigation Main Canal Pool," *Computers and Electronics in Agriculture*, vol. 69, no. 2, pp. 185–197, December 2009 (material found in Chapter 16).

Antonio J. Calderón, Blas M. Vinagre, and Vicente Feliu, "Fractional Order Control Strategies for Power Electronic Buck Converters," *Signal Processing*, vol. 86, 2006, pp. 2803–2819 (material found in Chapter 18).

Acknowledgement is given to Sage Publications to reproduce material from the following paper:

B. M. Vinagre, C. A. Monje, A. J. Calderón, and J. I. Suárez, "Fractional PID Controllers for Industry Application. A Brief Introduction." *Journal of Vibration and Control,* vol. 13, no. 9–10, pp. 1419–1429, 2007 (material found in Chapter 4).

Acknowledgement is given to Springer to reproduce material from the following papers:

Dingyü Xue and YangQuan Chen, "Suboptimum H_2 Pseudo-rational Approximations to Fractional order Linear Time Invariant Systems." *Advances in Fractional Calculus – Theoretical Developments and Applications in Physics and Engineering,* Jocelyn Sabatier, Om Prakash Agrawal and

J. A. Tenreiro Machado eds., pp. 61–75, 2007, DOI: 10.1007/978-1-4020-6042-7_5 (material found in Chapter 12).

B. M. Vinagre, I. Petras, I. Podlubny, and Y. Q. Chen, "Using Fractional Order Adjustment Rules and Fractional Order Reference Models in Model-Reference Adaptive Control," *Nonlinear Dynamics*, vol. 29, nos. 1-4, 2002, pp. 269–279, DOI:10.1023/A:1016504620249 (material found in Chapter 11).

Concepción A. Monje, Antonio J. Calderón, Blas M. Vinagre, YangQuan Chen, and Vicente Feliu, "On Fractional PI$^\lambda$ Controllers: Some Tuning Rules for Robustness to Plant Uncertainties." *Nonlinear Dynamics,* vol. 38, no. 1–4, pp. 369–381, 2004, DOI:10.1007/s11071-004-3767-3 (material found in Chapter 7).

Acknowledgement is given to the respective copyright holders to reproduce material from the following conference papers:

Blas M. Vinagre, Antonio J. Calderón, "On Fractional Sliding Mode Control." *Proceedings of the 7th Portuguese Conference On Automatic Control*, Lisbon (Portugal), September 2006 (material found in Chapter 11).

A. J. Calderón, C. A. Monje, and B. M. Vinagre, "Fractional Order Control of a Power Electronic Buck Converter." *Proceedings of the Portuguese Conference on Automatic Control* (CONTROLO 2002), Aveiro (Portugal), September 2002, pp. 365–370 (material found in Chapter 18).

Blas M. Vinagre, Concha A. Monje, and Inés Tejado, "Reset and Fractional Integrators in Control Applications." *Proceedings of the International Carpathian Control Conference,* Strbske Pleso (Slovak Republic), May 2007, pp. 754–757 (material found in Chapter 11).

A. J. Calderón, B. M. Vinagre, and V. Feliu, "Fractional Sliding Mode Control of a DC-DC Buck Converter with Application to DC Motor Drives." *Proceedings of the 11th International Conference on Advanced Robotics*, Coimbra (Portugal), June/July 2003, pp. 252–257 (material found in Chapter 18).

A. J. Calderón, B. M. Vinagre, and V. Feliu. "Linear Fractional Order Control of a DC-DC Buck Converter." *Proceedings of the European Control Conference,* Cambridge (UK), September 2003 (material found in Chapter 18).

Contents

Part II Fractional-order PID-Type Controllers

Part I
Fundamentals of Fractional-order Systems and Controls

Chapter 1
Introduction

Students of mathematics, sciences, and engineering encounter the differential operators d/dx, d^2/dx^2, *etc.*, but probably few of them ponder over whether it is necessary for the order of differentiation to be an integer. Why not be a rational, fractional, irrational, or even a complex number? At the very beginning of integral and differential calculus, in a letter to L'Hôpital in 1695, Leibniz himself raised the question: "Can the meaning of derivatives with integer order be generalized to derivatives with non-integer orders?" L'Hôpital was somewhat curious about that question and replied by another question to Leibniz: "What if the order will be 1/2?" Leibniz in a letter dated September 30, 1695 replied: "It will lead to a paradox, from which one day useful consequences will be drawn." The question raised by Leibniz for a non-integer-order derivative was an ongoing topic for more than 300 years, and now it is known as *fractional calculus*, a generalization of ordinary differentiation and integration to arbitrary (non-integer) order.

Before introducing fractional calculus and its applications to control in this book, it is important to remark that "fractional," or "fractional-order," are improperly used words. A more accurate term should be "non-integer-order," since the order itself can be irrational as well. However, a tremendous amount of work in the literature use "fractional" more generally to refer to the same concept. For this reason, we are using the term "fractional" in this book.

1.1 Why Fractional Order?

It is usual in undergraduate courses of feedback control to introduce the *basic control actions* and their effects in the controlled system behavior, in the frequency domain. So, we know that these actions are proportional, derivative, and integral, and their main effects over the controlled system behavior are [1]:

- to increase the speed of the response, and to decrease the steady-state error and relative stability, for proportional action;
- to increase the relative stability and the sensitivity to noise, for derivative action;
- to eliminate the steady-state error and to decrease the relative stability, for integral action.

The positive effects of the derivative action (increased relative stability) can be observed in the frequency domain by the $\pi/2$ phase lead introduced, and the negative ones (increased sensitivity to high-frequency noise) by the increasing gain with slope of $20\,\mathrm{dB/dec}$. For the integral action, the positive effects (elimination of steady-state errors) can be deduced by the infinite gain at zero frequency, and the negative ones (decreased relative stability) by the $\pi/2$ phase lag introduced. Considering this, it is quite natural to conclude that by introducing more general control actions of the form $s^n, 1/s^n, n \in \mathbb{R}^+$, we could achieve more satisfactory compromises between positive and negative effects, and combining the actions we could develop more powerful and flexible design methods to satisfy the controlled system specifications.

Let us turn now our attention from feedback control to systems modeling. Researchers in electrochemistry, biological systems, material science, viscoelasticity, and other fields in which electrochemical, mass transport, diffusion, or other memory phenomena appear [2–4], usually perform frequency domain experiments in order to obtain equivalent electrical circuits reflecting the dynamic behavior of the systems under study. It is quite usual in these fields to find behaviors that are far from the expected ones for common lumped elements such as resistors, inductors, and capacitors, and to define for operational purposes special impedances such as Warburg impedances, constant phase elements (CPEs), and others. All these special impedances have in common a frequency domain behavior of the form $k/(j\omega)^n, n \in \mathbb{R}$, and so, in the Laplace domain these elements should be modeled by $k/s^n, n \in \mathbb{R}$.

These operators in the frequency and Laplace domains give rise to the corresponding operators in the time domain. In what follows, by using standard definitions of repeated integrals and derivatives, we will try to show that these operators that arise in a quite natural way in the frequency domain, lead us to the definition of differential and integral operators of arbitrary order, the fundamental operators of the fractional calculus.

Let us assume for the time being zero initial conditions. If we define $F(s)$ as the Laplace transform of the function $f(t)$, $F(s) \equiv \mathscr{L}\left[f(t)\right]$, in the equation

$$\frac{1}{s^n}F(s), \tag{1.1}$$

we can recognize the Laplace domain equivalent for the n-fold integral of the function $f(t)$. Consider an antiderivative or primitive of the function $f(t)$, $\mathscr{D}^{-1}f(t)$, then

$$\mathscr{D}^{-1}f(t) = \int_0^t f(x)\mathrm{d}x. \qquad (1.2)$$

Now let us perform the repeated applications of the operator. For example,

$$\mathscr{D}^{-2}f(t) = \int_0^t \int_0^x f(y)\mathrm{d}y\mathrm{d}x. \qquad (1.3)$$

Equation 1.3 can be considered as a double integral, and taking into account the x-y plane over which it is integrated (see Figure 1.1), we can reverse the sequence of integrations by doing the proper changes in their limits. So, we obtain

$$\mathscr{D}^{-2}f(t) = \int_0^t \int_y^t f(y)\mathrm{d}x\mathrm{d}y. \qquad (1.4)$$

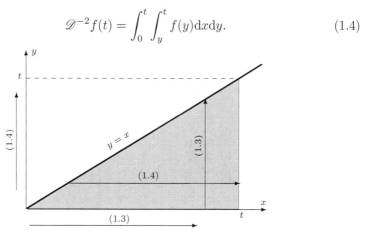

Figure 1.1 x-y plane for integration

As $f(y)$ is a constant with respect to x, we find that the inner integral is simply $(t-y)f(y)$, and we have

$$\mathscr{D}^{-2}f(t) = \int_0^t (t-y)f(y)\mathrm{d}y. \qquad (1.5)$$

Similarly, we can obtain

$$\mathscr{D}^{-3}f(t) = \frac{1}{2} \int_0^t (t-y)^2 f(y)\mathrm{d}y, \qquad (1.6)$$

and so on, giving the formula

$$\mathscr{D}^{-n}f(t) = \underbrace{\int \cdots \int_0^t}_{n} f(y) \underbrace{\mathrm{d}y \cdots \mathrm{d}y}_{n} = \int_0^t \frac{f(y)(t-y)^{n-1}}{(n-1)!}\mathrm{d}y. \qquad (1.7)$$

The last equation, in which we can see that an iterated integral may be expressed as a weighted single integral with a very simple weighting function, is known as the Cauchy's formula for iterated or repeated integral. If we generalize (1.7) for the case of $n \in \mathbb{R}^+$, we obtain

$$\mathscr{D}^{-n} f(t) = \frac{1}{\Gamma(n)} \int_0^t f(y)(t-y)^{n-1} dy, \tag{1.8}$$

which corresponds to the *Riemann–Liouville's definition* for the fractional-order integral of order $n \in \mathbb{R}^+$ [2,3].

We can obtain the same result by following a different path. By taking the inverse Laplace transform, the function corresponding to $1/s^n, n \in \mathbb{R}^+$ is

$$\mathscr{L}^{-1} \left[\frac{1}{s^n} \right] = \frac{t^{n-1}}{\Gamma(n)}. \tag{1.9}$$

So, if we consider (1.1) as the product of functions $1/s^n$ and $F(s)$ in the Laplace domain, it corresponds to the convolution product in the time domain, that is,

$$\mathscr{D}^{-n} f(t) = \frac{t^{n-1}}{\Gamma(n)} * f(t) = \frac{1}{\Gamma(n)} \int f(y)(t-y)^{n-1} dy. \tag{1.10}$$

Turning our attention from integrals to derivatives, the operator $s^n, n \in \mathbb{R}^+$ in the Laplace domain gives rise to an operator of the form d^n/dt^n in the time domain. According to the well known definition, the first-order derivative of the function $f(t)$, denoted by $\mathscr{D}^1 f(t)$, is defined by

$$\mathscr{D}^1 f(t) = \frac{df(t)}{dt} = \lim_{h \to 0} \frac{f(t) - f(t-h)}{h}, \tag{1.11}$$

that is, as the limit of a backward difference. Similarly,

$$\mathscr{D}^2 f(t) = \frac{d^2 f(t)}{dt^2} = \lim_{h \to 0} \frac{1}{h^2} [f(t) - 2f(t-h) + f(t-2h)] \tag{1.12}$$

and

$$\mathscr{D}^3 f(t) = \frac{d^3 f(t)}{dt^3} = \lim_{h \to 0} \frac{1}{h^3} \left[f(t) - 3f(t-h) + 3f(t-2h) - f(t-3h) \right]. \tag{1.13}$$

Iterating n-times, we can obtain

$$\mathscr{D}^n f(t) = \frac{d^n f(t)}{dt^n} = \lim_{h \to 0} \frac{1}{h^n} \sum_{k=0}^n (-1)^n \binom{n}{k} f(t-kh), \tag{1.14}$$

where

$$\binom{n}{k} = \frac{n(n-1)(n-2)\cdots(n-k+1)}{k!} \tag{1.15}$$

is the usual notation for the binomial coefficients. Equation 1.14 for $n \in \mathbb{R}^+$ leads us to the *Grünwald–Letnikov's definition* for the fractional-order derivative of order n [2,3].

We can wonder if (1.8) and (1.14) could be used for any $n \in \mathbb{R}$, in order to obtain unified definitions for generalized differential/integral operators. In fact, it is so, but considering some mathematical subtleties concerning the functions affected by the operators and the limits of the operation itself. Though avoiding unnecessary mathematical complexity, we will try to cover the essential of these aspects in Chapter 2, as well as the fundamental properties of the fractional-order operators and fractional-order systems (systems modeled by integro-differential equations involving fractional-order integro-differential operators).

For the time being, in the next section we give a brief historical overview of the development of the fractional calculus.

1.2 Brief Historical Overview

The earliest theoretical contributions to the field were made by Euler and Lagrange in the eighteenth century, and the first systematic studies seem to have been made at the beginning and middle of the nineteenth century by Liouville, Riemann, and Holmgren. It was Liouville who expanded functions in series of exponentials and defined the nth-order derivative of such a series by operating term-by-term as though n were a positive integer. Riemann proposed a different definition that involved a definite integral and was applicable to power series with non-integer exponents. It was Grünwald and Krug who first unified the results of Liouville and Riemann. Grünwald, by returning to the original sources and adopting as starting point the definition of a derivative as the limit of a difference quotient and arriving at definite-integral formulas for the nth-order derivative. Krug, working through Cauchy's integral formula for ordinary derivatives, showed that Riemann's definite integral had to be interpreted as having a finite lower limit while Liouville's definition corresponded to a lower limit $-\infty$.

The first application of the fractional calculus was made by Abel in 1823. He discovered that the solution of the integral equation for the tautochrone problem could be obtained via an integral in the form of derivative of order one half. Later in the nineteenth century, important stimuli to the use of fractional calculus were provided by the development by Boole of symbolic methods for solving linear differential equations of constant coefficients, or the operational calculus of Heaviside developed to solve certain problems in electromagnetic theory such as transmission lines. In the twentieth century contributions have been made to both the theory and applications of fractional calculus by very well known scientists such as Weyl and Hardy (properties of differintegrals), Erdély (integral equations), Riesz (functions of more than one variable), Scott Blair (rheology), or Oldham and Spanier (electrochemistry and general transport problems).

In the last decades of the last century there was continuing growth of the applications of fractional calculus mainly promoted by the engineering applications in the fields of feedback control, systems theory, and signals processing.

The interested reader can find good surveys of the history of fractional calculus in [2,5,6], and about particular applications related to the contents of this work, in the different chapters of the book.

1.3 Summary

In order to stimulate the interest of the reader, in this chapter we have tried to show how the necessity of defining and using the fractional-order differential and integral operators arises from very common and practical problems and applications (to extend the basic control actions or to model processes with memory phenomena), and how we can obtain a first approach to the definitions of these operators by using only mathematical tools well known by undergraduate students in science and engineering. Finally, a brief historical overview of the development of the fractional calculus has been presented, as well as the necessary references to satisfy the curiosity of the reader interested in this topic.

Chapter 2
Fundamentals of Fractional-order Systems

2.1 Fractional-order Operators: Definitions and Properties

2.1.1 Introduction

Essentially, the mathematical problem for defining fractional-order derivatives and integrals consists of the following [2,7]: to establish, for each function $f(z), z = x + jy$ of a general enough class, and for each number α (rational, irrational or complex), a correspondence with a function $g(z) = \mathscr{D}_c^\alpha f(z)$ fulfilling the following conditions:

- If $f(z)$ is an analytic function of the variable z, the derivative $g(z) = \mathscr{D}_c^\alpha f(z)$ is an analytic function of z and α.
- The operation $\mathscr{D}_c^\alpha$ and the usual derivative of order $n \in \mathbb{Z}^+$, $\alpha = n$ give the same result.
- The operation $\mathscr{D}_c^\alpha$ and the usual n-fold integral with $n \in \mathbb{Z}^-$, $\alpha = -n$ give the same result.
- $\mathscr{D}_c^\alpha f(z)$ and its first $(n-1)$th-order derivatives must vanish to zero at $z = c$.
- The operator of order $\alpha = 0$ is the identity operator.
- The operator must be linear: $\mathscr{D}_c^\alpha [af(z) + bh(z)] = a\mathscr{D}_c^\alpha f(z) + b\mathscr{D}_c^\alpha h(z)$.
- For the fractional-order integrals of arbitrary order, $\Re(\alpha) > 0, \Re(\beta) > 0$, it holds the additive law of exponents (semigroup property): $\mathscr{D}_c^\alpha \mathscr{D}_c^\beta f(z) = \mathscr{D}_c^{\alpha+\beta} f(z)$.

In the following sections the reader can find some of the more usual definitions of the fractional-order operators, fulfilling the above conditions, following mainly [3,7].

2.1.2 Fractional-order Integrals

In agreement with Riemann–Liouville's conception, the notion of fractional-order integral of order $\Re(\alpha) > 0$ is a natural consequence of Cauchy's formula for repeated integrals, which reduces the computation of the primitive corresponding to the n-fold integral of a function $f(t)$ to a simple convolution. This formula can be expressed as

$$\mathscr{I}_c^n f(t) \triangleq \mathscr{D}_c^{-n} f(t) = \frac{1}{(n-1)!} \int_c^t (t-\tau)^{n-1} f(\tau) d\tau, \quad t > c, \ n \in \mathbb{Z}^+. \quad (2.1)$$

In (2.1) we can see that $\mathscr{I}_c^n f(t)$ and its derivatives of orders $1, 2, 3, \cdots, n{-}1$ become zero for $t = c$.

In a natural way, we can extend the validity of (2.1) to $n \in \mathbb{R}^+$. Taking into account that $(n-1)! = \Gamma(n)$, and introducing the positive real number α, the Riemann–Liouville fractional-order integral is defined as

$$\mathscr{I}_c^\alpha f(t) \triangleq \frac{1}{\Gamma(\alpha)} \int_c^t (t-\tau)^{\alpha-1} f(\tau) d\tau, \quad t > c, \ \alpha \in \mathbb{R}^+. \quad (2.2)$$

It can be proved that this operator fulfils the aforementioned conditions.

When we deal with dynamic systems it is usual that $f(t)$ be a causal function of t, and so in what follows the definition for the fractional-order integral to be used is

$$\mathscr{I}^\alpha f(t) \triangleq \frac{1}{\Gamma(\alpha)} \int_0^t (t-\tau)^{\alpha-1} f(\tau) d\tau, \quad t > 0, \ \alpha \in \mathbb{R}^+. \quad (2.3)$$

In (2.3) we can see that the fractional-order integral can be expressed as a causal convolution of the form

$$\mathscr{I}^\alpha f(t) = \Phi_\alpha(t) * f(t), \quad \alpha \in \mathbb{R}^+, \quad (2.4)$$

with

$$\Phi_\alpha(t) = \frac{t_+^{\alpha-1}}{\Gamma(\alpha)}, \quad \alpha \in \mathbb{R}^+ \quad (2.5)$$

the causal kernel of the convolution and

$$t_+^{\alpha-1} = 0, \text{ for } t < 0; \ t_+^{\alpha-1} = t^{\alpha-1}, \text{ for } t \geqslant 0. \quad (2.6)$$

2.1.3 Fractional-order Derivatives

The definition (2.3) cannot be used for the fractional-order derivative by direct substitution of α by $-\alpha$, because we have to proceed carefully in order to guarantee the convergence of the integrals involved in the definition, and to preserve the properties of the ordinary derivative of integer-order.

Denoting the derivative operator of order $n \in \mathbb{N}$ by $\mathscr{D}^n$, and the identity operator by $\mathbb{I}$, we can verify that

$$\mathscr{D}^n \mathscr{I}^n = \mathbb{I}, \quad \mathscr{I}^n \mathscr{D}^n \neq \mathbb{I}, \quad n \in \mathbb{N}. \tag{2.7}$$

In other words, the operator $\mathscr{D}^n$ is only a left-inverse of the operator $\mathscr{I}^n$. In fact, from (2.1) we can deduce that

$$\mathscr{I}^n \mathscr{D}^n f(t) = f(t) - \sum_{k=0}^{n-1} f^{(k)}(0^+) \frac{t^k}{k!}, \quad t > 0, \tag{2.8}$$

where $f^{(k)}(\cdot)$ is the kth-order derivative of the function $f(\cdot)$. Consequently, it must be verified whether $\mathscr{D}^\alpha$ is a left-inverse of $\mathscr{I}^\alpha$ or not. For this purpose, introducing the positive integer m so that $m - 1 < \alpha < m$, the Riemann–Liouville definition for the fractional-order derivative of order $\alpha \in \mathbb{R}^+$ has the following form:

$$_R\mathscr{D}^\alpha f(t) \triangleq \mathscr{D}^m \mathscr{I}^{m-\alpha} f(t) = \frac{d^m}{dt^m} \left[\frac{1}{\Gamma(m-\alpha)} \int_0^t \frac{f(\tau)}{(t-\tau)^{\alpha-m+1}} d\tau \right], \tag{2.9}$$

where $m - 1 < \alpha < m$, $m \in \mathbb{N}$.

An alternative definition for the fractional-order derivative was introduced by Caputo as

$$_C\mathscr{D}^\alpha f(t) \triangleq \mathscr{I}^{m-\alpha} \mathscr{D}^m f(t) = \frac{1}{\Gamma(m-\alpha)} \int_0^t \frac{f^{(m)}(\tau)}{(t-\tau)^{\alpha-m+1}} d\tau, \tag{2.10}$$

where $m - 1 < \alpha < m$, $m \in \mathbb{N}$.

This definition is more restrictive than the Riemann–Liouville one because it requires the absolute integrability of the mth-order derivative of the function $f(t)$. It is clear that, in general,

$$_R\mathscr{D}^\alpha f(t) \triangleq \mathscr{D}^m \mathscr{I}^{m-\alpha} f(t) \neq \mathscr{I}^{m-\alpha} \mathscr{D}^m f(t) \triangleq {}_C \mathscr{D}^\alpha f(t), \tag{2.11}$$

except in the case of being zero at $t = 0^+$ for the function $f(t)$ and its first $(m-1)$th-order derivatives. In fact, between the two definitions there are the following relations:

$$_R\mathscr{D}^\alpha f(t) = {}_C \mathscr{D}^\alpha f(t) + \sum_{k=0}^{m-1} \frac{t^{k-\alpha}}{\Gamma(k-\alpha+1)} f^{(k)}(0^+), \tag{2.12}$$

$$_R\mathscr{D}^\alpha \left(f(t) - \sum_{k=0}^{m-1} f^{(k)}(0^+) \frac{t^k}{k!} \right) = {}_C \mathscr{D}^\alpha f(t). \tag{2.13}$$

Due to its importance in applications, we will consider here the Grünwald–Letnikov's definition, based on the generalization of the backward difference. This definition has the form

$$\mathscr{D}^\alpha f(t)|_{t=kh} = \lim_{h \to 0} \frac{1}{h^\alpha} \sum_{j=0}^k (-1)^j \binom{\alpha}{j} f(kh - jh). \tag{2.14}$$

An alternative definition of the Grünwald–Letnikov's derivative in integral form is [3]

$$_{\mathrm{L}}\mathscr{D}^{\alpha}f(t) = \sum_{k=0}^{m} \frac{f^{(k)}(0^{+})t^{k-\alpha}}{\Gamma(m+1-\alpha)} + \frac{1}{\Gamma(m+1-\alpha)} \int_{0}^{t} (t-\tau)^{m-\alpha} f^{(m+1)}(\tau)\mathrm{d}\tau,$$

(2.15)

where $m > \alpha - 1$.

2.1.4 Laplace and Fourier Transforms

Laplace and Fourier integral transforms are fundamental tools in systems and control engineering. For this reason, we will give here the equation of these transforms for the defined fractional-order operators. These equations are

$$\mathscr{L}\left[\mathscr{I}^{\alpha}f(t)\right] = s^{-\alpha}F(s), \tag{2.16}$$

$$\mathscr{L}\left[_{\mathrm{R}}\mathscr{D}^{\alpha}f(t)\right] = s^{\alpha}F(s) - \sum_{k=0}^{m-1} s^{k}\left[_{\mathrm{R}}\mathscr{D}^{\alpha-k-1}f(t)\right]_{t=0}, \tag{2.17}$$

$$\mathscr{L}\left[_{\mathrm{C}}\mathscr{D}^{\alpha}f(t)\right] = s^{\alpha}F(s) - \sum_{k=0}^{m-1} s^{\alpha-k-1}f^{(k)}(0), \tag{2.18}$$

$$\mathscr{L}\left[_{\mathrm{L}}\mathscr{D}^{\alpha}f(t)\right] = s^{\alpha}F(s), \tag{2.19}$$

$$\mathscr{F}\left[\mathscr{I}^{\alpha}f(t)\right] = \mathscr{F}\left[\frac{t_{+}^{\alpha-1}}{\Gamma(\alpha)}\right]\mathscr{F}\left\{f(t)\right\} = (\mathrm{j}\omega)^{-\alpha}F(\omega), \tag{2.20}$$

$$\mathscr{F}\left[\mathscr{D}^{\alpha}f(t)\right] = \mathscr{F}\left\{\mathscr{D}^{m}\mathscr{I}^{m-\alpha}f(t)\right\} = (\mathrm{j}\omega)^{\alpha}F(\omega), \tag{2.21}$$

where $(m-1 \leqslant \alpha < m)$. More on Laplace transform can be found in the Appendix.

2.2 Fractional-order Differential Equations

Once the basic definitions of the fractional calculus have been established, and as a prelude for the study of fractional-order linear time invariant systems, we will briefly review in this section the fundamentals of fractional-order ordinary differential equations. We will start with the two-term equations (relaxation and oscillation equations) and continue with the general equations for the solutions of n-term equations. A detailed study of fractional-order differential equations can be found in [3, 8].

2.2.1 Relaxation and Oscillation Equations

It is known that the classical problems of relaxation and oscillation are described by linear ordinary differential equations of orders 1 and 2, respectively (for control community, normal relaxation is equivalent to first-order dynamics). We can generalize the equations

$$\mathscr{D}u(t) + u(t) = q(t) \tag{2.22}$$

and

$$\mathscr{D}^2 u(t) + u(t) = q(t) \tag{2.23}$$

by simply substituting the integer-order derivatives by the fractional order α. If we want to preserve the usual initial conditions, we will use Caputo's fractional-order derivatives for obtaining

$$_C\mathscr{D}^\alpha u(t) + u(t) = {}_R\mathscr{D}^\alpha \left[u(t) - \sum_{k=0}^{m-1} \frac{t^k}{k!} u^{(k)}\left(0^+\right) \right] + u(t) = q(t), \ t > 0, \tag{2.24}$$

where m defined by $m - 1 < \alpha < m$ is a positive integer which determines the number of initial conditions $u^{(k)}(0^+) = b_k, k = 0, 1, 2, \cdots, m - 1$. It is obvious that in the case of $\alpha = m$, (2.24) becomes an ordinary differential equation, whose solution can be expressed as

$$u(t) = \sum_{k=0}^{m-1} b_k u_k(t) + \int_0^t q(t - \tau) u_\delta(\tau) \mathrm{d}\tau, \tag{2.25}$$

$$u_k(t) = \mathscr{I}^k u_0(t), \quad u_k^{(h)}(0^+) = \delta_{kh}, \quad u_\delta(t) = -\mathscr{D}u_0(t), \tag{2.26}$$

for $h, k = 0, 1, \cdots, m-1$, the m functions $u_k(t)$ are the fundamental solutions of the homogeneous differential equation, and the function $u_\delta(t)$ is the impulse response (the particular solution for $q(t) = \delta(t)$ under zero initial conditions).

It can be proved [3, 8] that the solution of (2.24) can be expressed in the same form as

$$u(t) = \sum_{k=0}^{m-1} b_k \mathscr{I}^k \mathscr{E}_\alpha(-t^\alpha) - \int_0^t q(t - \tau) \mathscr{E}'_\alpha(-\tau^\alpha) \mathrm{d}\tau, \tag{2.27}$$

where $\mathscr{E}_\alpha(-t^\alpha)$ is the *Mittag–Leffler function* defined by [3, 9, 10]

$$\mathscr{E}_\alpha(z) = \sum_{k=0}^{\infty} \frac{z^k}{\Gamma(\alpha k + 1)}. \tag{2.28}$$

Comparing (2.25) and (2.27) we can see that:

- The Mittag–Leffler function in (2.27) has the role of the exponential function in (2.25).

- When α is non-integer, $m - 1 < \alpha < m$, $m - 1$ is the integer part of α, $\left(m - 1 \triangleq [\alpha]\right)$ and m the number of initial conditions for uniqueness of the solution, $u(t)$.
- The m functions $\mathscr{I}^k \mathscr{E}_\alpha(-t^\alpha)$, with $k = 0, 1, \cdots, m - 1$ are the particular solutions of the homogeneous equation which satisfy the initial conditions, $i.e.$, the fundamental solutions of the homogenous equation.
- The function $\mathscr{E}'_\alpha(-t^\alpha)$, the first-order derivative of the function $\mathscr{E}_\alpha(-t^\alpha)$, is the impulse response.

It is clear that the form of the solutions is given by the properties of the Mittag–Leffler function. Figures 2.1 and 2.2 give the curves of the function for different values of α. As we can see, the behavior corresponds to an anomalous relaxation) (non-standard first-order decay) for $\alpha < 1$, is exponential for $\alpha = 1$, becomes a damped oscillation for $1 < \alpha < 2$, and oscillates for $\alpha = 2$.

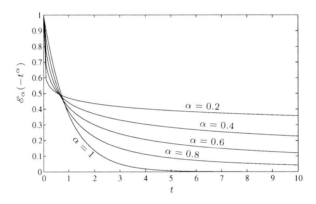

Figure 2.1 Mittag–Leffler functions $\mathscr{E}_\alpha(-t^\alpha)$ for $\alpha = 0.2, 0.4, 0.6, 0.8, 1$

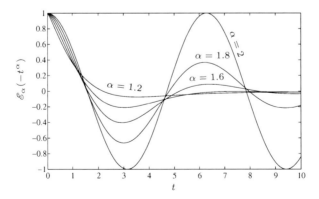

Figure 2.2 Mittag–Leffler functions $\mathscr{E}_\alpha(-t^\alpha)$ for $\alpha = 1.2, 1.4, 1.6, 1.8, 2$

For the general two-term equation with zero initial conditions

$$a\mathscr{D}^\alpha u(t) + bu(t) = q(t), \tag{2.29}$$

we can obtain the solution by applying the Laplace transform method. So, in the Laplace domain, the solution can be expressed as

$$U(s) = Q(s)\frac{1/a}{s^\alpha + b/a}, \tag{2.30}$$

and in the time domain as

$$u(t) = q(t) * \frac{1}{a}t^{\alpha-1}\mathscr{E}_{\alpha,\alpha}\left(-\frac{b}{a}t^\alpha\right), \tag{2.31}$$

where $t^{\alpha-1}\mathscr{E}_{\alpha,\alpha}\left(-bt^\alpha/a\right)/a$ is the impulse response, and $\mathscr{E}_{\alpha,\alpha}\left(-bt^\alpha/a\right)$ is the so-called *Mittag–Leffler function in two parameters* defined as [3]

$$\mathscr{E}_{\alpha,\beta}(z) = \sum_{k=0}^{\infty}\frac{z^k}{\Gamma\left(\alpha k + \beta\right)}, \quad \Re(\alpha) > 0, \; \Re(\beta) > 0. \tag{2.32}$$

2.2.2 Numerical Solutions

In order to obtain a numerical solution for the fractional-order differential equations, we can make use of the Grünwald–Letnikov's definition, and approximate

$$\mathscr{D}^\alpha f(t) \approx \Delta_h^\alpha f(t), \tag{2.33}$$

$$\Delta_h^\alpha f(t)\Big|_{t=kh} = h^{-\alpha}\sum_{j=0}^{k}(-1)^j\binom{\alpha}{j}f(kh - jh). \tag{2.34}$$

So, for the two-term equation (2.29) with $a = 1$ and zero initial conditions, this approximation leads to

$$h^{-\alpha}\sum_{j=0}^{k}w_j^{(\alpha)}y_{k-j} + by_k = q_k, \tag{2.35}$$

where $t_k = kh$, $y_k = y(t_k)$, $y_0 = 0$, $q_k = q(t_k)$, $k = 0, 1, 2, \cdots$, and

$$w_j^{(\alpha)} = (-1)^j\binom{\alpha}{j}, \tag{2.36}$$

for $j = 0, 1, 2, \cdots$, and the algorithm to obtain the numerical solution will be

$$y_i = 0, \quad i = 1, 2, \cdots, n - 1, \tag{2.37}$$

$$y_k = -bh^\alpha y_{k-1} - \sum_{j=1}^{k}w_j^{(\alpha)}y_{k-j} + h^\alpha q_k, \quad k = n, n+1, \cdots. \tag{2.38}$$

As we can see in former equations, as t grows, we need more and more terms to add for computing the solution, in other words, we need unlimited memory. To solve this problem, the *short memory principle* was proposed [11]. This principle is based on the observation that for large t the coefficients of the Grünwald–Letnikov's definition corresponding to values of the function near $t = 0$ (or any other point considered as initial) have little influence in the solution. This fact allows us to approximate the numerical solution by using the information of the "recent past," in other words, the interval $[t - L, t]$, L being the length of memory, a moving low limit to compute the derivatives. So, we will use

$$\mathscr{D}^\alpha f(t) \approx {}_{t-L}\mathscr{D}^\alpha f(t), \quad t > L, \tag{2.39}$$

and the number of terms to add is limited by the value of L/h. The error of the approximation when $\mid f(t) \mid \leqslant M, \ (0 < t \leqslant t_1)$ is bounded by

$$\varepsilon(t) = |\mathscr{D}^\alpha f(t) - {}_{t-L}\mathscr{D}^\alpha f(t)| \leqslant \frac{ML^{-\alpha}}{|\Gamma(1-\alpha)|}, \quad L \leqslant t \leqslant t_1, \tag{2.40}$$

which can be used to determine the necessary memory length, L, to obtain a certain error bound, as

$$\varepsilon(t) < \epsilon, \ \ L \leqslant t \leqslant t_1 \Rightarrow L \geqslant \left(\frac{M}{\epsilon\,|\Gamma(1-\alpha)|}\right)^{1/\alpha}. \tag{2.41}$$

For the computation of the coefficients to obtain the numerical solution, several methods can be used. For the case of a fixed value of derivative order α, we can use the following recursive formula:

$$w_0^{(\alpha)} = 1; \quad w_k^{(\alpha)} = \left(1 - \frac{\alpha+1}{k}\right) w_{k-1}^{(\alpha)}, \quad k = 1, 2, \cdots. \tag{2.42}$$

For a non-fixed α (for example, if we need to identify a system and α is a parameter to be determined) it is more convenient to use the fast Fourier transform (FFT). In such a case, it should be noted that the coefficients can be considered as the coefficients of the series expansion for the function $(1-z)^\alpha$

$$(1-z)^\alpha = \sum_{k=0}^{\infty} (-1)^k \binom{\alpha}{k} z^k = \sum_{k=0}^{\infty} w_k^{(\alpha)} z^k, \tag{2.43}$$

and with $z = \mathrm{e}^{-\mathrm{j}\omega}$

$$\left(1 - \mathrm{e}^{-\mathrm{j}\omega}\right)^\alpha = \sum_{k=0}^{\infty} w_k^{(\alpha)} \mathrm{e}^{-\mathrm{j}k\omega}, \tag{2.44}$$

we can express the coefficients in terms of the inverse Fourier transform

$$w_k^{(\alpha)} = \frac{1}{2\pi} \int_0^{2\pi} \left(1 - \mathrm{e}^{-\mathrm{j}\omega}\right)^\alpha \mathrm{e}^{\mathrm{j}k\omega} \mathrm{d}\omega \tag{2.45}$$

that can be computed by using FFT algorithms.

2.3 Fractional-order Systems

After establishing the fundamental definitions of the fractional calculus in the previous sections and determining the kind of solutions of the differential equations of fractional order, this section will deal with the analysis of the systems described by such equations. This analysis, as is usual for the systems of integer-order in the classical control theory framework, will start with the input-output models or representations of these systems in different domains (*e.g.*, time, Laplace, and Z) to study their performance, both transient and steady state, discussing the conditions and criteria for stability, and determining the steady-state error coefficients.

2.3.1 Models and Representations

The equations for a continuous-time dynamic system of fractional-order can be written as follows:

$$H\left(\mathscr{D}^{\alpha_0\alpha_1\alpha_2\cdots\alpha_m}\right)(y_1, y_2, \cdots, y_l) = G\left(\mathscr{D}^{\beta_0\beta_1\beta_2\cdots\beta_n}\right)(u_1, u_2, \cdots, u_k), \quad (2.46)$$

where y_i, u_i are functions of time and $H(\cdot), G(\cdot)$ are the combination laws of the fractional-order derivative operator. For the linear time-invariant single-variable case, the following equation would be obtained:

$$H\left(\mathscr{D}^{\alpha_0\alpha_1\alpha_2\cdots\alpha_n}\right)y(t) = G\left(\mathscr{D}^{\beta_0\beta_1\beta_2\cdots\beta_m}\right)u(t), \quad (2.47)$$

with

$$H\left(\mathscr{D}^{\alpha_0\alpha_1\alpha_2\cdots\alpha_n}\right) = \sum_{k=0}^{n} a_k \mathscr{D}^{\alpha_k}; \quad G\left(\mathscr{D}^{\beta_0\beta_1\beta_2\cdots\beta_m}\right) = \sum_{k=0}^{m} b_k \mathscr{D}^{\beta_k}, \quad (2.48)$$

where $a_k, b_k \in \mathbb{R}$. Or, explicitly,

$$\begin{aligned} a_n \mathscr{D}^{\alpha_n} y(t) + a_{n-1} \mathscr{D}^{\alpha_{n-1}} y(t) + \cdots + a_0 \mathscr{D}^{\alpha_0} y(t) \\ = b_m \mathscr{D}^{\beta_m} u(t) + b_{m-1} \mathscr{D}^{\beta_{m-1}} u(t) + \cdots + b_0 \mathscr{D}^{\beta_0} u(t). \end{aligned} \quad (2.49)$$

If in the previous equation all the orders of derivation are integer multiples of a base order, α, that is, $\alpha_k, \beta_k = k\alpha$, $\alpha \in \mathbb{R}^+$, the system will be of *commensurate-order*, and (2.49) becomes

$$\sum_{k=0}^{n} a_k \mathscr{D}^{k\alpha} y(t) = \sum_{k=0}^{m} b_k \mathscr{D}^{k\alpha} u(t). \quad (2.50)$$

If in (2.50) $\alpha = 1/q$, $q \in \mathbb{Z}^+$, the system will be of *rational-order*.

This way, linear time-invariant systems can be classified as follows:

$$\text{LTI Systems} \begin{cases} \text{Non-integer} \begin{cases} \text{Commensurate} \begin{cases} \text{Rational} \\ \text{Irrational} \end{cases} \\ \text{Non-commensurate} \end{cases} \\ \text{Integer} \end{cases}$$

In the case of discrete-time systems (or discrete equivalents of continuous-time systems) we can use (2.33) and (2.34) to obtain models of the form

$$a_n \Delta_h^{\alpha_n} y(t) + a_{n-1} \Delta_h^{\alpha_{n-1}} y(t) + \cdots + a_0 \Delta_h^{\alpha_0} y(t)$$
$$= b_m \Delta_h^{\beta_m} u(t) + b_{m-1} \Delta_h^{\beta_{m-1}} u(t) + \cdots + b_0 \Delta_h^{\beta_0} u(t). \tag{2.51}$$

Applying the Laplace transform to (2.49) with zero initial conditions, or the Z transform to (2.51), the input-output representations of fractional-order systems can be obtained. In the case of continuous models, a fractional-order system will be given by a transfer function of the form

$$G(s) = \frac{Y(s)}{U(s)} = \frac{b_m s^{\beta_m} + b_{m-1} s^{\beta_{m-1}} + \cdots + b_0 s^{\beta_0}}{a_n s^{\alpha_n} + a_{n-1} s^{\alpha_{n-1}} + \cdots + a_0 s^{\alpha_0}}. \tag{2.52}$$

In the case of discrete-time systems, the discrete-time transfer function will be of the form

$$G(z) = \frac{b_m \left(w\left(z^{-1}\right)\right)^{\beta_m} + b_{m-1} \left(w\left(z^{-1}\right)\right)^{\beta_{m-1}} + \cdots + b_0 \left(w\left(z^{-1}\right)\right)^{\beta_0}}{a_n \left(w\left(z^{-1}\right)\right)^{\alpha_n} + a_{n-1} \left(w\left(z^{-1}\right)\right)^{\alpha_{n-1}} + \cdots + a_0 \left(w\left(z^{-1}\right)\right)^{\alpha_0}}, \tag{2.53}$$

where $\left(w\left(z^{-1}\right)\right)$ is the Z transform of the operator Δ_h^1, or, in other words, the discrete equivalent of Laplace's operator, s.

As can be seen in the previous equations, a fractional-order system has an irrational-order transfer function in Laplace's domain or a discrete transfer function of unlimited order in the Z domain, since only in the case of $\alpha_k \in \mathbb{Z}$ will there be a limited number of coefficients $(-1)^l \binom{\alpha_k}{l}$ different from zero. Because of this, it can be said that a fractional-order system has an unlimited memory or is infinite-dimensional, and obviously the systems of integer-order are just particular cases.

In the case of a commensurate-order system, the continuous-time transfer function is given by

$$G(s) = \frac{\sum_{k=0}^{m} b_k (s^\alpha)^k}{\sum_{k=0}^{n} a_k (s^\alpha)^k}, \tag{2.54}$$

which can be considered as a pseudo-rational function, $H(\lambda)$, of the variable

$$\lambda = s^\alpha,$$

$$H(\lambda) = \frac{\sum\limits_{k=0}^{m} b_k \lambda^k}{\sum\limits_{k=0}^{n} a_k \lambda^k}. \tag{2.55}$$

2.3.2 Stability

2.3.2.1 Previous Considerations

In a general way, the study of the stability of fractional-order systems can be carried out by studying the solutions of the differential equations that characterize them. An alternative way is the study of the transfer function of the system (2.52). To carry out this study it is necessary to remember that a function of the type

$$a_n s^{\alpha_n} + a_{n-1} s^{\alpha_{n-1}} + \dots + a_0 s^{\alpha_0}, \tag{2.56}$$

with $\alpha_i \in \mathbb{R}^+$, is a multi-valued function of the complex variable s whose domain can be seen as a Riemann surface [12, 13] of a number of sheets which is finite only in the case of $\forall i$, $\alpha_i \in \mathbb{Q}^+$, being the principal sheet defined by $-\pi < \arg(s) < \pi$. In the case of $\alpha_i \in \mathbb{Q}^+$, that is, $\alpha = 1/q$, q being a positive integer, the q sheets of the Riemann surface are determined by

$$s = |s| e^{j\phi}, \quad (2k+1)\pi < \phi < (2k+3)\pi, \quad k = -1, 0, \cdots, q-2. \tag{2.57}$$

Correspondingly, the case of $k = -1$ is the *principal sheet*. For the mapping $w = s^\alpha$, these sheets become the regions of the plane w defined by

$$w = |w| e^{j\theta}, \quad \alpha(2k+1)\pi < \theta < \alpha(2k+3)\pi. \tag{2.58}$$

This mapping is illustrated in Figures 2.3 and 2.4 for the case of $w = s^{1/3}$. Figure 2.3 represents the Riemann surface that corresponds to the transformation introduced above, and Figure 2.4 represents the regions of the complex plane w that correspond to each sheet of the Riemann surface. These three sheets correspond to

$$k = \begin{cases} -1, & -\pi < \arg(s) < \pi, \text{ (the principal sheet)} \\ 0, & \pi < \arg(s) < 3\pi, \text{ (sheet 2)} \\ 1 \, (= 3 - 2), & 3\pi < \arg(s) < 5\pi, \text{ (sheet 3)} \end{cases}$$

Thus, an equation of the type

$$a_n s^{\alpha_n} + a_{n-1} s^{\alpha_{n-1}} + \cdots + a_0 s^{\alpha_0} = 0, \tag{2.59}$$

which in general is not a polynomial, will have an infinite number of roots, among which only a finite number of roots will be on the principal sheet of

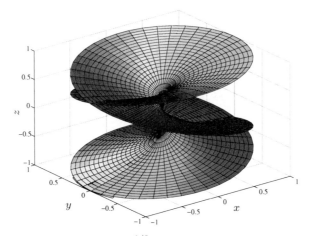

Figure 2.3 Riemann surface for $w = s^{1/3}$

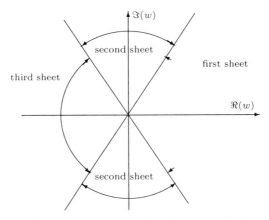

Figure 2.4 w-plane regions corresponding to the Riemann surface for $w = s^{1/3}$

the Riemann surface. It can be said that the roots which are in the secondary sheets of the Riemann surface are related to solutions that are always monotonically decreasing functions (they go to zero without oscillations when $t \to \infty$) and only the roots that are in the principal sheet of the Riemann surface are responsible for a different dynamics: damped oscillation, oscillation of constant amplitude, oscillation of increasing amplitude with monotonic growth.

For example, for equation

$$s^\alpha + b = 0, \tag{2.60}$$

the solutions are given by

$$s = (-b)^{1/\alpha} = |b|^{1/\alpha} \angle \frac{\arg(b) + 2l\pi}{\alpha}, \quad l = 0, \pm 1, \pm 2, \cdots,$$

and only the roots satisfying the condition

$$\left| \frac{\arg(b) + 2l\pi}{\alpha} \right| < \pi$$

will be on the principal sheet.

This definition of the principal sheet, which assumes a cut along $\mathbb{R}^-$, corresponds to the Cauchy principal value of the integral corresponding to the inverse transformation of Laplace, that is, to that obtained by direct application of the residue theorem. The roots which are in this sheet are called *structural roots* [14] or *relevant roots* [8].

For example, for the function

$$f(s) = \frac{1}{s^\alpha + b}, \quad \alpha = \frac{1}{\pi}, \quad b \in \mathbb{R}^+,$$

the roots of the denominator are given by the equation

$$s_l = (-b)^\pi|_l = |b|^\pi \angle \pi \left(\pi + 2l\pi \right), \quad l = 0, \pm 1, \pm 2, \cdots.$$

For the roots to be in the principal sheet, they must fulfil the following condition:

$$|\arg(s_l)| < \pi \implies |\pi \left(\pi + 2l\pi \right)| < \pi \implies |\pi \left(1 + 2l \right)| < 1.$$

As can be seen, there is no value of l to fulfil this condition, so there are no structural roots of this function. This fact can be observed in Figure 2.5 for $b = 1$: the function is analytical for every s, $|\arg(s)| < \pi$ with a maximum at $s = 0 + j0$, and the point $s^\alpha = -1$ is not a pole, but a branch point.

Studying the function for $\alpha = 4/\pi$, the condition becomes

$$\frac{|(\pi + 2l\pi)|}{4} < 1,$$

and it is fulfilled for $l = 0$ and $l = -1$, being the corresponding arguments $\angle s_0 = \pi^2/4$, and $\angle s_{-1} = -\pi^2/4$, respectively. In Figure 2.6 it can be observed that the function has poles at s_0 and s_{-1}.

2.3.2.2 Stability Conditions

After the previous considerations, the stability conditions of the fractional-order systems can be established.

In general, it can be said that a fractional-order system, with an irrational-order transfer function $G(s) = P(s)/Q(s)$, is bounded-input bounded-output stable (BIBO stable) if and only if the following condition is fulfilled (see [14] for more details):

$$\exists M, \quad |G(s)| \leqslant M, \quad \forall s \; \Re(s) \geqslant 0. \tag{2.61}$$

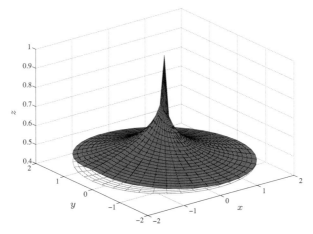

Figure 2.5 Magnitude of the function $f(s), \alpha = 1/\pi, b = 1$

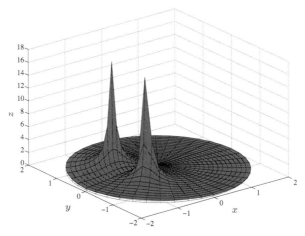

Figure 2.6 Magnitude of the function $f(s), \alpha = 4/\pi, b = 1$

The previous condition is satisfied if all the roots of $Q(s) = 0$ in the principal Riemann sheet, not being roots of $P(s) = 0$, have negative real parts.

For the case of commensurate-order systems, whose characteristic equation is a polynomial of the complex variable $\lambda = s^{\alpha}$, the stability condition is expressed as

$$|\arg(\lambda_i)| > \alpha \frac{\pi}{2}, \qquad (2.62)$$

where λ_i are the roots of the characteristic polynomial in λ. For the particular case of $\alpha = 1$ the well known stability condition for linear time-invariant systems of integer-order is recovered:

$$|\arg(\lambda_i)| > \frac{\pi}{2}, \quad \forall \lambda_i / Q(\lambda_i) = 0. \qquad (2.63)$$

An equivalent result was previously obtained by Mittag–Leffler in [9].

2.3.2.3 Stability Criteria

Nowadays there are no polynomial techniques, either Routh or Jury type, to analyze the stability of fractional-order systems. Only the geometrical techniques of complex analysis based on the Cauchy's argument principle can be applied, since they are techniques that inform about the number of singularities of the function within a rectifiable curve by observing the evolution of the function's argument through this curve.

In this way, applying the argument principle to the curve generally known as the Nyquist path (a curve that encloses the right half-plane of the Riemann principal sheet), the stability of the system can be determined by determining the number of revolutions of the resultant curve around the origin. To determine the closed-loop stability, it will be enough to check whether the evaluation curve encloses the critical point $(-1, j0)$ or not.

For the case of rational-order systems, a similar procedure can be applied. Given a system defined by the transfer function

$$G(s) = \frac{1}{a_n s^{n\alpha} + a_{n-1} s^{(n-1)\alpha} + \cdots + a_1 s^\alpha + a_0}, \qquad (2.64)$$

where $\alpha = 1/q$, $q, n \in \mathbb{Z}^+$, $a_k \in \mathbb{R}$, we can introduce the mapping $\lambda = s^\alpha$ to obtain the function $G(\lambda)$, and applying the condition (2.62) the stability of the system can be studied by evaluating the function $G(\lambda)$ along the curve Γ defined in the λ-plane in Figure 2.7:

$$\Gamma = \Gamma_1 \cup \Gamma_2 \cup \Gamma_3, \qquad (2.65)$$

with

$$\Gamma_1 : \lambda \diagup \arg(\lambda) = -\alpha \frac{\pi}{2}, \quad |\lambda| \in [0, \infty),$$

$$\Gamma_2 : \lambda = \lim_{R \to \infty} R^{j\phi}, \quad \phi \in \left(-\alpha \frac{\pi}{2}, \alpha \frac{\pi}{2}\right),$$

$$\Gamma_3 : \lambda \diagup \arg(\lambda) = \alpha \frac{\pi}{2}, \quad |\lambda| \in (\infty, 0).$$

If the transfer function has the form

$$H(s) = \frac{1}{a_n s^{p_n/q_n} + a_{n-1} s^{p_{n-1}/q_{n-1}} + \cdots + a_1 s^{p_1/q_1} + a_0}, \qquad (2.66)$$

where p_i, $q_i \in \mathbb{Z}^+$, and $p_n/q_n > p_{n-1}/q_{n-1} > \cdots > p_1/q_1$, the same procedure can be applied by using the function

$$H(\lambda), \quad \lambda = \frac{1}{q}, \quad q = \mathrm{lcm}\,(q_n, q_{n-1}, \cdots, q_1), \qquad (2.67)$$

where lcm(·) stands for the least common multiplier.

To illustrate the application of this procedure, two examples are given.

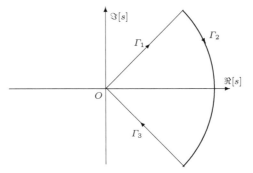

Figure 2.7 The Γ evaluation contour

Example 1 Given a system with transfer function

$$G(s) = \frac{1}{s^{2/3} - s^{1/2} + 1/2},$$

in a unity negative feedback structure with gain K, its evaluation along the Nyquist path defined in the Riemann principal sheet gives the result shown in Figure 2.8. It can be observed that the evaluation curve does not enclose the point $(-1, \mathrm{j}0)$. So, we can conclude that the closed-loop system is stable for any $K > 0$.

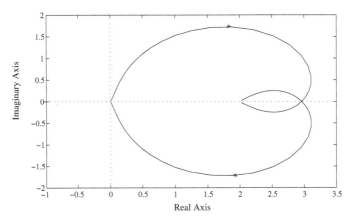

Figure 2.8 Evaluation result for Example 1

An equivalent result can be obtained by evaluating the function

$$G(\lambda) = \frac{1}{\lambda^4 - \lambda^3 + 1/2}, \quad \lambda = s^{1/6},$$

along the curve Γ of the complex plane λ defined by

$$\Gamma = \Gamma_1 \cup \Gamma_2 \cup \Gamma_3, \tag{2.68}$$

with

$$\Gamma_1: \quad \lambda \diagup \arg(\lambda) - \frac{\pi}{12}, \quad |\lambda| \in [0, \infty),$$

$$\Gamma_2: \quad \lambda = \lim_{R \to \infty} R^{j\phi}, \quad \phi \in \left(-\frac{\pi}{12}, \frac{\pi}{12}\right),$$

$$\Gamma_3: \quad \lambda \diagup \arg(\lambda) = \frac{\pi}{12}, \quad |\lambda| \in (\infty, 0).$$

Effectively, the roots of the characteristic equation can be obtained from the roots of the polynomial

$$Q(\lambda) = \lambda^4 - \lambda^3 + \frac{3}{2},$$

which are $\lambda_{1,2} = 1.0891 \pm j0.6923 = 1.2905 \angle \pm 0.5662$, $\lambda_{3,4} = -0.5891 \pm j0.7441 = 0.9491 \angle \pm 2.2404$, being the roots of the denominator of $G(s)$:

$$s_{1,2} = (\lambda_{1,2})^6 = 4.6183 \angle \pm 3.3975, \quad |\arg(s_{1,2})| > \pi,$$

$$s_{3,4} = (\lambda_{3,4})^6 = 0.7308 \angle \pm 13.4423, \quad |\arg(s_{3,4})| > \pi.$$

As can be observed, there are no structural roots (roots on the Riemann principal sheet defined by $|\arg(s)| < \pi$), which indicates the closed-loop system stability. □

Example 2 If now we deal with the stability of the closed-loop system whose transfer function is

$$G(s) = \frac{1}{s - 2s^{1/2} + 1.25},$$

evaluating the function

$$G(\lambda) = \frac{1}{\lambda^2 - 2\lambda + 1.25}, \quad \lambda = s^{1/2},$$

along the curve defined by

$$\Gamma = \Gamma_1 \cup \Gamma_2 \cup \Gamma_3, \quad (2.69)$$

with

$$\Gamma_1: \quad \lambda \diagup \arg(\lambda) = -\frac{\pi}{4}, \quad |\lambda| \in [0, \infty),$$

$$\Gamma_2: \quad \lambda = \lim_{R \to \infty} R^{j\phi}, \quad \phi \in \left(-\frac{\pi}{4}, \frac{\pi}{4}\right) 2$$

$$\Gamma_3: \quad \lambda \diagup \arg(\lambda) = \frac{\pi}{4}, \quad |\lambda| \in (\infty, 0),$$

the result shown in Figure 2.9 is obtained.

As can be observed, the resultant curve encloses twice the critical point $(-1, j0)$ in the negative direction for $K > 0.75$. Taking into account that $G(\lambda)$ has two unstable poles, $\lambda_{1,2} = 1 \pm j0.5, \arg(\lambda_{1,2}) < \pi/4$, it can be concluded that the system is stable for $K > 0.75$. For $K < 0.75$ the closed-loop system

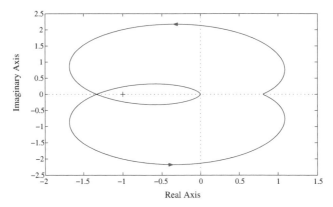

Figure 2.9 Evaluation result for Example 2

is unstable, with two structural roots in the right half-plane of the Riemann principal sheet.

Effectively, the roots of the polynomial

$$Q(\lambda) = \lambda^2 - 2\lambda + 1.25 + K$$

are, for $K = 1$,

$$\lambda_{1,2} = 1.0 \pm j1.1180,$$

and the structural poles of $G(s)$ are

$$s_{1,2} = (\lambda_{1,2})^2 = 2.25\angle \pm 0.9273, \quad |\arg(s_{1,2})| < \frac{\pi}{2} = 1.5708. \qquad \square$$

It is also important to note that the root locus technique can be applied to a commensurate-order system as easily as it can be applied to an integer-order one. Only the interpretation changes, that is, the relation of the complex plane points $\lambda = s^\alpha$ with the dynamic characteristics of the system.

2.3.3 Analysis of Time and Frequency Domain Responses

2.3.3.1 Transient Response

The general equation for the response of a fractional-order system in the time domain can be determined by using the analytical methods described previously.

The response will depend on the roots of the characteristic equation, having six different cases:

- There are no roots in the Riemann principal sheet. In this case the response will be a monotonically decreasing function.
- There are roots in the Riemann principal sheet, located in $\Re(s)<0$, $\Im(s) = 0$. In this case the response will be a monotonically decreasing function.
- There are roots in the Riemann principal sheet, located in $\Re(s)<0$, $\Im(s) \neq 0$. In this case the response will be a function with damped oscillations.
- There are roots in the Riemann principal sheet, located in $\Re(s)=0$, $\Im(s) \neq 0$. In this case the response will be a function with oscillations of constant amplitude.
- There are roots in the Riemann principal sheet, located in $\Re(s)>0$, $\Im(s) \neq 0$. In this case the response will be a function with oscillations of increasing amplitude.
- There are roots in the Riemann principal sheet, located in $\Re(s)>0$, $\Im(s) = 0$. In this case the response will be a monotonically increasing function.

In the particular case of commensurate-order systems, the impulse response can be written as follows:

$$\mathcal{L}^{-1}\left\{H(\lambda),\lambda=s^{\alpha}\right\}=\mathcal{L}^{-1}\left\{\frac{\displaystyle\sum_{k=0}^{m}a_{k}\lambda^{k}}{\displaystyle\sum_{k=0}^{n}b_{k}\lambda^{k}}\right\}=\mathcal{L}^{-1}\left\{\sum_{k=0}^{n}\frac{r_{k}}{\lambda-\lambda_{k}}\right\}.$$

Taking into account the general equation

$$\mathcal{L}^{-1}\left\{\frac{s^{\alpha-\beta}}{s^{\alpha}-\lambda_{k}}\right\}=t^{\beta-1}\mathcal{E}_{\alpha,\beta}\left(\lambda_{k}t^{\alpha}\right), \tag{2.70}$$

the impulse response, $g(t)$, can be obtained by setting $\alpha=\beta$ in the previous equation as follows:

$$g(t)=\sum_{k=0}^{n}r_{k}t^{\alpha-1}\mathcal{E}_{\alpha,\alpha}\left(\lambda_{k}t^{\alpha}\right). \tag{2.71}$$

The step response, given by the equation

$$y(t)=\mathcal{L}^{-1}\left\{\sum_{k=0}^{n}\frac{r_{k}s^{-1}}{\left(s^{\alpha}-\lambda_{k}\right)}\right\}, \tag{2.72}$$

can be obtained setting $\alpha=\beta-1$ in (2.71), in the following form:

$$y(t)=\sum_{k=0}^{n}r_{k}t^{\alpha}\mathcal{E}_{\alpha,\alpha+1}\left(\lambda_{k}t^{\alpha}\right). \tag{2.73}$$

The form of these responses will be:

- monotonically decreasing if $|\arg(\lambda_{k})| \geqslant \alpha\pi$;
- oscillatory with decreasing amplitude if $\alpha\pi/2 < |\arg(\lambda_{k})| < \alpha\pi$;

- oscillatory with constant amplitude if $|\arg(\lambda_k)| = \alpha\pi/2$;
- oscillatory with increasing amplitude if $|\arg(\lambda_k)| < \alpha\pi/2$, $|\arg(\lambda_k)| \neq 0$;
- monotonically increasing if $|\arg(\lambda_k)| = 0$.

These responses are depicted in Figure 2.10.

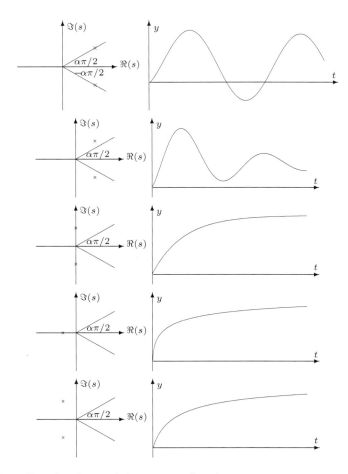

Figure 2.10 Root locations and the corresponding time responses

2.3.3.2 Frequency Domain Response

In general, the frequency response has to be obtained by the direct evaluation of the irrational-order transfer function of the fractional-order system along the imaginary axis for $s = j\omega$, $\omega \in (0, \infty)$. However, for the commensurate-order systems we can obtain Bode-like plots, in other words, the frequency

response can be obtained by the addition of the individual contributions of
the terms of order α resulting from the factorization of the function

$$G(s) = \frac{P(s^\alpha)}{Q(s^\alpha)} = \frac{\prod\limits_{k=0}^{m} (s^\alpha + z_k)}{\prod\limits_{k=0}^{n} (s^\alpha + \lambda_k)}, \quad z_k, P(z_k) = 0, \quad \lambda_k, Q(\lambda_k) = 0, \quad z_k \neq \lambda_k.$$

For each of these terms, referred to as $(s^\alpha + \gamma)^{\pm 1}$, the magnitude curve
will have a slope which starts at zero and tends to $\pm\alpha 20\,\mathrm{dB/dec}$ for higher
frequencies, and the phase plot will go from 0 to $\pm\alpha\pi/2$. Besides, there will
be resonances for $\alpha > 1$. To illustrate this, Figure 2.11 shows the frequency
response of the system whose transfer function is

$$G(s) = \frac{1}{s^{3/2} + 1}, \tag{2.74}$$

in which can be observed a slope which goes from 0 to $-30\,\mathrm{dB/dec}$, a phase
which starts at 0 and tends to $-3\pi/4$, and a resonant frequency of $\omega \approx$
$0.8\,\mathrm{rad/sec}$.

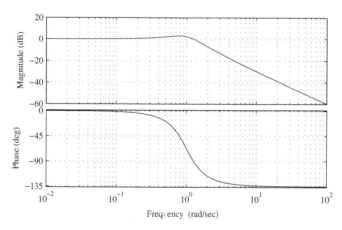

Figure 2.11 Bode plot of the system (2.74)

2.3.3.3 Steady-state Response

To finish this brief analysis of the response, the behavior of the fractional-
order systems in steady-state will be considered now, starting from the typical
system with unity negative feedback and using the usual definitions of steady-
state error coefficients. These definitions are:

- Position error coefficient:

$$K_p = \lim_{s \to 0} G(s). \tag{2.75}$$

- Velocity error coefficient:

$$K_v = \lim_{s \to 0} sG(s). \tag{2.76}$$

- Acceleration error coefficient:

$$K_a = \lim_{s \to 0} s^2 G(s). \tag{2.77}$$

For a fractional-order system whose transfer function can be expressed by

$$G(s) = \frac{K \left(a_m s^{\beta_m} + a_{m-1} s^{\beta_{m-1}} + \cdots + 1\right)}{s^\gamma \left(b_n s^{\alpha_n} + b_{n-1} s^{\alpha_{n-1}} + \cdots + 1\right)}, \tag{2.78}$$

the following relations are obtained:

$$K_p = \lim_{s \to 0} \frac{K}{s^\gamma} = \lim_{s \to 0} K s^{-\gamma}, \quad e_p = \frac{1}{1 + K_p}, \tag{2.79}$$

$$K_v = \lim_{s \to 0} K \frac{s}{s^\gamma} = \lim_{s \to 0} K s^{1-\gamma}, \quad e_v = \frac{1}{K_v}, \tag{2.80}$$

$$K_a = \lim_{s \to 0} K \frac{s^2}{s^\gamma} = \lim_{s \to 0} K s^{2-\gamma}, \quad e_a = \frac{1}{K_a}. \tag{2.81}$$

In Table 2.1, the steady-state errors and steady-state error coefficients are summarized for different values of γ. As can be observed, the fractional-order systems always have steady-state error coefficients 0 or ∞, and this shows that their behavior, also in steady-state, has to do with the behavior of the integer-order systems of higher or lower order than the fractional order. These systems will have finite coefficients only for inputs whose temporal dependence is of the form

$$r(t) = At^\gamma, \tag{2.82}$$

or, in Laplace domain,

$$\mathcal{L}\{r(t)\} = R(s) = A \frac{\Gamma(\gamma + 1)}{s^{\gamma+1}}. \tag{2.83}$$

Table 2.1 Steady-state error coefficients

	Steady state					Steady state			
γ	K_p, e_p	K_v, e_v	K_a, e_a	Type	γ	K_p, e_p	K_v, e_v	K_a, e_a	Type
0	K, $1/(1+K)$	0, ∞	0, ∞	0	(0,1)	∞, 0	0, ∞	0, ∞	0/1
1	∞, 0	K, $1/K$	0, ∞	1	(1,2)	∞, 0	∞, 0	0, ∞	1/2
2	∞, 0	∞, 0	K, $1/K$	2	(2,3)	∞, 0	∞, 0	∞, 0	2/3

2.3.4 Bode's Ideal Loop Transfer Function as Reference System

2.3.4.1 Introduction

In the previous sections it has been shown that the system

$$G(s) = \frac{A}{s^\alpha + A}, \quad 0 < \alpha < 2, \tag{2.84}$$

can exhibit behaviors that range from relaxation to oscillation, including the behaviors corresponding to first- and second-order systems as particular cases. For this reason, it is interesting to take this system as the reference system. It was first proposed in [15] and is the starting point for the CRONE control [16, 17]. This function can be considered as the result of the closed-loop connection of a fractional-order integrator with gain A and order α (see Figure 2.12), that is, a system whose open-loop transfer function is given by

$$F(s) = \frac{A}{s^\alpha}, \quad 0 < \alpha < 2. \tag{2.85}$$

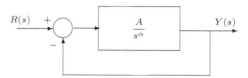

Figure 2.12 Reference system

Bode called this transfer function, $F(s)$, *the ideal open-loop transfer function* [3, 18].

2.3.4.2 General Characteristics

This ideal transfer function has the following characteristics:

1. Open-loop:

- The magnitude curve has a constant slope of $-20\alpha \, \text{dB/dec}$.
- The gain crossover frequency depends on A.
- The phase plot is a horizontal line of value $-\alpha\pi/2$.
- The Nyquist plot is a straight line which starts from the origin with argument $-\alpha\pi/2$.

2. Closed-loop with unity negative feedback:

- The gain margin is infinite.
- The phase margin is constant with value $\varphi_m = \pi(1 - \alpha/2)$, only depending on α.
- The step response is of the form

$$y(t) = At^\alpha \mathscr{E}_{\alpha,\alpha+1}(-At^\alpha).\qquad(2.86)$$

2.3.4.3 Step Response and Characteristic Parameters

In Figure 2.13 the step responses of the system $F(s)$ for $A = 1$ and different values of α are represented.

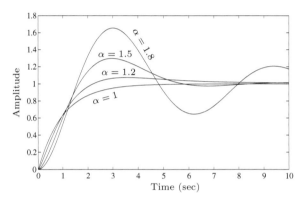

Figure 2.13 Step response (2.86) with $A = 1$

These curves correspond to the damping ratios and the natural frequency that can be obtained from the structural roots of the denominator of $F(s)$. These roots are

$$s_{1,2} = A^{1/\alpha}e^{j\pi/\alpha} = A^{1/\alpha}\left(\cos\frac{\pi}{\alpha} + j\sin\frac{\pi}{\alpha}\right).\qquad(2.87)$$

By using the well known relations

$$\omega_n = |s_{1,2}|,\quad -\delta\omega_n = \Re(s_{1,2}),\quad \omega_p = \omega_n\sqrt{1 - \delta^2},\qquad(2.88)$$

for the natural frequency ω_n, the damping ratio δ, and the damped natural frequency of the system ω_p as functions of the position of the poles, these characteristic parameters can be determined by

$$\delta = -\cos\frac{\pi}{\alpha},\quad \omega_n = A^{1/\alpha},\quad \omega_p = A^{1/\alpha}\sqrt{1 - \left(-\cos\frac{\pi}{\alpha}\right)^2} = A^{1/\alpha}\sin\frac{\pi}{\alpha}.\qquad(2.89)$$

Other alternative definitions can be found in [15, 16].

2.3.4.4 Frequency Response and Characteristic Parameters

To complete the characterization of the system as done for the integer-order ones, the frequency and the resonant peak can be determined. For this purpose, we will set $s = j\omega$ to obtain

$$F(j\omega) = \frac{A}{(j\omega)^\alpha + A} = \frac{A}{(\omega^\alpha \cos \alpha\pi/2 + A) + j\omega^\alpha \sin \alpha\pi/2}. \tag{2.90}$$

The magnitude of this function is given by

$$|F(j\omega)| = \frac{A}{\sqrt{\omega^{2\alpha} + 2A\omega^\alpha \cos \alpha\pi/2 + A^2}}, \tag{2.91}$$

having the maximum at

$$\omega^\alpha = -A \cos \alpha\frac{\pi}{2} \implies \omega_r = \left(-A \cos \alpha\frac{\pi}{2}\right)^{1/\alpha}, \quad \alpha > 1. \tag{2.92}$$

By substituting the equation obtained for the resonant frequency ω_r in the equation of the magnitude, the equation for the resonant peak is

$$M_r = \frac{1}{\sin \alpha\pi/2}. \tag{2.93}$$

As can be seen, the resonant peak, like the damping ratio, only depends on α. Figure 2.14 shows the magnitude of the frequency responses for $A = 1, \alpha = 1, 1.2, 1.5, 1.8$.

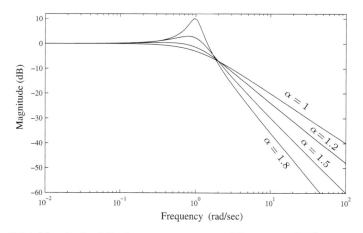

Figure 2.14 Magnitude of the frequency responses of the system (2.84)

2.4 Summary

The aim of this chapter has been to provide the reader with the essentials of input-output models (external representations) for fractional-order linear time invariant systems, as well as the dynamical properties (stability, time transient and steady-state responses, and frequency response) usually considered in classical control theory. With this aim, we have introduced two preliminary sections, the first devoted to the fundamental definitions of fractional-order operators in both time and Laplace domain, and the second to the analytical and numerical solutions of the fractional-order ordinary differential equations. As an introduction to the fractional-order control, a brief study of the so-called *Bode's ideal function* has been included. The necessary tools for using modern control theory (state-space or internal representations) will be given in the following chapter. Numerical implementations of the content of this chapter can be found in Chapter 13.

Chapter 3
State-space Representation and Analysis

The principles presented in Chapter 2 for studying fractional-order dynamic systems expressed in the input-output representation are extended in this chapter to the state-space representation. This chapter considers only systems that exhibit the following features, which are similar to those considered in the previous chapter: (1) systems are linear, (2) systems are time invariant, and (3) systems are of commensurate-order. We will show that this last property enables models and properties that are a quite straightforward generalization of well known results for integer-order LTI (linear time invariant) systems to be obtained. This chapter is divided into two parts: the first is devoted to the study of the representation and analysis of continuous systems, and the second studies discrete systems. Similar results are obtained for both kinds of systems in modeling, solution of the differential (or difference) equations, stability analysis, controllability, and observability.

3.1 Continuous-time LTI State-space Models

Consider a multivariable LTI system, that is, a system with multiple inputs and/or outputs. A state-space representation can be obtained for these systems of the form

$$\mathscr{D}^{\alpha} \boldsymbol{x} = \boldsymbol{A}\boldsymbol{x} + \boldsymbol{B}\boldsymbol{u} \tag{3.1}$$

$$\boldsymbol{y} = \boldsymbol{C}\boldsymbol{x} + \boldsymbol{D}\boldsymbol{u}, \tag{3.2}$$

where $\boldsymbol{\alpha} = [\alpha_1, \alpha_2, \cdots, \alpha_n]$, $\boldsymbol{u} \in \mathbb{R}^l$ is the input column vector, $\boldsymbol{x} \in \mathbb{R}^n$ is the state column vector, $\boldsymbol{y} \in \mathbb{R}^p$ is the output column vector, $\boldsymbol{A} \in \mathbb{R}^{n \times n}$ is the state matrix, $\boldsymbol{B} \in \mathbb{R}^{n \times l}$ is the input matrix, $\boldsymbol{C} \in \mathbb{R}^{p \times n}$ is the output matrix, and $\boldsymbol{D} \in \mathbb{R}^{p \times l}$ is the direct transmission matrix. The above notation establishes that fractional-order differentiation $\mathscr{D}^{\alpha_i}$ is applied only to the element x_i of state $\boldsymbol{x}$ in (3.1), where (3.1) is referred to as the *fractional-order state equation*, and (3.2) the *output equation*.

The above fractional-order model can be simplified in the particular case when $\alpha_i \equiv \alpha$, $1 \leqslant i \leqslant n$, and the state-space equation (3.1) becomes

$$\mathscr{D}^\alpha \boldsymbol{x} = \boldsymbol{A}\boldsymbol{x} + \boldsymbol{B}\boldsymbol{u}, \tag{3.3}$$

where now operator $\mathscr{D}^\alpha$ means that all the states are α-differentiated. Moreover, assume that $0 < \alpha \leqslant 1$.

Using Laplace transform in (3.3), taking into account the Caputo's definition for the fractional-order derivatives in (2.10), and applying property (2.18) in the case that $0 < \alpha \leqslant 1$, yields

$$s^\alpha \boldsymbol{X}(s) - s^{\alpha-1}\boldsymbol{x}(0) = \boldsymbol{A}\boldsymbol{X}(s) + \boldsymbol{B}\boldsymbol{U}(s) \implies$$
$$\boldsymbol{X}(s) = (s^\alpha \boldsymbol{I} - \boldsymbol{A})^{-1}\boldsymbol{B}\boldsymbol{U}(s) + (s^\alpha \boldsymbol{I} - \boldsymbol{A})^{-1}s^{\alpha-1}\boldsymbol{x}(0), \tag{3.4}$$

$$\boldsymbol{Y}(s) = \boldsymbol{C}\boldsymbol{X}(s) + \boldsymbol{D}\boldsymbol{U}(s). \tag{3.5}$$

Note that Caputo's definition is needed if we want initial conditions to be expressed directly as the values of the states at $t = 0$. In the case of null initial conditions ($\boldsymbol{x}(0) = \boldsymbol{0}$), (3.4) becomes

$$\boldsymbol{X}(s) = (s^\alpha \boldsymbol{I} - \boldsymbol{A})^{-1}\boldsymbol{B}\boldsymbol{U}(s), \tag{3.6}$$

and combining this with (3.5) yields

$$\boldsymbol{Y}(s) = \boldsymbol{G}(s)\boldsymbol{U}(s), \quad \boldsymbol{G}(s) = \boldsymbol{C}(s^\alpha \boldsymbol{I} - \boldsymbol{A})^{-1}\boldsymbol{B} + \boldsymbol{D}, \tag{3.7}$$

where $\boldsymbol{I}$ is the identity matrix of dimension $n \times n$, and matrix $\boldsymbol{G}(s)$ has p rows and l columns. $\boldsymbol{G}(s)$ represents a transfer function matrix whose numerator and denominator polynomials are expressed in terms of integer powers of s^α. Then as a consequence of having imposed that $\alpha_i = \alpha$, the dynamics has become that of a commensurate-order system.

3.1.1 Stability Analysis

Elements of $\boldsymbol{G}(s)$ are transfer functions with a common polynomial denominator given by

$$\phi(s^\alpha) = \det(s^\alpha \boldsymbol{I} - \boldsymbol{A}). \tag{3.8}$$

Let us denote as p_i the poles of system (3.7). They are defined as the solution of the equation $\phi(s^\alpha) = 0$. Then it is easily derived from (3.8) that system poles can be obtained from

$$p_i = \lambda_i^{1/\alpha}, \tag{3.9}$$

where λ_i, $1 \leqslant i \leqslant n$ are the eigenvalues of matrix $\boldsymbol{A}$.

The stability condition in the BIBO sense is attained if the poles of the system lie in the negative half-plane of the s-complex plane ($|\arg(p_i)| > \pi/2$). Then, taking into account (3.9), and the considerations given in Section 2.3.2, the stability condition for commensurate-order systems is obtained [14]:

$$|\arg(\lambda_i)| > \alpha\frac{\pi}{2}, \quad 1 \leqslant i \leqslant n. \tag{3.10}$$

In the case of rational commensurate-order systems ($\alpha = 1/q$) this condition becomes

$$|\arg(\lambda_i)| > \frac{\pi}{2q}, \quad 1 \leqslant i \leqslant n, \tag{3.11}$$

and the maximum number of poles p_i is n.

3.1.2 State-space Realizations

The results presented from now on in this chapter are limited only to single-input single-output (SISO) systems ($l = p = 1$).

Consider again the commensurate-order transfer function defined in (2.54). For convenience, it is presented here:

$$G(s) = \frac{Y(s)}{U(s)} = \frac{\sum\limits_{k=0}^{m} b_k(s^\alpha)^k}{\sum\limits_{k=0}^{n} a_k(s^\alpha)^k}, \quad a_n = 1, \quad m \leqslant n, \tag{3.12}$$

which can also be obtained from any state-space model of a SISO fractional-order system by using (3.7).

Associated with this transfer function, three canonical state-space representations can be proposed, which are similar to the classical ones developed for integer-order differential equation systems.

3.1.2.1 Controllable Canonical Form

Defining the first state in terms of its Laplace transform as

$$X_1(s) = \frac{1}{\sum\limits_{k=0}^{n} a_k(s^\alpha)^k}U(s), \tag{3.13}$$

and the remaining elements of the state vector in a recursive way from this one as $x_{i\,|\,1} = \mathscr{D}^\alpha x_l, i = 1, 2, \cdots, n-1$, the state representation, expressed in the controllable canonical form, is given by the matrix equations

$$
\begin{bmatrix} \mathscr{D}^{\alpha}x_1 \\ \mathscr{D}^{\alpha}x_2 \\ \vdots \\ \mathscr{D}^{\alpha}x_{n-2} \\ \mathscr{D}^{\alpha}x_{n-1} \\ \mathscr{D}^{\alpha}x_n \end{bmatrix} = \begin{bmatrix} 0 & 1 & 0 & \cdots & 0 & 0 \\ 0 & 0 & 1 & \cdots & 0 & 0 \\ \vdots & \vdots & \vdots & \ddots & \vdots & \vdots \\ 0 & 0 & 0 & \cdots & 1 & 0 \\ 0 & 0 & 0 & \cdots & 0 & 1 \\ -a_0 & -a_1 & -a_2 & \cdots & -a_{n-2} & -a_{n-1} \end{bmatrix} \begin{bmatrix} x_1 \\ x_2 \\ \vdots \\ x_{n-2} \\ x_{n-1} \\ x_n \end{bmatrix} + \begin{bmatrix} 0 \\ 0 \\ \vdots \\ 0 \\ 0 \\ 1 \end{bmatrix} u, \quad (3.14)
$$

$$
y = [b_0 - b_n a_0, \; b_1 - b_n a_1, \; \cdots, \; b_{n-1} - b_n a_{n-1}] \begin{bmatrix} x_1 \\ x_2 \\ \vdots \\ x_n \end{bmatrix} + b_n u, \qquad (3.15)
$$

where $b_i = 0$, for $m < i \leqslant n$.

3.1.2.2 Observable Canonical Form

From (3.12), collecting in the left hand side of the equation the terms that contain the powers of s and in the right hand side the terms that do not, we have

$$
s^{\alpha} \underbrace{\left[((s^{\alpha})^{n-1} + a_{n-1}(s^{\alpha})^{n-2} + \cdots + a_1)Y(s) - (b_n(s^{\alpha})^{n-1} + \cdots + b_1)U(s) \right]}_{X_1(s)}
$$
$$
= b_0 U(s) - a_0 Y(s),
$$
$$(3.16)$$

and, from the definition of x_1 given in the above expression, the first state equation and the equation used to define the following states are obtained:

$$
\mathscr{D}^{\alpha}x_1 = b_0 u - a_0 y,
$$
$$
X_1(s) = ((s^{\alpha})^{n-1} + \cdots + a_1)Y(s) - (b_n(s^{\alpha})^{n-1} + \cdots + b_1)U(s). \tag{3.17}
$$

Repeating in the second equation of (3.17) the rearrangement carried out in (3.16) gives

$$
s^{\alpha} \underbrace{\left[((s^{\alpha})^{n-2} + a_{n-1}(s^{\alpha})^{n-3} + \cdots + a_2)Y(s) - (b_n(s^{\alpha})^{n-2} + \cdots + b_2)U(s) \right]}_{X_2(s)}
$$
$$
= X_1(s) + b_1 U(s) - a_1 Y(s),
$$
$$(3.18)$$

and, from the definition of x_2 given in the above expression, the second state equation, and the equation used to define the following states are obtained:

$$
\mathscr{D}^{\alpha}x_2 = x_1 + b_1 u - a_1 y,
$$
$$
X_2(s) = ((s^{\alpha})^{n-2} + \cdots + a_2)Y(s) - (b_n(s^{\alpha})^{n-2} + \cdots + b_2)U(s). \tag{3.19}
$$

Repeating this procedure until x_n is defined, we get finally that

$$\mathscr{D}^\alpha x_n = x_{n-1} + b_{n-1}u - a_{n-1}y,$$
$$X_n(s) = Y(s) - b_n U(s). \tag{3.20}$$

Note that the last expression of (3.20) is the output equation.

Substituting y by the output equation in the first equations of (3.17), (3.19), (3.20), and rearranging the above fractional-order state equations, the observable canonical form follows in matrix form:

$$
\begin{bmatrix}
\mathscr{D}^\alpha x_1 \\
\mathscr{D}^\alpha x_2 \\
\mathscr{D}^\alpha x_3 \\
\vdots \\
\mathscr{D}^\alpha x_n \\
\mathscr{D}^\alpha x_n
\end{bmatrix}
=
\begin{bmatrix}
0 & 0 & \cdots & 0 & 0 & -a_0 \\
1 & 0 & \cdots & 0 & 0 & -a_1 \\
0 & 1 & \cdots & 0 & 0 & -a_2 \\
\vdots & \vdots & \ddots & \vdots & \vdots & \vdots \\
0 & 0 & \cdots & 1 & 0 & -a_{n-2} \\
0 & 0 & \cdots & 0 & 1 & -a_{n-1}
\end{bmatrix}
\begin{bmatrix}
x_1 \\
x_2 \\
x_3 \\
\vdots \\
x_{n-1} \\
x_n
\end{bmatrix}
+
\begin{bmatrix}
b_0 - b_n a_0 \\
b_1 - b_n a_1 \\
b_2 - b_n a_2 \\
\vdots \\
b_{n-2} - b_n a_{n-2} \\
b_{n-1} - b_n a_{n-1}
\end{bmatrix}
u, \tag{3.21}
$$

$$
y = [0, \cdots, 0, 1]
\begin{bmatrix}
x_1 \\
\vdots \\
x_{n-1} \\
x_n
\end{bmatrix}
+ b_n u, \tag{3.22}
$$

where $b_i = 0$, for $m < i \leqslant n$.

3.1.2.3 Modal Canonical Form

Carrying out the partial fraction expansion of (3.12) in functions of s^α, and assuming that the first pole λ_1 is of r multiplicity, and the remaining $n - r$ poles are of multiplicity 1, yields

$$
Y(s) = \Big[b_n + \frac{\rho_1}{(s^\alpha - \lambda_1)^r} + \cdots + \frac{\rho_{r-1}}{(s^\alpha - \lambda_1)^2} + \frac{\rho_r}{s^\alpha - \lambda_1} \\
+ \frac{\rho_{r+1}}{s^\alpha - \lambda_{r+1}} + \cdots \frac{\rho_n}{s^\alpha - \lambda_n} \Big] U(s). \tag{3.23}
$$

The state variables are defined as

$$X_1(s) = \frac{1}{(s^\alpha - \lambda_1)^r} U(s) = \frac{1}{s^\alpha - \lambda_1} X_2(s),$$

$$X_2(s) = \frac{1}{(s^\alpha - \lambda_1)^{r-1}} U(s) = \frac{1}{s^\alpha - \lambda_1} X_3(s),$$

$$\vdots$$

$$X_{r-1}(s) = \frac{1}{(s^\alpha - \lambda_1)^2} U(s) = \frac{1}{s^\alpha - \lambda_1} X_r(s),$$

$$X_r(s) = \frac{1}{s^\alpha - \lambda_1} U(s),$$

$$X_{r+1}(s) = \frac{1}{s^\alpha - \lambda_{r+1}} U(s),$$

$$\vdots \tag{3.24}$$

$$X_n(s) = \frac{1}{s^\alpha - \lambda_n} U(s).$$

In the above set of equations, the equations of the first $(r-1)$ states yield that $\mathscr{D}^\alpha x_i = \lambda_1 x_i + x_{i+1}$, $1 \leqslant i < r$, and the equations of the remaining states yield that $\mathscr{D}^\alpha x_i = \lambda_i x_i + u$, $r \leqslant i \leqslant n$, (where it is assumed $\lambda_r = \lambda_1$). Taking into account these relations and (3.23), the modal canonical form follows, given by the matrix equations

$$\begin{bmatrix} \mathscr{D}^\alpha x_1 \\ \mathscr{D}^\alpha x_2 \\ \vdots \\ \mathscr{D}^\alpha x_r \\ \mathscr{D}^\alpha x_{r+1} \\ \vdots \\ \mathscr{D}^\alpha x_n \end{bmatrix} = \begin{bmatrix} \lambda_1 & 1 & \cdots & 0 & 0 & \cdots & 0 \\ 0 & \lambda_1 & \cdots & 0 & 0 & \cdots & 0 \\ \vdots & \vdots & \ddots & \vdots & \vdots & \ddots & \vdots \\ 0 & 0 & \cdots & \lambda_1 & 0 & \cdots & 0 \\ 0 & 0 & \cdots & 0 & \lambda_{r+1} & \cdots & 0 \\ \vdots & \vdots & \ddots & \vdots & \vdots & \ddots & \vdots \\ 0 & 0 & \cdots & 0 & 0 & \cdots & \lambda_n \end{bmatrix} \begin{bmatrix} x_1 \\ x_2 \\ \vdots \\ x_r \\ x_{r+1} \\ \vdots \\ x_n \end{bmatrix} + \begin{bmatrix} 0 \\ 0 \\ \vdots \\ 1 \\ 1 \\ \vdots \\ 1 \end{bmatrix} u, \tag{3.25}$$

$$y = \begin{bmatrix} \rho_1, & \rho_2, & \cdots, & \rho_r, & \rho_{r+1}, & \cdots, & \rho_n \end{bmatrix} \begin{bmatrix} x_1 \\ \vdots \\ x_n \end{bmatrix} + b_n u, \tag{3.26}$$

where $b_n = 0$ if $m < n$.

Example 1 We examine the dynamic model of an immersed plate.
Consider a thin rigid plate of mass M and area S immersed in a Newtonian fluid of infinite extent and connected by a massless spring of stiffness K to a fixed point as shown in Figure 3.1. A force $f(t)$ is applied to the plate. It is assumed that the spring does not disturb the fluid and that the area of

the plate is sufficiently large to produce in the fluid adjacent to the plate the velocity $v(t, z)$ and stresses $\sigma(t, z)$ described by [19], which are represented by the time fractional-order derivative relation

$$\sigma(t, z) = \sqrt{\mu\rho}\mathscr{D}^{0.5}v(t, z), \tag{3.27}$$

where z is the distance of a point in the fluid to the submerged plate. Assuming that the plate-fluid system is initially in an equilibrium state, and displacement and velocities are initially zero, the dynamics of the system of Figure 3.1 is given by

$$M\mathscr{D}^2y(t) = f(t) - Ky(t) - 2S\sigma(t, 0). \tag{3.28}$$

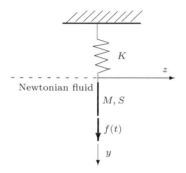

Figure 3.1 Thin rigid plate immersed in a Newtonian fluid

Substituting (3.27) in (3.28) and taking into account that $v(t, 0) = \mathscr{D}^1y(t)$, it follows that [20]

$$A_B\mathscr{D}^2y(t) + B_B\mathscr{D}^{1.5}y(t) + C_By(t) = f(t), \tag{3.29}$$

where $A_B = M$, $B_B = 2S\sqrt{\mu\rho}$, $C_B = K$, and the equilibrium initial state of the system is

$$y(0) = 0, \quad \dot{y}(0) = 0. \tag{3.30}$$

In the proposed example the following values are given (parameter values were taken from [3]):

$$A_B = 1, \quad B_B = 0.5, \quad C_B = 0.5, \quad f(t) = \begin{cases} 1, & 0 \leqslant t \leqslant 1, \\ 0, & t > 1. \end{cases} \tag{3.31}$$

This system is clearly of commensurate-order with $\alpha = 0.5$. Moreover it is of rational-order with $q = 2$. Taking into account that the maximum order of this differential equation is $n_d = 2$, then the number of states to be considered is $n = n_d q = 4$, the Laplace transform of (3.29) is

$$G(s) = \frac{1}{(s^{0.5})^4 + 0.5(s^{0.5})^3 + 0.5},\tag{3.32}$$

and the fractional-order state-space models are as follows:

1. Controllable canonical form given by (3.14), (3.15):

$$\mathscr{D}^{0.5}\boldsymbol{x} = \begin{bmatrix} 0 & 1 & 0 & 0 \\ 0 & 0 & 1 & 0 \\ 0 & 0 & 0 & 1 \\ -0.5 & 0 & 0 & -0.5 \end{bmatrix} \boldsymbol{x} + \begin{bmatrix} 0 \\ 0 \\ 0 \\ 1 \end{bmatrix} f,\tag{3.33}$$

$$y = \begin{bmatrix} 1 & 0 & 0 & 0 \end{bmatrix} \boldsymbol{x}.\tag{3.34}$$

2. Observable canonical form given by (3.21), (3.22):

$$\mathscr{D}^{0.5}\boldsymbol{x} = \begin{bmatrix} 0 & 0 & 0 & -0.5 \\ 1 & 0 & 0 & 0 \\ 0 & 1 & 0 & 0 \\ 0 & 0 & 1 & -0.5 \end{bmatrix} \boldsymbol{x} + \begin{bmatrix} 1 \\ 0 \\ 0 \\ 0 \end{bmatrix} f,\tag{3.35}$$

$$y = \begin{bmatrix} 0 & 0 & 0 & 1 \end{bmatrix} \boldsymbol{x}.\tag{3.36}$$

3. Modal canonical form given by (3.25), (3.26).

First the roots of the characteristic equation $\phi(\lambda) = 0$, $\lambda = s^{0.5}$ of (3.32) are calculated: $\lambda_{1,2} = -0.7388 \pm j0.5688$, $\lambda_{3,4} = 0.4888 \pm j0.5799$, and their respective residues: $\rho_{1,2} = 0.2882 \mp j0.3136$, $\rho_{3,4} = -0.2882 \mp j0.3025$. Then the matrix equations are obtained:

$$\mathscr{D}^{0.5}\boldsymbol{x} = \begin{bmatrix} \lambda_1 & 0 & 0 & 0 \\ 0 & \lambda_1^* & 0 & 0 \\ 0 & 0 & \lambda_3 & 0 \\ 0 & 0 & 0 & \lambda_3^* \end{bmatrix} \boldsymbol{x} + \begin{bmatrix} 1 \\ 1 \\ 1 \\ 1 \end{bmatrix} f,\tag{3.37}$$

$$y = \begin{bmatrix} \rho_1 & \rho_1^* & \rho_3 & \rho_3^* \end{bmatrix} \boldsymbol{x},\tag{3.38}$$

where an upper $(*)$ denotes complex conjugate.

Finally, the stability of this system is studied. The phase of the roots λ_i are: $|\arg(\lambda_{1,2})| = 2.4855\,\mathrm{rad}$, $|\arg(\lambda_{3,4})| = 0.8704\,\mathrm{rad}$. As all these values are larger than $\alpha\pi/2 = 0.7854\,\mathrm{rad}$, stability condition (3.10) is always satisfied and the system of this example is stable.   □

3.2 Solution of the State Equation of Continuous LTI Commensurate-order Systems

Given a SISO system of commensurate-order, its states can be obtained from the inverse Laplace transform of (3.4):

$$x(t) = \mathscr{L}^{-1}\left\{X(s)\right\} = \mathscr{L}^{-1}\left\{(s^{\alpha}I - A)^{-1}BU(s) + (s^{\alpha}I - A)^{-1}s^{\alpha-1}x(0)\right\}.$$
(3.39)

Define $\hat{\boldsymbol{\Phi}}(t) = \mathscr{L}^{-1}\left\{(s^{\alpha}I - A)^{-1}\right\}$ and $\boldsymbol{\Phi}(t) = \mathscr{L}^{-1}\left\{(s^{\alpha}I - A)^{-1}s^{\alpha-1}\right\}$, both for $t \geqslant 0$, so that $\boldsymbol{\Phi}(t) = \hat{\boldsymbol{\Phi}}(t) * \chi_{\alpha-1}(t)$ being (see Table A.2)

$$\chi_{\alpha-1}(t) = \mathscr{L}^{-1}\left(s^{\alpha-1}\right) = \begin{cases} \dfrac{t^{-\alpha}}{\Gamma(1-\alpha)}, & \alpha < 1, \\ \delta(t), & \alpha = 1, \end{cases}$$
(3.40)

where $\delta(t)$ is the Dirac's delta function. Applying the property of the Laplace transform related to the convolution product, the following expression is obtained:

$$x(t) = \boldsymbol{\Phi}(t)x(0) + \hat{\boldsymbol{\Phi}}(t) * [Bu(t)]$$

$$= \boldsymbol{\Phi}(t)x(0) + \int_0^t \hat{\boldsymbol{\Phi}}(t-\tau)Bu(\tau)\mathrm{d}\tau, \quad \alpha \leqslant 1.$$
(3.41)

As can be seen in (3.41), $\boldsymbol{\Phi}(t)$ is the matrix usually known as the state transition matrix.

Using a procedure similar to that used for linear systems of integer-order, or as proposed in [20], the form of the state transition matrix can be determined. For that purpose the following expression will be used:

$$_C\mathscr{D}^{\alpha}x(t) = Ax(t), \quad x(0) = x_0,$$
(3.42)

where we recall that the Caputo's definition is being used, and it is being made explicit in the notation of the differential operator. Then we assume that the solution is of the form

$$x(t) = A_0 + A_1 t^{\alpha} + A_2 t^{2\alpha} + \cdots + A_k t^{k\alpha} + \cdots.$$
(3.43)

At the initial instant $t = 0$, (3.43) yields

$$x(0) = A_0.$$

Performing the Caputo's α-fractional-order differentiation of (3.43), and using the property [21]

$$_C\mathscr{D}^{\alpha}t^{\gamma} = \frac{\Gamma(\gamma+1)}{\Gamma(\gamma+1-\alpha)}t^{\gamma-\alpha},$$
(3.44)

it is found that

$$
{}_C\mathscr{D}^{\alpha}\boldsymbol{x}(t) = \boldsymbol{0} + \boldsymbol{A}_1\Gamma\left(1+\alpha\right) + \frac{\boldsymbol{A}_2\Gamma\left(1+2\alpha\right)}{\Gamma\left(1+\alpha\right)}t^{\alpha} + \cdots
$$
$$
+ \frac{\boldsymbol{A}_k\Gamma\left(1+k\alpha\right)}{\Gamma\left(1+(k-1)\alpha\right)}t^{(k-1)\alpha} + \cdots = \boldsymbol{A}\boldsymbol{x}(t). \tag{3.45}
$$

Setting $t = 0$ in (3.45) it follows that

$$
\boldsymbol{A}_1 = \frac{\boldsymbol{A}\boldsymbol{x}(0)}{\Gamma\left(1+\alpha\right)}.
$$

Applying successive derivatives of order α to (3.42), and carrying out the following iterative procedure:

$$
{}_C\mathscr{D}^{k\alpha}\boldsymbol{x}(t) = \boldsymbol{A}_k\Gamma\left(1+k\alpha\right) = {}_C\mathscr{D}^{\alpha}({}_C\mathscr{D}^{(k-1)\alpha}\boldsymbol{x}(t)) = \boldsymbol{A}(\boldsymbol{A}^{k-1}\boldsymbol{x}(t)), \tag{3.46}
$$

matrixes of coefficients $\boldsymbol{A}_k$ can be obtained:

$$
\boldsymbol{A}_k = \frac{\boldsymbol{A}^k\boldsymbol{x}(0)}{\Gamma\left(1+k\alpha\right)},
$$

and the solution (3.43) is given by

$$
\boldsymbol{x}(t) = \boldsymbol{x}(\boldsymbol{0}) + \frac{\boldsymbol{A}\boldsymbol{x}(0)}{\Gamma\left(1+\alpha\right)}t^{\alpha} + \frac{\boldsymbol{A}^2\boldsymbol{x}(0)}{\Gamma\left(1+2\alpha\right)}t^{2\alpha} + \cdots + \frac{\boldsymbol{A}^k\boldsymbol{x}(0)}{\Gamma\left(1+k\alpha\right)}t^{k\alpha} + \cdots
$$
$$
= \left(\sum_{k=0}^{\infty}\frac{\boldsymbol{A}^k t^{k\alpha}}{\Gamma\left(1+k\alpha\right)}\right)\boldsymbol{x}(0) = \mathscr{E}_{\alpha,1}\left(\boldsymbol{A}t^{\alpha}\right)\boldsymbol{x}(0) = \boldsymbol{\Phi}(t)\boldsymbol{x}(0). \tag{3.47}
$$

It is clear that the Mittag–Leffler function here performs the same role as that performed by the exponential function for the integer-order systems. The well known exponential matrix, $e^{\boldsymbol{A}t}$ is just a particular case of the generalized exponential matrix, $\mathscr{E}_{\alpha,1}\left(\boldsymbol{A}t^{\alpha}\right)$, which can be called *Mittag–Leffler matrix function*.

Moreover, two important remarks follow on the previous results.

Remark 3.1. As the fractional-order derivative of $\boldsymbol{x}$ depends on the "history" of $\boldsymbol{x}$ from the lower limit of the integral that defines this operator (usually, we made this limit equal to 0 in this chapter, only for simplification purposes) until the present instant, the state defined in this chapter is not a state in the classical sense of integer derivative models. In standard state models the state at instant t includes all the information needed to calculate the future behavior of the system provided that future inputs were known. In a fractional-order model it is not true as the fractional-order derivative has "memory" from t up to the lower limit that defines the fractional-order operator. Then, in order to determine the future behavior of a fractional-order system, not only is the value of the state at instant t needed but also all the values of the state in the interval $[0, t]$. This condition is reduced to the knowledge of only $\boldsymbol{x}(t)$ in the case of integer-order derivatives because the fractional-order differentiation operator collapses into a local operator that only depends on the value of the state in the immediate vecinity of t. *Then we should denote $\boldsymbol{x}(t)$ as a pseudo-state. However, we call this vector a state throughout the remainder of this chapter only for lexical simplifying purposes.*   □

Remark 3.2. Matrix $\boldsymbol{\Phi}(t)$ in (3.41) is not a state transition matrix in the usual sense. It really means $\boldsymbol{\Phi}(t,0)$, the second argument being the lower limit of the integral that defines the fractional-order derivative used in the state model. Then this argument is fixed and cannot be changed to any τ. For the same reason, the semi-group property verified in standard state transition matrices $(\boldsymbol{\Phi}(t,t_0) = \boldsymbol{\Phi}(t,\tau)\boldsymbol{\Phi}(\tau,t_0),\ t_0 < \tau < t)$ is not satisfied here. *We denote $\boldsymbol{\Phi}(t)$ as the state pseudo-transition matrix throughout the remainder of this chapter.*   □

In the following, three methods are proposed to calculate the state pseudo-transition matrix, which are the generalizations of the well known methods used for integer-order LTI systems.

It will be assumed that matrix $\boldsymbol{A}$ has n distinct eigenvalues λ_i, $1 \leqslant i \leqslant n$. The case of multiple eigenvalues can also be considered using the same modifications reported in the state methods literature for these three methods in integer-order LTI systems.

3.2.1 Inverse Laplace Transform Method

This method is based on computing the expression

$$\boldsymbol{\Phi}(t) = \mathscr{L}^{-1}\left\{ (s^\alpha \boldsymbol{I} - \boldsymbol{A})^{-1} s^{\alpha-1} \right\}, \quad t \geqslant 0. \tag{3.48}$$

The procedure is as follows:

1. Operate $(s^\alpha \boldsymbol{I} - \boldsymbol{A})^{-1}$. This yields a matrix of transfer functions which are rational functions in s^α. Each element of this matrix is denoted as $\hat{\phi}_{i_1,i_2}(s^\alpha)$, $1 \leqslant i_1, i_2 \leqslant n$.
2. Find the partial fraction expansion of each $\hat{\phi}_{i_1,i_2}(s^\alpha)$ in terms of s^α:

$$\hat{\phi}_{i_1,i_2}(s) = \sum_{i_3=1}^{n} \frac{r(i_1,i_2,i_3)}{s^\alpha - \lambda_{i_3}}. \tag{3.49}$$

3. Carry out the inverse Laplace transform of (3.49) multiplied by $s^{\alpha-1}$:

$$\phi_{i_1,i_2}(t) = \mathscr{L}^{-1}\left\{ \sum_{i_3=1}^{n} \frac{r(i_1,i_2,i_3)s^{\alpha-1}}{s^\alpha - \lambda_{i_3}} \right\}, \tag{3.50}$$

where the partial fraction expansion of the previous step has been used. Note that inverse Laplace transforms of the terms of this expansion are the Mittag–Leffler functions $r(i_1, i_2, i_3)\mathscr{E}_{\alpha,1}(\lambda_{i_3}t^\alpha)$, which can be obtained using the numerical definition of these functions given in (2.28), (2.32).

Repeating the second and third steps of this procedure for all the elements of matrix (3.48) yields the state pseudo-transition matrix $\boldsymbol{\Phi}(t)$.

3.2.2 Jordan Matrix Decomposition Method

Performing the Jordan matrix decomposition $\boldsymbol{A} = \boldsymbol{V}\boldsymbol{\Lambda}\boldsymbol{V}^{-1}$, where $\boldsymbol{\Lambda} = \text{diag}\,(\lambda_1, \lambda_2, \cdots \lambda_n)$, and substituting this in (3.47), it follows that

$$\boldsymbol{\Phi}(t) = \boldsymbol{V}\left(\sum_{k=0}^{\infty}\frac{\boldsymbol{\Lambda}^k t^{k\alpha}}{\Gamma\,(1+k\alpha)}\right)\boldsymbol{V}^{-1}, \qquad (3.51)$$

and using the definition of the scalar Mittag–Leffler function it is found that

$$\boldsymbol{\Phi}(t) = \boldsymbol{V}\,\text{diag}\,(\mathscr{E}_{\alpha,1}(\lambda_1 t^\alpha), \mathscr{E}_{\alpha,1}(\lambda_2 t^\alpha), \cdots, \mathscr{E}_{\alpha,1}(\lambda_n t^\alpha))\,\boldsymbol{V}^{-1}. \qquad (3.52)$$

In this expression, functions $\mathscr{E}_{\alpha,1}(\lambda_i t^\alpha)$ have to be calculated numerically, similar to the previous method.

3.2.3 Cayley–Hamilton Method

This method is based on the well known Cayley–Hamilton Theorem which states that a matrix $\boldsymbol{A}$ satisfies its own characteristic equation (3.8):

$$\phi(\boldsymbol{A}) = \boldsymbol{0}. \qquad (3.53)$$

Given an arbitrary function $f(\boldsymbol{A}, t)$, that can be expanded in a Taylor's matrix series, and applying (3.53) reduces this infinite series to a polynomial of order $n-1$ of the matrix $\boldsymbol{A}$ with time varying coefficients.

Taking into account that $\phi(\lambda_i) = 0$ is also verified (λ_i being the eigenvalues of $\boldsymbol{A}$), it follows that $f(\lambda_i, t)$ can also be expressed as a polynomial of order $n-1$ of λ_i with the same time varying coefficients as before.

All these results can be expressed in a compact form by the Sylvester's interpolation formula [22], which states that

$$
\begin{vmatrix}
1 & \lambda_1 & \lambda_1^2 & \cdots & \lambda_1^{n-1} & f(\lambda_1, t) \\
1 & \lambda_2 & \lambda_2^2 & \cdots & \lambda_2^{n-1} & f(\lambda_2, t) \\
\cdot & & & & & \\
\cdot & & & & & \\
1 & \lambda_n & \lambda_n^2 & \cdots & \lambda_n^{n-1} & f(\lambda_n, t) \\
\boldsymbol{I} & \boldsymbol{A} & \boldsymbol{A}^2 & & \boldsymbol{A}^{n-1} & f(\boldsymbol{A}, t)
\end{vmatrix} = \boldsymbol{0}
$$

in the case of distinct eigenvalues (a version of this formula also exists for the multiple eigenvalue case). Solving the previous equation for $f(\boldsymbol{A}, t)$ yields

$$
f(\boldsymbol{A}, t) = a_0(t) \boldsymbol{I} + a_1(t) \boldsymbol{A} + \cdots + a_{n-1}(t) \boldsymbol{A}^{n-1}, \qquad (3.54)
$$

where coefficients a_i are to be determined. Solving for the other n equations associated with the eigenvalues of $\boldsymbol{A}$, a system of equations is obtained that allows for calculating these n functions $a_i(t), (i = 0, 1, \cdots, n-1)$:

$$
\begin{bmatrix}
f(\lambda_1, t) \\
f(\lambda_2, t) \\
\vdots \\
f(\lambda_n, t)
\end{bmatrix}
=
\underbrace{
\begin{bmatrix}
1 & \lambda_1 & \cdots & \lambda_1^{n-1} \\
1 & \lambda_2 & \cdots & \lambda_2^{n-1} \\
\vdots & \vdots & \ddots & \vdots \\
1 & \lambda_n & \cdots & \lambda_n^{n-1}
\end{bmatrix}
}_{H}
\begin{bmatrix}
a_0(t) \\
a_1(t) \\
\vdots \\
a_{n-1}(t)
\end{bmatrix}.
\qquad (3.55)
$$

Functions $a_i(t)$ could be calculated from this expression if matrix $\boldsymbol{H}$ were invertible. As it is a Vandermonde matrix, it would be full-rank if all the λ_i were distinct [23], which is the case, and, then, functions $a_i(t)$ can be determined from

$$
\begin{bmatrix}
a_0(t) \\
a_1(t) \\
\vdots \\
a_{n-1}(t)
\end{bmatrix}
= \boldsymbol{H}^{-1}
\begin{bmatrix}
f(\lambda_1, t) \\
f(\lambda_2, t) \\
\vdots \\
f(\lambda_n, t)
\end{bmatrix}.
\qquad (3.56)
$$

Defining $f(\boldsymbol{A}, t) = \boldsymbol{\Phi}(t) = \mathscr{E}_{\alpha,1}(\boldsymbol{A}t^\alpha)$ in (3.54), it is shown that

$$
\mathscr{E}_{\alpha,1}(\boldsymbol{A}t^\alpha) = \sum_{i=0}^{n-1} a_i(t) \boldsymbol{A}^i, \qquad (3.57)
$$

where functions $a_i(t)$ can be obtained from (3.56) making $f(\lambda_i, t) = \mathscr{E}_{\alpha,1}(\lambda_i t^\alpha)$, $1 \leqslant i \leqslant n$. And the state pseudo-transition matrix is then obtained.

Next some lemmas will be proposed about the expansion of $\mathscr{E}_{\alpha,1}$ shown in (3.57), that will be used later to obtain the controllability and observability conditions.

Lemma 3.3. The set of fractional power functions $\{1, t^\alpha, t^{2\alpha}, \cdots t^{(k-1)\alpha}\}$, where $k \in N^+$, $t \geqslant 0$, and t^α accounts for the real positive root of that function, is linearly independent.

Proof. It is demonstrated by contradiction. Assume that these k functions are linearly dependent. Then there should exist a set of real numbers $\{v_1, v_2, \cdots, v_i, \cdots, v_k\}$ not all trivially null such that

$$\sum_{i=1}^{k} v_i t^{(i-1)\alpha} = 0, \tag{3.58}$$

where t^α means the real positive root of this fractional power function. Choosing a set of increasing time instants $\{t_1, t_2, t_3, \cdots, t_k\}$, condition (3.58) must be verified for all the elements of this set. This condition can be expressed in matrix form as

$$\begin{bmatrix} 1 & t_1^\alpha & (t_1^\alpha)^2 & \cdots & (t_1^\alpha)^{k-1} \\ 1 & t_2^\alpha & (t_2^\alpha)^2 & \cdots & (t_2^\alpha)^{k-1} \\ 1 & t_3^\alpha & (t_3^\alpha)^2 & \cdots & (t_3^\alpha)^{k-1} \\ \vdots & \vdots & \vdots & \ddots & \vdots \\ 1 & t_k^\alpha & (t_k^\alpha)^2 & \cdots & (t_k^\alpha)^{k-1} \end{bmatrix} \begin{bmatrix} v_1 \\ v_2 \\ \vdots \\ v_k \end{bmatrix} = \mathbf{0}. \tag{3.59}$$

The coefficient matrix of (3.59) is a Vandermonde matrix which is full-rank if $t_{i_1}^\alpha \neq t_{i_2}^\alpha, 1 \leqslant i_1, i_2 \leqslant k$. Taking into account that t^α is an increasing function if $\alpha > 0$, this means that as all t_i in the previous set are different, then all the t_i^α will also be different, the Vandermonde matrix will be full-rank and invertible, and this will imply that column vector containing the scalars v_i in (3.59) will be zero, in contradiction with the assumption. Then the lemma is proven. $\square$

Lemma 3.4. Given any arbitrary set $\{z_1, z_2, \cdots z_i, \cdots z_n\}$ where z_i ($1 \leqslant i \leqslant n$) are distinct complex numbers, and n is the order of matrix $\boldsymbol{A}$, then the set of functions $\{\mathscr{E}_{\alpha,1}(z_1 t^\alpha), \mathscr{E}_{\alpha,1}(z_2 t^\alpha), \cdots \mathscr{E}_{\alpha,1}(z_i t^\alpha), \cdots, \mathscr{E}_{\alpha,1}(z_n t^\alpha)\}$ is linearly independent.

Proof. It is demonstrated by contradiction. Assume that these functions are linearly dependent. Then there should exist a set of complex numbers $\{v_1, v_2, \cdots, v_i, \cdots, v_n\}$ not all trivially null such that

$$\sum_{i=1}^{n} v_i \mathscr{E}_{\alpha,1}(z_i t^\alpha) = 0. \tag{3.60}$$

Taking into account the definition of the Mittag–Leffler function in (2.28), (2.32), (3.60) becomes

$$\sum_{k=0}^{\infty} \frac{t^{\alpha k}}{\Gamma(1 + k\alpha)} \sum_{i=1}^{n} v_i z_i^k = 0. \tag{3.61}$$

This expression is a summation of fractional powers in t; Lemma 3.3 states that these fractional power functions are linearly independent and, then, the only possibility of this being zero for any $t \in [0, \infty)$ is that all the coefficients that multiply the powers of t in that series are zero. This implies that $\sum_{i=1}^{n} v_i z_i^k = 0$, $\forall k$. If we consider the first n terms of series (3.61), that condition yields $(k = 0, \cdots, n-1)$

$$
\begin{bmatrix}
1 & 1 & \cdots & 1 \\
z_1 & z_2 & \cdots & z_n \\
\vdots & \vdots & \ddots & \vdots \\
z_1^{n-1} & z_2^{n-1} & \cdots & z_n^{n-1}
\end{bmatrix}
\begin{bmatrix}
v_1 \\
v_2 \\
\vdots \\
v_n
\end{bmatrix}
= \mathbf{0}.
\tag{3.62}
$$

The coefficient matrix in (3.62) is the transpose of a Vandermonde matrix. Then if all the z_i were distinct this matrix would be full-rank and the only possibility that (3.62) be verified would be that all the $v_i = 0$, in contradiction with the initial assumption. □

Lemma 3.5. If all the eigenvalues of matrix $\mathbf{A}$, $\lambda_i, 1 \leqslant i \leqslant n$ are distinct, then functions $a_i(t), 0 \leqslant i \leqslant n-1$ of (3.57) are linearly independent.
Proof. It is also demonstrated by contradiction. Assume that these functions were linearly dependent. Then there should exist a vector of complex numbers $\mathbf{v} = [v_1, v_2, \cdots, v_i, \cdots, v_n]^T$ not all trivially null such that

$$
\mathbf{v}^T
\begin{bmatrix}
a_0(t) \\
a_1(t) \\
\vdots \\
a_{n-1}(t)
\end{bmatrix}
= 0.
\tag{3.63}
$$

Equation 3.56 holds if all the eigenvalues λ_i were distinct, which is the case. Then substituting this expression into (3.63), and taking into account that $f(\lambda_i, t) = \mathscr{E}_{\alpha,1}(\lambda_i t^\alpha)$, it is obtained that

$$
\mathbf{v}^T \mathbf{H}^{-1}
\begin{bmatrix}
\mathscr{E}_{\alpha,1}(\lambda_1, t) \\
\mathscr{E}_{\alpha,1}(\lambda_2, t) \\
\vdots \\
\mathscr{E}_{\alpha,1}(\lambda_n, t)
\end{bmatrix}
= 0.
\tag{3.64}
$$

As the product of $\mathbf{v}^T \mathbf{H}^{-1}$ is a constant complex row vector, (3.64) implies that the n functions $\mathscr{E}_{\alpha,1}(\lambda_i t^\alpha)$ are linearly dependent which is in contradiction with the statement of Lemma 3.4 (just make $z_i = \lambda_i$ in that lemma). □

Next an example is developed which illustrates the general solution and applies one of the proposed methods to calculate the state pseudo-transition matrix.

Example 2 Consider again the example of the previous section. It is desired to study the response of the plate when the force given by (3.31) is applied at the bottom of the plate. It is also assumed that initially the position is $0.1\,\mathrm{m}$ and the velocity $-0.01\,\mathrm{m/sec}$.

We are interested in monitoring the derivatives of the output (position). Looking at the output equation of the controllable canonical form (3.34), the output coincides with the first state x_1 in this representation, the other states being the successive fractional-order derivatives of the output. Moreover the initial state in this representation is straightforward: $\boldsymbol{x}^c(0) = [0.1, 0, -0.01, 0]^{\mathrm{T}}$, when the third state is the velocity and the initial states of x_2 and x_4 are supposed to be zero. Therefore the controllable canonical form is adequate for this problem, and the state vector is denoted $\boldsymbol{x}^c(t)$ in this representation.

However the calculation of the Mittag–Leffler matrix function (3.47) in this representation is not easy. For the sake of simplicity, the modal canonical form (3.37), (3.38) will be used, and then the obtained states will be transformed to the controllable canonical form. It has the advantage of a diagonal matrix $\boldsymbol{A}$, which allows the use of (3.52) of the second method proposed to obtain $\boldsymbol{\Phi}(t)$, but has the drawback of having to handle complex numbers.

We need to determine the matrix transformation of states between the controllable and modal canonical forms. Then the Jordan decomposition of the matrix $\boldsymbol{A}$, expressed in the controllable canonical form (see (3.33)), is carried out first:

$$\begin{bmatrix} 0 & 1 & 0 & 0 \\ 0 & 0 & 1 & 0 \\ 0 & 0 & 0 & 1 \\ -0.5 & 0 & 0 & -0.5 \end{bmatrix} = \boldsymbol{V}_j \begin{bmatrix} \lambda_1 & 0 & 0 & 0 \\ 0 & \lambda_1^* & 0 & 0 \\ 0 & 0 & \lambda_3 & 0 \\ 0 & 0 & 0 & \lambda_3^* \end{bmatrix} \boldsymbol{V}_j^{-1}, \tag{3.65}$$

where λ_1 and λ_3 are the values given in (3.37). The transformation vector $\boldsymbol{V}$ is calculated combining the above diagonalization with the condition

$$\begin{bmatrix} 1, 1, 1, 1 \end{bmatrix}^{\mathrm{T}} = \boldsymbol{V}_j^{-1} \begin{bmatrix} 0, 0, 0, 1 \end{bmatrix}^{\mathrm{T}}, \tag{3.66}$$

needed to guarantee the transformation between matrices $\boldsymbol{B}$ in these canonical forms. Assume that a matrix $\boldsymbol{V}_j$ has been obtained that verifies (3.65), *e.g.*, using the `eig()` function of MATLAB™. Then any matrix $\boldsymbol{V} = \boldsymbol{V}_j\boldsymbol{\Pi}$, $\boldsymbol{\Pi}$ being an arbitrary diagonal matrix, would also verify (3.65). Therefore (3.66) is the condition that allows for the evaluation of the particular matrix that transforms the controllable into the modal canonical form. If the above matrix $\boldsymbol{V}$ is used in (3.66) instead of $\boldsymbol{V}_j$ it yields

$$\boldsymbol{\Pi} \begin{bmatrix} 1, 1, 1, 1 \end{bmatrix}^{\mathrm{T}} = \boldsymbol{V}_j^{-1} \begin{bmatrix} 0, 0, 0, 1 \end{bmatrix}^{\mathrm{T}}, \tag{3.67}$$

from which diagonal elements of matrix $\boldsymbol{\Pi}$ can easily be obtained:

$$[\Pi_{1,1}, \Pi_{2,2}, \Pi_{3,3}, \Pi_{4,4}]^T = V_j^{-1}[0, 0, 0, 1]^T, \qquad (3.68)$$

and the transform matrix is calculated:

$$V = V_j \boldsymbol{\Pi}$$

$$= \begin{bmatrix} 0.288-j0.314 & 0.288+j0.314 & -0.288-j0.303 & -0.288+j0.303 \\ -0.035+j0.396 & -0.035-j0.396 & 0.0345-j0.315 & 0.0345+j0.315 \\ -0.2-j0.312 & -0.2+j0.312 & 0.2-j0.134 & 0.2+j0.134 \\ 0.325+j0.117 & 0.325-j0.117 & 0.175+j0.05 & 0.175-j0.05 \end{bmatrix}.$$

Then the transformation between states of the controllable and modal canonical forms is given by [22]

$$\boldsymbol{x}^c(t) = V\boldsymbol{x}^j(t), \qquad (3.69)$$

being $\boldsymbol{x}^j(t)$ the state in the modal canonical form.

Next we determine the initial state in the modal canonical form. Thus it is found that

$$\boldsymbol{x}^j(0) = V^{-1}\boldsymbol{x}^c(0)$$

$$= 0.1 \times [0.45+j0.27, 0.45-j0.27, -0.524+j0.446, -0.524-j0.446]^T.$$

Free state evolution, which is given by (3.47), can be more easily calculated using the modal canonical form than using the other canonical forms. This is because the evaluation of the Mittag–Leffler matrix function from the series expansion given in that expression is transformed into the evaluation of two Mittag–Leffler scalar functions, according to (3.51) and (3.52). This yields

$$\boldsymbol{\Phi}^j(t) = \mathscr{E}_{\alpha,1}(At^\alpha) = \begin{bmatrix} \mathscr{E}_{\alpha,1}(\lambda_1 t^\alpha) & 0 & 0 & 0 \\ 0 & \mathscr{E}_{\alpha,1}^*(\lambda_1 t^\alpha) & 0 & 0 \\ 0 & 0 & \mathscr{E}_{\alpha,1}(\lambda_3 t^\alpha) & 0 \\ 0 & 0 & 0 & \mathscr{E}_{\alpha,1}^*(\lambda_3 t^\alpha) \end{bmatrix}, \qquad (3.70)$$

where $\boldsymbol{\Phi}^j$ is the pseudo-state transition matrix given in the modal canonical form. Note also that, from the series expansion definition of the Mittag–Leffler function given in Chapter 2, it easily follows that $\mathscr{E}_{\alpha,1}(\lambda^* t^\alpha) = \mathscr{E}_{\alpha,1}^*(\lambda t^\alpha)$, a result that has also been used in the evaluation of (3.70).

In this example $\alpha = 0.5$. This is a particular case of the Mittag–Leffler function which involves the well known *erfc function* [3]:

$$\mathscr{E}_{0.5,1}(z) = e^{z^2}\text{erfc}(-z), \quad \text{erfc}(z) = \frac{2}{\sqrt{\pi}}\int_z^\infty e^{-t^2}\,dt. \qquad (3.71)$$

Function $\mathscr{E}_{0.5,1}(z)$ could easily be calculated if argument z were real by using the erfc function. MATLAB includes `erfc(z)` as one of its basic functions.

But in our example the argument z is complex. In this case the Mittag–Leffler function can be calculated for z values of moderate absolute value by summing the series $\mathscr{E}_{0.5,1}(z) = \sum_{k=0}^{\infty} \dfrac{z^k}{\Gamma(1+0.5k)}$, which quickly converges. If the modulus of z were large then summation of the previous series would exhibit convergence problems. In this case the Mittag–Leffler function can be approximated by using some rational formulas [24].

States are expressed in the controllable canonical form using (3.69). In this representation, the state evolution in free motion from the initial state is given by

$$\boldsymbol{x}^c(t) = \boldsymbol{V}\operatorname{diag}\left(\mathscr{E}_{0.5,1}(\lambda_1 t^\alpha), \mathscr{E}^*_{0.5,1}(\lambda_1 t^\alpha), \mathscr{E}_{0.5,1}(\lambda_3 t^\alpha), \mathscr{E}^*_{0.5,1}(\lambda_3 t^\alpha)\right)\boldsymbol{x}^j(0), \tag{3.72}$$

and is plotted in Figure 3.2.

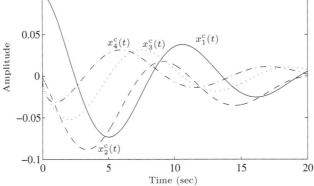

Figure 3.2 Free evolution of the controllable canonical form states of the immersed plate

Evolution of the system from a zero initial state as consequence of an external input can be obtained from (3.39), taking into account the structure of the modal canonical form matrices:

$$\boldsymbol{x}^c(t) = \boldsymbol{V}\mathscr{L}^{-1}\left\{\begin{bmatrix} \dfrac{1}{s^\alpha - \lambda_1} \\ \dfrac{1}{s^\alpha - \lambda_1^*} \\ \dfrac{1}{s^\alpha - \lambda_3} \\ \dfrac{1}{s^\alpha - \lambda_3^*} \end{bmatrix} U(s)\right\}. \tag{3.73}$$

Moreover the form of the input is given by (3.31), which can be expressed in terms of delayed step functions, $u(t) = f(t) = u_o(t) - u_o(t-1)$, $u_o(t)$

being the unit step function. Then the Laplace transform of the input is $U(s) = (1 - e^{-s})/s$, and the inverse Laplace transform

$$\mathscr{L}^{-1}\left\{\frac{1}{s(s^\alpha - \lambda)}\right\} = \frac{e^{\lambda^2 t}\mathrm{erfc}(-\lambda\sqrt{t}) - 1}{\lambda} = \frac{\mathscr{E}_{0.5,1}\left(\lambda\sqrt{t}\right) - 1}{\lambda}, \qquad (3.74)$$

which was obtained from Table A.2, is needed in order to solve for (3.73). Substituting (3.74) into (3.73) it follows that

$$\boldsymbol{x}^c(t) = \boldsymbol{V}\begin{bmatrix} \dfrac{\mathscr{E}_{0.5,1}\left(\lambda_1\sqrt{t}\right)u_o(t) - \mathscr{E}_{0.5,1}\left(\lambda_1\sqrt{t-1}\right)u_o(t-1)}{\lambda_1} \\[2mm] \dfrac{\mathscr{E}_{0.5,1}\left(\lambda_1^*\sqrt{t}\right)u_o(t) - \mathscr{E}_{0.5,1}\left(\lambda_1^*\sqrt{t-1}\right)u_o(t-1)}{\lambda_1^*} \\[2mm] \dfrac{\mathscr{E}_{0.5,1}\left(\lambda_3\sqrt{t}\right)u_o(t) - \mathscr{E}_{0.5,1}\left(\lambda_3\sqrt{t-1}\right)u_o(t-1)}{\lambda_3} \\[2mm] \dfrac{\mathscr{E}_{0.5,1}\left(\lambda_3^*\sqrt{t}\right)u_o(t) - \mathscr{E}_{0.5,1}\left(\lambda_3^*\sqrt{t-1}\right)u_o(t-1)}{\lambda_3^*} \end{bmatrix}, \qquad (3.75)$$

where functions $\mathscr{E}_{0.5,1}(\cdot)$ are calculated as before.

State evolution under the combined effects of non-zero initial conditions (3.72) and the external input (3.75) in the controllable canonical form is obtained by summing these two expressions, being plotted in Figure 3.3. In Figures 3.2 and 3.3 it can be observed that state x_3 is the first-order derivative of x_1, while states x_2 and x_4 are respectively the 0.5th- and 1.5th-order derivatives of x_1. □

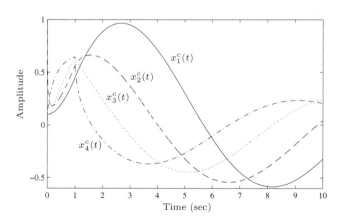

Figure 3.3 Evolution of the controllable canonical form states of the immersed plate

3.3 Controllability of Continuous LTI Commensurate-order Systems

The controllability and observability conditions for commensurate-order systems can be seen in [25] without proof. Here the proofs are given following a method similar to that used for integer-order systems [22, 26].

To obtain the controllability conditions the following definition will be taken into account.

Definition 3.6. A system is controllable if it is possible to establish a non-restricted control vector which can lead the system from an initial state, $x(t_0)$, to another final state, $x(t_f)$, in a finite time $t_0 \leqslant t \leqslant t_f$. ☐

For LTI systems, it can be supposed without loss of generality that the desired final state is $x(t_f) = 0$, i.e., the origin of the state-space, that $x(t_0) \neq 0$, and that $t_0 = 0$. Moreover as we are considering only SISO systems, the control vector reduces to the scalar input $u(t)$.

The solution of the state equation given by (3.41) can be modified as follows:

$$x(t) = \Phi(t)x(0) + \int_0^t \Phi(t - \tau)B\hat{u}(\tau)\mathrm{d}\tau, \quad t \geqslant 0, \tag{3.76}$$

where

$$\hat{u}(t) = \mathscr{L}^{-1}\left\{U(s)s^{1-\alpha}\right\} \tag{3.77}$$

is a fictitious input.

From (3.76) it is shown that

$$x(t_f) = 0 = \Phi(t_f)x(0) + \int_0^{t_f} \Phi(t_f - \tau)B\hat{u}(\tau)\mathrm{d}\tau \Rightarrow$$

$$\Phi(t_f)x(0) = -\int_0^{t_f} \Phi(t_f - \tau)B\hat{u}(\tau)\mathrm{d}\tau. \tag{3.78}$$

Recalling equation (3.57) for $\Phi(t)$, and substituting this into (3.78), gives

$$\Phi(t_f)x(0) = -\int_0^{t_f} \sum_{i=0}^{n-1} a_i \left(t_f - \tau\right) A^i B\hat{u}(\tau)\mathrm{d}\tau$$

$$= -\sum_{i=0}^{n-1} A^i B \int_0^{t_f} a_i \left(t_f - \tau\right) \hat{u}(\tau)\mathrm{d}\tau, \tag{3.79}$$

which can be written in a matrix form as follows:

$$\Phi(t_f)x(0) = -\left[B, AB, A^2B, \cdots, A^{n-1}B\right] \begin{bmatrix} \psi_0 \\ \psi_1 \\ \vdots \\ \psi_{n-1} \end{bmatrix} = -\mathcal{C}\Psi, \tag{3.80}$$

where $\boldsymbol{\Psi} = [\psi_0, \psi_1, \cdots, \psi_{n-1}]^{\mathrm{T}}$,

$$\psi_i = \int_0^{t_f} a_i(t_f - \tau)\,\hat{u}(\tau)\mathrm{d}\tau, \quad 0 \leqslant i \leqslant n - 1, \tag{3.81}$$

and

$$\mathcal{C} = \left[\boldsymbol{B}, \boldsymbol{AB}, \boldsymbol{A}^2\boldsymbol{B}, \cdots, \boldsymbol{A}^{n-1}\boldsymbol{B} \right]. \tag{3.82}$$

Equation 3.80 has a unique solution $\boldsymbol{\Psi}$ if and only if the rank of $\mathcal{C}$ is n, *i.e.*, its determinant is not zero. If this were true, then it would always be possible to find at least one function $\hat{u}(t)$, defined in the interval $[0, t_f]$, such that all conditions (3.81) are fulfilled. This last assertion is proven next.

Lemma 3.7. A control input of the form

$$\hat{u}(t) = \boldsymbol{\xi}(t_f - t)\boldsymbol{\Xi}^{-1}(t_f)\boldsymbol{\Psi} \tag{3.83}$$

where

$$\boldsymbol{\Xi}(t_f) = \int_0^{t_f} \boldsymbol{\xi}^{\mathrm{T}}(t_f - \tau_1)\boldsymbol{\xi}(t_f - \tau_1)\mathrm{d}\tau_1, \tag{3.84}$$

and

$$\boldsymbol{\xi}(t) = \left[a_0(t), a_1(t), \cdots, a_{n-1}(t) \right], \tag{3.85}$$

satisfies the set of conditions (3.81).

Proof. Conditions (3.81) can be expressed in a compact matrix form as

$$\boldsymbol{\Psi} = \int_0^{t_f} \boldsymbol{\xi}^{\mathrm{T}}(t_f - \tau)\,\hat{u}(\tau)\mathrm{d}\tau. \tag{3.86}$$

Input $\hat{u}(t)$ given by (3.83) exists if and only if matrix $\boldsymbol{\Xi}(t_f)$ can be inverted. Then it is demonstrated that this matrix is full-rank, which is the condition to be invertible. This is proven by contradiction. Assuming that this matrix were not invertible, then it would have at least a zero eigenvalue. Denoting by $\boldsymbol{v}$ the eigenvector corresponding to such eigenvalue, it is verified from the eigenvalues definition that

$$\boldsymbol{\Xi}(t_f)\boldsymbol{v} = \boldsymbol{0}, \tag{3.87}$$

and left multiplying $\boldsymbol{\Xi}(t_f)\boldsymbol{v}$ by $\boldsymbol{v}^{\mathrm{T}}$ and rearranging terms it follows that

$$\int_0^{t_f} \left(\boldsymbol{\xi}(t_f - \tau_1)\,\boldsymbol{v} \right)^{\mathrm{T}} \left(\boldsymbol{\xi}(t_f - \tau_1)\,\boldsymbol{v} \right) \mathrm{d}\tau_1 = 0, \tag{3.88}$$

which expresses the integral of the square of the scalar $\boldsymbol{\xi}(t_f - \tau_1)\,\boldsymbol{v}$. This integral is zero only if $\boldsymbol{\xi}(t_f - \tau_1)\,\boldsymbol{v} = 0$, $\forall \tau_1 \in [0, t_f]$, which means that the n functions $a_i(t_f - \tau_1)$ are linearly dependent in that time interval. As this is in contradiction with the statement of Lemma 3.5, then matrix $\boldsymbol{\Xi}(t_f)$ is full-rank, it is invertible, and the $\hat{u}(t)$ given by (3.83) exists.

Finally, substituting the $\hat{u}(t)$ given by (3.83) in (3.86), the lemma easily follows.   □

Once a fictitious input signal $\hat{u}(t)$ has been obtained that verifies (3.79), the real input $u(t)$ can be determined from (3.77). This states that $\hat{U}(s) = U(s)s^{1-\alpha}$ from which it is shown that

$$u(t) = \mathcal{L}^{-1}\left\{\hat{U}(s)\frac{1}{s^{1-\alpha}}\right\} = \hat{u}(t) * \chi_{\alpha-1}(t) = \int_0^t \hat{u}(t-\tau)\chi_{\alpha-1}(\tau)d\tau. \quad (3.89)$$

Expression (3.83) shows that $\hat{u}(t)$ is bounded. Assume that $|\hat{u}(t)| \leqslant \hat{u}_{max}$, then from (3.89) it is shown (taking into account that $\chi_{\alpha-1}(t) \geqslant 0, t \geqslant 0$) that

$$|u(t)| \leqslant \int_0^t |\hat{u}(t-\tau)|\chi_{\alpha-1}(\tau) \leqslant \hat{u}_{max}\int_0^t \chi_{\alpha-1}(\tau)d\tau, \quad (3.90)$$

which, considering the form of $\chi_{\alpha-1}(t)$ given by (3.40), and integrating, yields

$$|u(t)| \leqslant \begin{cases} \hat{u}_{max}\dfrac{t^{1-\alpha}}{\Gamma(2-\alpha)}, & \alpha < 1 \\ \hat{u}_{max}, & \alpha = 1 \end{cases}. \quad (3.91)$$

Then it is concluded that if (3.82) is verified, a bounded control signal $u(t)$ exists that, according to Definition 3.6, leads the system to the zero state. And the controllability condition has been established.

Matrix $\mathcal{C}$ of (3.82) is denoted *the controllability matrix*, and the following criterion can be defined.

Controllability Criterion. The system given by (3.3) and (3.2) is controllable if and only if matrix $\mathcal{C}$ defined by (3.82) is full-rank. □

It is concluded that this controllability condition (3.82) for a commensurate-order system is the same as for an integer-order system, constructing its state description in relation to the operator $\mathscr{D}^\alpha$ or its equivalence in the Laplace domain, $\lambda = s^\alpha$.

Example 3 Consider the immersed plate studied previously. Its state equation is given by (3.33) in the controllable canonical form. Its controllability is easily determined by calculating the rank of matrix $\mathcal{C}$ given by (3.82). Then it is shown from (3.33) that

$$\mathcal{C} = \begin{bmatrix} \boldsymbol{B}, & \boldsymbol{AB}, & \boldsymbol{A}^2\boldsymbol{B}, & \boldsymbol{A}^3\boldsymbol{B} \end{bmatrix} = \begin{bmatrix} 0 & 0 & 0 & 1 \\ 0 & 0 & 1 & -0.5 \\ 0 & 1 & -0.5 & 0.25 \\ 1 & -0.5 & 0.25 & -0.125 \end{bmatrix}, \quad (3.92)$$

whose determinant is 1. Therefore matrix $\mathcal{C}$ is full-rank and the system is controllable. □

3.4 Observability of Continuous LTI Commensurate-order Systems

To obtain the observability conditions the following definition will be taken into account.

Definition 3.8. A system is observable if any state, $x(t_0)$, can be determined from the observation of $y(t)$ and the knowledge of the input $u(t)$ in a finite interval of time $t_0 \leqslant t \leqslant t_f$. □

Remark 3.9. In this section this definition is interpreted in a special manner as the fractional-order derivative according to Caputo's definition is considered, and the lower limit of this operator is made to coincide with t_0. □

Taking into account the above remark, the solution of the state equation (3.41) can be modified to consider an initial non-zero time t_0. And combining this with the output equation of the state description (3.2) it can be written that

$$y(t) = C\boldsymbol{\Phi}(t - t_0)\boldsymbol{x}(t_0) + C\int_{t_0}^{t} \hat{\boldsymbol{\Phi}}(t - \tau)\boldsymbol{B}u(\tau)\mathrm{d}\tau + \boldsymbol{D}u(t), \quad t \geqslant t_0. \quad (3.93)$$

The last two terms on the right hand side of the previous equation are known since they are defined by $\boldsymbol{C}, \boldsymbol{B}, \boldsymbol{D}, \hat{\boldsymbol{\Phi}}(t)$, and $u(t)$. Then these terms can be subtracted from the observed value $y(t)$, and an equivalent observed output,

$$\hat{y}(t) = y(t) - C\int_{t_0}^{t} \hat{\boldsymbol{\Phi}}(t - \tau)\boldsymbol{B}u(\tau)\mathrm{d}\tau - \boldsymbol{D}u(t), \quad t \geqslant t_0, \quad (3.94)$$

is defined. Therefore the observability condition can be derived from

$$\hat{y}(t) = C\boldsymbol{\Phi}(t - t_0)\boldsymbol{x}(t_0), \quad (3.95)$$

which is obtained from (3.93), and allows for the estimation of $\boldsymbol{x}(t_0)$.

Substituting again $\boldsymbol{\Phi}(t) = \mathscr{E}_{\alpha,1}(\boldsymbol{A}t^{\alpha})$ in (3.95) by its expansion (3.57), it is shown that

$$\hat{y}(t) = C\sum_{i=0}^{n-1} a_i(t - t_0)\boldsymbol{A}^i\boldsymbol{x}(t_0) = \sum_{i=0}^{n-1} a_i(t - t_0)\psi_i, \quad (3.96)$$

where now

$$\psi_i = \boldsymbol{C}\boldsymbol{A}^i\boldsymbol{x}(t_0), \quad 0 \leqslant i \leqslant n - 1. \quad (3.97)$$

Multiplying (3.96) by functions $a_i(t - t_0), 0 \leqslant i \leqslant n - 1$, integrating in the interval $[t_0, t_f]$ where t_f is any time value larger than t_0, and expressing the resulting set of n equations in a compact matrix form, it follows that

$$\int_{t_0}^{t_f} \boldsymbol{\xi}^{\mathrm{T}}(t - t_0)\hat{y}(t)\mathrm{d}t = \left[\int_{t_0}^{t_f} \boldsymbol{\xi}^{\mathrm{T}}(t - t_0)\boldsymbol{\xi}(t - t_0)\mathrm{d}t\right]\boldsymbol{\Psi}, \quad (3.98)$$

where $\boldsymbol{\xi}(t)$ is given again by (3.85), and $\boldsymbol{\Psi} = [\psi_0 \quad \psi_1 \cdots \psi_{n-1}]^{\mathrm{T}}$.

The column vector $\boldsymbol{\Psi}$ could be determined from $\hat{y}(t)$ and (3.98), if the square matrix $\int_{t_0}^{t_f} \boldsymbol{\xi}^{\mathrm{T}}(t - t_0) \boldsymbol{\xi}(t - t_0) \, \mathrm{d}t$ were invertible, i.e., full-rank. This fact has already been proven in Lemma 3.7 of the previous section. Therefore $\boldsymbol{\Psi}$ can be determined from $\hat{y}(t)$ and (3.98) if all the eigenvalues of $\boldsymbol{A}$ are distinct.

Taking into account the form of ψ_i given in (3.97), the following matrix equation follows:

$$\boldsymbol{\Psi} = \mathcal{O}\boldsymbol{x}(t_0), \tag{3.99}$$

where

$$\mathcal{O} = \begin{bmatrix} \boldsymbol{C} \\ \boldsymbol{C}\boldsymbol{A} \\ \vdots \\ \boldsymbol{C}\boldsymbol{A}^{n-1} \end{bmatrix}. \tag{3.100}$$

If matrix $\mathcal{O}$ is full-rank then it can be inverted, and $\boldsymbol{x}(t_0)$ can be calculated from (3.99). Then the observability property depends on the rank of matrix $\mathcal{O}$, which has to be n in order for the system be observable. This matrix is called *the observability matrix*, and the next criterion can be defined.

Observability Criterion. The system given by (3.3) and (3.2) is observable if and only if matrix $\mathcal{O}$ defined by (3.100) is full-rank. □

As can be seen, the observability condition for commensurate-order LTI systems coincides with the well known one for integer-order LTI systems, considering the state representation introduced previously in Section 3.1.

Example 4 Consider again the immersed plate system, and its state-space model given by (3.33) and (3.34) in the controllable canonical form. Its observability is easily determined by calculating the rank of matrix $\mathcal{O}$ given by expression (3.100). Then it is shown from (3.33) and (3.34) that

$$\mathcal{O} = \begin{bmatrix} \boldsymbol{C} \\ \boldsymbol{C}\boldsymbol{A} \\ \boldsymbol{C}\boldsymbol{A}^2 \\ \boldsymbol{C}\boldsymbol{A}^3 \end{bmatrix} = \begin{bmatrix} 1 & 0 & 0 & 0 \\ 0 & 1 & 0 & 0 \\ 0 & 0 & 1 & 0 \\ 0 & 0 & 0 & 1 \end{bmatrix}, \tag{3.101}$$

whose determinant is 1. Therefore matrix $\mathcal{O}$ is full-rank and the system is observable. □

3.5 Discrete-time LTI State-space Models

In this section the state-space model defined in Section 3.1 for continuous-time systems is extended to discrete-time systems, according to the definitions given in books [3, 27] and journal papers [28–30].

3.5.1 Discrete-time State-space Equivalent Model of a Continuous LTI System

Consider a fractional-order LTI continuous system given by the state equation (3.3) and the output equation (3.2). In this section the discrete-time state-space model equivalent to the series connection of a zero-order hold, system (3.3)-(3.2), and a sampler, is studied. It is assumed that the zero-order hold and the sampler work at the same frequency f_s or, equivalently, with a sampling period $T = 1/f_s$. The discrete model to be obtained should behave exactly as the before series connection set which is shown in Figure 3.4 at sampling instants $t = kT, \ k = 0, 1, 2, \cdots$.

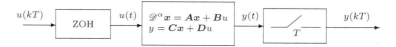

Figure 3.4 Discrete-time state-space model equivalent to the series connection of a zero-order hold, system (3.3)-(3.2), and a sampler

First recall how this equivalent discrete-time model is obtained in the standard case of integer-order LTI continuous systems [31]:

$$\dot{x} = Ax + Bu. \qquad (3.102)$$

The solution of this state equation is

$$x(t) = \Phi(t, t_0)x(t_0) + \int_{t_0}^{t} \Phi(t, \tau)Bu(\tau)d\tau, \quad t \geqslant t_0, \qquad (3.103)$$

which is based on the state transition matrix Φ. The equivalent discrete-time model is obtained from this equation by making $t_0 = kT$ and $t = (k + 1)T$, and taking into account that in LTI systems $\Phi(t, \tau) = \Phi(t - \tau)$.

Consider the solution of the fractional-order state equation given by (3.41). The above procedure cannot be applied using this expression because of the two remarks stated in Section 3.2:

- As the fractional-order derivative of $\boldsymbol{x}$ depends on the "history" of $\boldsymbol{x}$ from the lower limit of the integral that defines this operator until the present instant, in order to determine the future behavior of a fractional-order system from instant t not only is the value of the state at instant t needed but also all the values of the state in the interval $[0, t]$. Consequently, it does not make sense to obtain an expression of $\boldsymbol{x}((k+1)T)$ as function only of $\boldsymbol{x}(kT)$ and $u(kT)$ because previous state values also influence the value of $\boldsymbol{x}((k+1)T)$.
- As was stated in Remark 3.2, matrix $\boldsymbol{\Phi}(t)$ in (3.41) really means $\boldsymbol{\Phi}(t, 0)$, the second argument being the lower limit of the integral that defines the fractional-order derivative used in the state-space model. Then this argument is fixed and cannot be changed to kT as was done in the integer-order derivative model discretization.

Therefore discrete-time space state models must include the history of the state starting from the lower limit time value used in the definition of the fractional-order operator. A model that takes this into account is proposed in the next section.

3.5.2 Discrete-time State-space Model Based on Finite Differences

A different approach is proposed in this section in order to develop a discrete-time state-space model. It is based on the Grünwald–Letnikov's definition of fractional-order operators [3, 27, 28]. This operator was applied in [29] to develop the discrete-time model studied here. Moreover, a MATLAB toolbox has been built [32, 33] to analyze and simulate these systems.

In order to introduce a discrete-time fractional-order state-space system, let us assume that the standard integer-order model (3.102) is discretized with sampling period T by a numerical method. The simplest way of approximating the first-order derivative is calculating its first-order forward difference:

$$\dot{\boldsymbol{x}}(kT) \approx \frac{\boldsymbol{x}((k+1)T) - \boldsymbol{x}(kT)}{T}. \tag{3.104}$$

Substituting this in (3.102) gives

$$\frac{\boldsymbol{x}((k+1)T) - \boldsymbol{x}(kT)}{T} \approx \boldsymbol{A}\boldsymbol{x}(kT) + \boldsymbol{B}u(kT), \tag{3.105}$$

which yields the approximate equivalent discrete state-space model

$$\boldsymbol{x}((k+1)T) = (\boldsymbol{A}T + \boldsymbol{I})\boldsymbol{x}(kT) + \boldsymbol{B}Tu(kT), \tag{3.106}$$

$$y(kT) = \boldsymbol{C}\boldsymbol{x}(kT) + \boldsymbol{D}u(kT). \tag{3.107}$$

Note that the first-order backward difference could also have been used to approximate the first-order derivative, but it would not yield a discrete-time model like (3.106). This algorithm is not adequate to obtain discrete-time models that relate $\boldsymbol{x}((k+1)T)$ with previous states and inputs. This consideration will be taken into account in the next discretization method.

Let us generalize the previous discretization method to a fractional-order differentiation operator by using the Grünwald–Letnikov definition, and leading the operator one sample in order to get a model where $\boldsymbol{x}((k+1)T)$ appears, similar to the one sample forward difference used in the standard case. Then it is shown that

$$\mathscr{D}^\alpha \boldsymbol{x}(kT) \approx \frac{1}{T^\alpha} \sum_{i=0}^{k+1} (-1)^i \binom{\alpha}{i} \boldsymbol{x}((k+1-i)T)$$

$$= \frac{1}{T^\alpha} \left(\boldsymbol{x}((k+1)T) - \binom{\alpha}{1}\boldsymbol{x}(kT) + \sum_{i=2}^{k+1} (-1)^i \binom{\alpha}{i} \boldsymbol{x}((k+1-i)T) \right), \tag{3.108}$$

which, substituted in (3.3), yields for $k \geqslant 1$

$$\boldsymbol{x}((k+1)T) = (\boldsymbol{A}T^\alpha + \alpha\boldsymbol{I})\boldsymbol{x}(kT)$$

$$- \sum_{i=2}^{k+1} (-1)^i \binom{\alpha}{i} \boldsymbol{x}((k+1-i)T) + \boldsymbol{B}T^\alpha u(kT), \tag{3.109}$$

and for $k = 0$

$$\boldsymbol{x}((k+1)T) = (\boldsymbol{A}T^\alpha + \alpha\boldsymbol{I})\boldsymbol{x}(kT) + \boldsymbol{B}T^\alpha u(kT). \tag{3.110}$$

Equations (3.109), (3.110) combined with (3.107) constitute the discrete-time LTI state-space model. In this model the state at instant $(k+1)T$ depends on all the previous states until $kT = 0$ in a direct way (as could be expected from the considerations of the previous subsection about the "memory" of the fractional-order derivative operators). From now on, and without loss of generality, the sampling period T is removed from the sampled signal sequences, i.e., $\boldsymbol{x}(k) \equiv \boldsymbol{x}(kT)$ and $u(k) \equiv u(kT)$.

3.5.3 Discrete-time State-space Model Based on the Expanded State

An alternative way of expressing the system given by (3.109), (3.110), and (3.107) was proposed in [29, 34]. By using an expanded state formed by the actual state and all the past states, i.e., the "history" of the system from the initial instant, an infinite-dimensional system is constructed that takes the form

$$\begin{bmatrix} \boldsymbol{x}((k+1)T) \\ \boldsymbol{x}(kT) \\ \boldsymbol{x}((k-1)T) \\ \vdots \end{bmatrix} = \tilde{\boldsymbol{A}} \begin{bmatrix} \boldsymbol{x}(kT) \\ \boldsymbol{x}((k-1)T) \\ \boldsymbol{x}((k-2)T) \\ \vdots \end{bmatrix} + \tilde{\boldsymbol{B}}u(kT),$$

$$y(kT) = \tilde{\boldsymbol{C}} \begin{bmatrix} \boldsymbol{x}(kT) \\ \boldsymbol{x}((k-1)T) \\ \boldsymbol{x}((k-2)T) \\ \vdots \end{bmatrix} + \boldsymbol{D}u(kT),$$

(3.111)

where

$$\tilde{\boldsymbol{A}} = \begin{bmatrix} (\boldsymbol{A}T^\alpha + \alpha\boldsymbol{I}) & -\boldsymbol{I}(-1)^2\binom{\alpha}{2} & -\boldsymbol{I}(-1)^3\binom{\alpha}{3} & \cdots \\ \boldsymbol{I} & \boldsymbol{0} & \boldsymbol{0} & \cdots \\ \boldsymbol{0} & \boldsymbol{I} & \boldsymbol{0} & \cdots \\ \vdots & \vdots & \vdots & \ddots \end{bmatrix}, \quad \tilde{\boldsymbol{B}} = \begin{bmatrix} \boldsymbol{B}T^\alpha \\ \boldsymbol{0} \\ \boldsymbol{0} \\ \vdots \end{bmatrix}, \quad (3.112)$$

and

$$\tilde{\boldsymbol{C}} = [\boldsymbol{C}, \ \boldsymbol{0}, \ \boldsymbol{0}, \ \cdots], \quad (3.113)$$

where $\boldsymbol{0}$ is the null matrix of dimension $n \times n$.

3.6 Solution of the Discrete-time LTI Commensurate-order State Equation

In this section the solution of the state-space difference equation

$$\boldsymbol{x}(k+1) = (\boldsymbol{A}T^\alpha + \alpha\boldsymbol{I})\boldsymbol{x}(k) - \sum_{i=2}^{k+1}(-1)^i\binom{\alpha}{i}\boldsymbol{x}(k+1-i) + \boldsymbol{B}T^\alpha u(k), \quad (3.114)$$

for $k \geqslant 1$, and

$$\boldsymbol{x}(1) = (\boldsymbol{A}T^\alpha + \alpha\boldsymbol{I})\boldsymbol{x}(0) + \boldsymbol{B}T^\alpha u(0), \quad (3.115)$$

$$y(k) = \boldsymbol{C}\boldsymbol{x}(k) + \boldsymbol{D}u(k), \quad (3.116)$$

equivalent to (3.109), (3.110), and (3.107) (in accordance with the last paragraph of Section 3.5.2), is obtained. Before presenting the solution, we note that this equation has a varying number of terms depending on the value of k. This makes it different from standard difference equations.

3.6.1 Solution of the Homogeneous Discrete-time State Equation

Consider the homogeneous difference equation derived from (3.114), (3.115):

$$x(k+1) = (AT^\alpha + \alpha I)x(k) - \sum_{i=2}^{k+1}(-1)^i \binom{\alpha}{i}x(k+1-i), \qquad (3.117)$$
$$x(1) = (AT^\alpha + \alpha I)x(0).$$

As this is an LTI system, it is expected that the solution of this equation will be of the form $x(k) = \Phi(k)x(0)$, where $\Phi(k)$ is a discrete-time state pseudo-transition matrix. Applying the recursive equation (3.117) in a sequential way, and starting from an initial state $x(0)$ it follows that

$$x(0) = Ix(0) \Longrightarrow \Phi(0) = I, \qquad (3.118)$$

for $k = 0$, we have

$$x(1) = (AT^\alpha + \alpha I)x(0) \Longrightarrow \Phi(1) = (AT^\alpha + \alpha I), \qquad (3.119)$$

for $k = 1$, we have

$$x(2) = (AT^\alpha + \alpha I)x(1) - \binom{\alpha}{2}x(0) = (AT^\alpha + \alpha I)\Phi(1)x(0) - \binom{\alpha}{2}x(0)$$
$$\Phi(2) = (AT^\alpha + \alpha I)^2 - I\binom{\alpha}{2},$$
$$\qquad (3.120)$$

for $k = 2$, we have

$$x(3) = (AT^\alpha + \alpha I)x(2) - \binom{\alpha}{2}x(1) + \binom{\alpha}{3}x(0)$$
$$= (AT^\alpha + \alpha I)\Phi(2)x(0) - \binom{\alpha}{2}\Phi(1)x(0) + \binom{\alpha}{3}x(0), \qquad (3.121)$$
$$\Phi(3) = (AT^\alpha + \alpha I)^3 - (AT^\alpha + \alpha I)2\binom{\alpha}{2} + I\binom{\alpha}{3}$$

and so on. Substituting $x(k) = \Phi(k)x(0)$ in (3.117), and taking into account that this expression must be verified for any initial state value $x(0)$, it follows, for $k \geq 1$, that

$$\Phi(k+1) = (AT^\alpha + \alpha I)\Phi(k) - \sum_{i=2}^{k+1}(-1)^i \binom{\alpha}{i}\Phi(k+1-i), \qquad (3.122)$$
$$\Phi(1) = (AT^\alpha + \alpha I)\Phi(0), \quad \Phi(0) = I,$$

which allows the calculation of matrix function $\Phi(k)$ in a recursive way, and then consequently the computation of $x(k)$.

This state pseudo-transition matrix really means $\Phi(k, 0)$, i.e., has two arguments, the second one being the lower limit of the Grünwald–Letnikov's operator. Then this argument is fixed and cannot be changed, similar to the state pseudo-transition matrix of fractional-order continuous systems. For this reason this matrix is expressed as $\Phi(k)$ with a single argument.

Remark 3.10. The semi-group property is not satisfied in the discretized state pseudo-transition matrix, in the same way as the continuous state pseudo-state matrix (see Remark 3.2): $\boldsymbol{\Phi}(k_2, 0) \neq \boldsymbol{\Phi}(k_2, k_1)\boldsymbol{\Phi}(k_1, 0)$, $0 < k_1 < k_2$. A demonstration of this can be found in [30]. □

3.6.2 Solution of the Complete Discrete-time State Equation

The solution of the complete difference equation (3.114), (3.115) is given by

$$x(k) = \boldsymbol{\Phi}(k)x(0) + \sum_{i=0}^{k-1} \boldsymbol{\Phi}(k-1-i)BT^\alpha u(i), \quad k \geqslant 1. \tag{3.123}$$

The demonstration of this is carried out by induction [30], and is developed in the following.

Proof. Step 1: for $k = 0$, using (3.115) and taking into account (3.118) and (3.119), the state is

$$x(1) = (AT^\alpha + \alpha I)x(0) + BT^\alpha u(0) = \boldsymbol{\Phi}(1)x(0) + \boldsymbol{\Phi}(0)BT^\alpha u(0), \tag{3.124}$$

which is according to (3.123).

Step 2: for $k = 1$, now using (3.114) and from (3.118) to (3.120), and substituting $x(1)$ by (3.124), the state is

$$x(2) = (AT^\alpha + \alpha I)x(1) - \binom{\alpha}{2}x(0) + BT^\alpha u(1)$$
$$= \boldsymbol{\Phi}(2)x(0) + BT^\alpha \left[\boldsymbol{\Phi}(1)u(0) + \boldsymbol{\Phi}(0)u(1)\right], \tag{3.125}$$

also in accordance with (3.123).

Assume that it is true in Step k, *i.e.*, that (3.123) is satisfied. Then it has to be proven that it is true in Step $k + 1$. From (3.114) it is shown that

$$x(k+1) = (AT^\alpha + \alpha I)x(k) - \sum_{i=2}^{k}(-1)^i\binom{\alpha}{i}x(k+1-i)$$
$$- (-1)^{k+1}\binom{\alpha}{k+1}x(0) + BT^\alpha u(k), \tag{3.126}$$

and, since (3.123) is verified by all $l \leqslant k$, then states $x(k+1-i)$ and $x(k)$ in (3.126) can be expressed in terms of the initial state $x(0)$ yielding

$$x(k+1) = (AT^\alpha + \alpha I) \left[\boldsymbol{\Phi}(k)x(0) + \sum_{l=0}^{k-1} \boldsymbol{\Phi}(k-1-l)BT^\alpha u(l) \right]$$

$$- \sum_{i=2}^{k}(-1)^i \binom{\alpha}{i} \left[\boldsymbol{\Phi}(k+1-i)x(0) + \sum_{l=0}^{k-i} \boldsymbol{\Phi}(k-i-l)BT^\alpha u(l) \right] \quad (3.127)$$

$$- (-1)^{k+1} \binom{\alpha}{k+1} x(0) + BT^\alpha u(k).$$

This equation is rearranged as

$$x(k+1) = \left[(AT^\alpha + \alpha I)\boldsymbol{\Phi}(k) - \sum_{i=2}^{k+1}(-1)^i \binom{\alpha}{i}\boldsymbol{\Phi}(k+1-i) \right] x(0) + BT^\alpha u(k)$$

$$+ BT^\alpha \left[(AT^\alpha + \alpha I) \sum_{l=0}^{k-1} \boldsymbol{\Phi}(k-1-l)u(l) - \sum_{i=2}^{k} (-1)^i \binom{\alpha}{i} \sum_{l=0}^{k-i} \boldsymbol{\Phi}(k-i-l)u(l) \right].$$

Then substituting (3.122) in the term inside the square bracket that multiplies $x(0)$, rearranging terms inside the second square bracket, and permutating indexes i and l in the two sums, it is shown that

$$x(k+1) = \boldsymbol{\Phi}(k+1)x(0) + BT^\alpha u(k)$$

$$+ BT^\alpha \left[\sum_{l=0}^{k-1}(AT^\alpha + \alpha I)\boldsymbol{\Phi}(k-1-l)u(l) - \sum_{l=0}^{k-2}\sum_{i=2}^{k-l}(-1)^i \binom{\alpha}{i}\boldsymbol{\Phi}(k-i-l)u(l) \right].$$

Decomposing the term $k-1$ of the first sum inside the square bracket

$$x(k+1) = \boldsymbol{\Phi}(k+1)x(0) + BT^\alpha \left[u(k) + (AT^\alpha + \alpha I)\boldsymbol{\Phi}(0)u(k-1) \right]$$

$$+ BT^\alpha \sum_{l=0}^{k-2} \left[(AT^\alpha + \alpha I)\boldsymbol{\Phi}(k-1-l) - \sum_{i=2}^{k-l}(-1)^i \binom{\alpha}{i}\boldsymbol{\Phi}(k-i-l) \right] u(l),$$

and again substituting (3.122) in the term inside the square bracket it follows that

$$x(k+1) = \boldsymbol{\Phi}(k+1)x(0) + BT^\alpha \left[u(k) + (AT^\alpha + \alpha I)\boldsymbol{\Phi}(0)u(k-1) \right]$$

$$+ BT^\alpha \sum_{l=0}^{k-2} \boldsymbol{\Phi}(k-l)u(l). \quad (3.128)$$

Taking into account (3.118) and (3.119) we have that

$$BT^\alpha = BT^\alpha \boldsymbol{\Phi}(0), \text{ and } BT^\alpha(AT^\alpha + \alpha I)\boldsymbol{\Phi}(0) = BT^\alpha \boldsymbol{\Phi}(1). \quad (3.129)$$

Substituting this in (3.128) and rearranging yields

$$x(k+1) = \boldsymbol{\Phi}(k+1)x(0) + \sum_{l=0}^{k} \boldsymbol{\Phi}(k-l)BT^\alpha u(l), \quad (3.130)$$

which completes the proof. $\qquad \Box$

3.7 Stability of Discrete-time LTI Commensurate-order Systems

Let us now turn our attention to the stability issues related to the state-space models defined in Section 3.5. The approach presented here is based on the stability analysis originally developed in [29].

First a stability definition is proposed which is adopted from the stability definition of infinite dimensional systems presented in [35].

Definition 3.11. The linear, discrete-time, infinite-dimensional system given by (3.114), (3.115), and (3.116), or alternatively by (3.111), (3.112), and (3.113), is finite-time stable with respect to $\{\alpha, \beta, N, M, || \cdot ||\}$, $\alpha < \beta$, $\alpha, \beta \in \Re^+$, if and only if

$$||\boldsymbol{x}(i)|| < \alpha, \quad i = 0, -1, \cdots, -N \tag{3.131}$$

implies

$$||\boldsymbol{x}(i)|| < \beta, \quad i = 0, 1, 2, \cdots, M. \tag{3.132}$$

This stability condition is similar to that for traditional, integer-order, infinite-dimensional state-space systems.   □

Stability Criterion 1. The system given by (3.114), (3.115), and (3.116), or alternatively by (3.111), (3.112), and (3.113), is asymptotically stable if and only if

$$||\widetilde{\boldsymbol{A}}|| < 1 \tag{3.133}$$

where $|| \cdot ||$ denotes the matrix norm defined as $\max |\lambda_i|$ where λ_i is the ith eigenvalue of the matrix $\widetilde{\boldsymbol{A}}$.

A drawback of this result is that it is exact for a matrix $\widetilde{\boldsymbol{A}}$ of infinite dimension. In practice the number of factors $-\boldsymbol{I}(-1)^i \binom{\alpha}{i}$ in this matrix has to be limited. This reduction may cause a decrease in the accuracy of the stability determination, however, especially when the system is close to the stability margin.

Next a sufficient stability condition is presented, which was also developed in [35], and can be applied more easily than the above.

Stability Criterion 2. The system given by (3.114), (3.115), and (3.116), or alternatively by (3.111), (3.112), and (3.113), is asymptotically stable if

$$||\boldsymbol{A}T^{\alpha} + \alpha \boldsymbol{I}|| + \sum_{i=2}^{k} \left\| -\boldsymbol{I}(-1)^i \binom{\alpha}{i} \right\| < 1, \quad \forall k > 1. \tag{3.134}$$

This criterion implies that the state pseudo-transition matrix $\boldsymbol{\Phi}$ satisfies the condition

$$||\boldsymbol{\Phi}(k)|| < \sum_{i=1}^{k} ||\boldsymbol{\Phi}(k-i)|| \leq \epsilon. \tag{3.135}$$

In order to make this criterion more useful, the next lemma is used.

Lemma 3.12. For $i \geq 2$, the factors $(-1)^i \binom{\alpha}{i}$ are

$$(-1)^i \binom{\alpha}{i} = \begin{cases} < 0, & \text{for } 0 < \alpha < 1, \\ > 0, & \text{for } 1 < \alpha < 2, \\ = 0, & \text{for } \alpha = 1, 2. \end{cases} \tag{3.136}$$

Proof. With (1.15) for $\binom{\alpha}{i}$, it is easy to see that, for $i \geq 2$ and $0 < \alpha < 1$,

$$\binom{\alpha}{i} < 0, \text{ for } i = 2, 4, 6, \cdots, \qquad \binom{\alpha}{i} > 0, \text{ for } i = 3, 5, 7, \cdots. \tag{3.137}$$

Moreover factors $(-1)^i$ are equal to 1 for even values of i, and equal to -1 for its odd values. Then the signs of the two factors $\binom{\alpha}{i}$ and $(-1)^i$ are thus opposite $\forall i \geq 2$ which means that their product is always negative for $i \geq 2$ if $0 < \alpha < 1$.

For $i \geq 2$ and $1 < \alpha < 2$, the situation is the opposite:

$$\binom{\alpha}{i} > 0, \text{ for } i = 2, 4, 6, \cdots; \qquad \binom{\alpha}{i} < 0, \text{ for } i = 3, 5, 7, \cdots \tag{3.138}$$

In this case the signs of the two factors $\binom{\alpha}{i}$ and $(-1)^i$ are thus identical $\forall i \geq 2$ which means that their product is always positive for $i \geq 2$ if $1 < \alpha < 2$.

$\square$

Using this lemma, the next relation is obtained:

$$\sum_{i=2}^{k} \left| \left| -I(-1)^i \binom{\alpha}{i} \right| \right| = \sum_{i=2}^{k} \left| (-1)^i \binom{\alpha}{i} \right| = \begin{cases} -\sum_{i=2}^{k} (-1)^i \binom{\alpha}{i}, & \text{for } 0 < \alpha < 1 \\ \sum_{i=2}^{k} (-1)^i \binom{\alpha}{i}, & \text{for } 1 < \alpha < 2. \end{cases}$$

And the property [36] $\tag{3.139}$

$$\sum_{i=0}^{k} (-1)^i \binom{\alpha}{i} = \frac{\Gamma(k+1-\alpha)}{\Gamma(1-\alpha)\Gamma(k+1)}, \tag{3.140}$$

leads to the relation

$$\sum_{i=2}^{k} (-1)^i \binom{\alpha}{i} = \frac{\Gamma(k+1-\alpha)}{\Gamma(1-\alpha)\Gamma(k+1)} - 1 + \alpha, \tag{3.141}$$

which permits a rewrite of Stability Criterion 2 as follows.

Stability Criterion 3. The system given by (3.114), (3.115), and (3.116), or alternatively by (3.111), (3.112), and (3.113), is asymptotically stable if

$$||\boldsymbol{A}\boldsymbol{T}^\alpha + \alpha\boldsymbol{I}|| < r(k, \alpha), \tag{3.142}$$

where

$$r(k, \alpha) = \begin{cases} \alpha + \dfrac{\Gamma(k+1-\alpha)}{\Gamma(1-\alpha)\Gamma(k+1)}, & \text{for } 0 \leqslant \alpha < 1, \\[2ex] 2 - \alpha - \dfrac{\Gamma(k+1-\alpha)}{\Gamma(1-\alpha)\Gamma(k+1)}, & \text{for } 1 \leqslant \alpha \leqslant 2. \end{cases} \tag{3.143}$$

Some comments are made to this result.

Remark 3.13.

- $r(k, \alpha)$ is the stability radius of the system, *i.e.*, it is the radius of a circle within which stable eigenvalues (in the sense of fulfilment of (3.135)) of the system must be located.
- The exact stability radius will be given for the limit $k \longrightarrow \infty$.
- It is worth noting that the stability radius for $\alpha = 0$ has the same value as that for $\alpha = 1$, because the system given by (3.114), (3.115) for $\alpha = 0$ is in fact the first-order system.   □

3.8 Controllability of Discrete-time LTI Commensurate-order Systems

In this section the concept of controllability is extended to fractional-order discrete-time LTI systems of the form given by (3.114), (3.115), and (3.116). First some definitions are presented.

Definition 3.14. The linear, discrete-time, fractional-order system modeled by (3.114), (3.115), and (3.116) is *reachable* if and only if for an arbitrary final state $\boldsymbol{x}_\mathrm{f}$ there exists a number N and an input sequence $\{u(0), u(1), \ldots, u(N-1)\}$ which carries the system from the initial state $\boldsymbol{x}_0 = \boldsymbol{0}$ to the desired final state $\boldsymbol{x}_\mathrm{f}$.   □

Definition 3.15. The system modeled by (3.114), (3.115), and (3.116) is *controllable* if and only if for any initial state $\boldsymbol{x}_0$ there exists a number N and an input sequence $\{u(0), u(1), \ldots, u(N-1)\}$ which carries the system from the initial state $\boldsymbol{x}_0$ to zero final state $\boldsymbol{x}_\mathrm{f} = \boldsymbol{0}$.   □

These two definitions extend the concepts of state reachability (or controllability from the origin) and controllability (or controllability to the origin) to our discrete-time fractional-order systems. The next results and criteria for reachability and controllability were originally developed in [37].

We mention that an alternative approach to these problems based on a generalized discrete-time controllability Gramian can be found in [30]. The controllability condition obtained in [30] from this Gramian is more complex than that presented in [37]. However, [30] provides, besides its condition, the control sequence that allows one to reach the final state.

3.8.1 Reachability Conditions

Reachability Criterion 1. The system given by (3.114) and (3.115) is reachable (in N steps) if and only if matrix $\boldsymbol{\mathcal{C}}$ defined as

$$\boldsymbol{\mathcal{C}} = [\boldsymbol{\Phi}(0)\boldsymbol{B}, \ \boldsymbol{\Phi}(1)\boldsymbol{B}, \ \boldsymbol{\Phi}(2)\boldsymbol{B}, \ \cdots, \ \boldsymbol{\Phi}(N-1)\boldsymbol{B}] \tag{3.144}$$

is full-rank.

Proof. The solution of the system (3.114) and (3.115) is given by (3.123), which for zero initial condition $\boldsymbol{x}_0 = \boldsymbol{x}(0) = \boldsymbol{0}$ and final state $\boldsymbol{x}_{\mathrm{f}}$ yields

$$
\begin{aligned}
\boldsymbol{x}_{\mathrm{f}} &= \sum_{i=0}^{N-1} \boldsymbol{\Phi}(N-1-i)\boldsymbol{B}T^{\alpha}u(i) \\
&= T^{\alpha}\left[\boldsymbol{\Phi}(0)\boldsymbol{B}, \boldsymbol{\Phi}(1)\boldsymbol{B}, \boldsymbol{\Phi}(2)\boldsymbol{B}, \cdots, \boldsymbol{\Phi}(N-1)\boldsymbol{B}\right]
\begin{bmatrix}
u(N-1) \\
u(N-2) \\
\vdots \\
u(0)
\end{bmatrix} .
\end{aligned}
\tag{3.145}
$$

This forms a set of n equations with N unknowns, which are the components of the input sequence. This set of equations has a solution if and only if matrix (3.144) is full-rank (note that $T^{\alpha} \neq 0$), and the criterion is proven. □

A first consequence of this criterion is that $N \geqslant n$ must be verified.

Moreover the above criterion can be rewritten in an easier form as follows.

Reachability Criterion 2. The system given by (3.114) and (3.115) is reachable (in N steps) if and only if matrix $\boldsymbol{\mathcal{C}}_1$ defined as

$$\boldsymbol{\mathcal{C}}_1 = \left[\boldsymbol{B}, \ \boldsymbol{AB}, \ \boldsymbol{A}^2\boldsymbol{B}, \ \cdots, \ \boldsymbol{A}^{N-1}\boldsymbol{B}\right] \tag{3.146}$$

is full-rank.

Proof. Right multiplying (3.122) by $\boldsymbol{B}$, substituting $k+1$ by k, splitting the first term on the right hand side of this expression, and then rearranging terms, it is shown that

$$\boldsymbol{\Phi}(k)\boldsymbol{B} = T^{\alpha}\boldsymbol{A}\boldsymbol{\Phi}(k-1)\boldsymbol{B} \sum_{i=1}^{k}(-1)^i\binom{\alpha}{i}\boldsymbol{\Phi}(k-i)\boldsymbol{B}, \quad k \geqslant 1, \tag{3.147}$$

which shows that $\boldsymbol{\Phi}(k)\boldsymbol{B}$ can be expressed as the sum of $T^{\alpha}\boldsymbol{A}\boldsymbol{\Phi}(k-1)\boldsymbol{B}$ and a linear combination of past terms $\boldsymbol{\Phi}(k-i)\boldsymbol{B}$, $1 \leqslant i \leqslant k$. Use (3.147) to define $\boldsymbol{\Phi}(k-i)\boldsymbol{B}$ on the left hand side of the equation. Then, substituting this in the first term on the right hand side yields

$$\boldsymbol{\Phi}(k)\boldsymbol{B} = (T^{\alpha}\boldsymbol{A})^2\boldsymbol{\Phi}(k-2)\boldsymbol{B} + \sum_{i=0}^{k-1}\hat{\nu}_i\boldsymbol{\Phi}(i)\boldsymbol{B}, \qquad (3.148)$$

where $\hat{\nu}_i$ are scalars whose particular values are not of interest. And repeating this process recursively until $\boldsymbol{\Phi}(0)$ appears in the first term of the right hand side of the expression, it is shown that

$$\boldsymbol{\Phi}(k)\boldsymbol{B} = (T^{\alpha}\boldsymbol{A})^k\boldsymbol{\Phi}(0)\boldsymbol{B} + \sum_{i=0}^{k-1}\nu_i\boldsymbol{\Phi}(i)\boldsymbol{B}, \qquad (3.149)$$

which shows that $\boldsymbol{\Phi}(k)\boldsymbol{B}$ can be expressed as the sum of $T^{k\alpha}\boldsymbol{A}^k\boldsymbol{B}$ and a linear combination of past terms $\boldsymbol{\Phi}(i)\boldsymbol{B}$, $0 \leqslant i < k$ (again values of scalars ν_i are not of interest), for $k \geqslant 1$.

Then column vector $\boldsymbol{\Phi}(k)\boldsymbol{B}$ of matrix $\boldsymbol{\mathcal{C}}$ can be expressed as a linear combination of $\boldsymbol{A}^k\boldsymbol{B}$ and previous column vectors $\boldsymbol{\Phi}(i)\boldsymbol{B}$, $0 \leqslant i < k$.

Matrix $\boldsymbol{\mathcal{C}}$ would be full-rank if it had n column vectors linearly independent. According to the previous result (3.149), $\boldsymbol{\Phi}(k)\boldsymbol{B}$ would be linearly independent of the previous column vectors $\boldsymbol{\Phi}(i)\boldsymbol{B}$, $0 \leqslant i < k$ (columns located at its left in matrix $\boldsymbol{\mathcal{C}}$), if and only if $\boldsymbol{A}^k\boldsymbol{B}$ were linearly independent of these previous column vectors (or equivalently, linearly independent of previous $\boldsymbol{A}^i\boldsymbol{B}$). Then, taking into account that $\boldsymbol{\Phi}(0)\boldsymbol{B} = \boldsymbol{B}$, matrix $\boldsymbol{\mathcal{C}}_1$ defined in (3.146) has the same rank as $\boldsymbol{\mathcal{C}}$, and the criterion is proven.   □

3.8.2 Controllability Conditions

Controllability Criterion 1. The system given by (3.114) and (3.115) is controllable (in N steps) if and only if matrix $\boldsymbol{\mathcal{C}}$ defined as

$$\boldsymbol{\mathcal{C}} = [\boldsymbol{\Phi}(0)\boldsymbol{B} \quad \boldsymbol{\Phi}(1)\boldsymbol{B} \quad \boldsymbol{\Phi}(2)\boldsymbol{B} \ldots \boldsymbol{\Phi}(N-1)\boldsymbol{B}], \qquad (3.150)$$

is full-rank.

Proof. The solution of the system (3.114) and (3.115) is given by (3.123) which, for an arbitrary initial condition $\boldsymbol{x}_0 = \boldsymbol{x}(0)$ and final state $\boldsymbol{x}_{\mathrm{f}} = \boldsymbol{0}$, yields

$$\boldsymbol{0} = \boldsymbol{\Phi}(N)\boldsymbol{x}_0 + \sum_{i=0}^{N-1}\boldsymbol{\Phi}(N-1-i)\boldsymbol{B}T^{\alpha}u(i) \implies$$

$$-\boldsymbol{\Phi}(N)\boldsymbol{x}_0 = T^{\alpha}\left[\boldsymbol{\Phi}(0)\boldsymbol{B}, \boldsymbol{\Phi}(1)\boldsymbol{B}, \cdots, \boldsymbol{\Phi}(N-1)\boldsymbol{B}\right]\begin{bmatrix} u(N-1) \\ u(N-2) \\ \vdots \\ u(0) \end{bmatrix}. \qquad (3.151)$$

This forms a set of n equations with N unknowns, which are the components of the input sequence. This set of equations has a solution if and only if matrix (3.150) is full-rank (note that $T^\alpha \neq 0$), and the criterion is proven.

$\square$

As before, a consequence of this criterion is that $N \geqslant n$ must be verified.

The above criterion can also be expressed more easily, adopting the same form of the second reachability criterion.

Controllability Criterion 2. The system given by (3.114) and (3.115) is controllable (in N steps) if and only if matrix $\boldsymbol{C}_1$ defined as

$$\boldsymbol{C}_1 = \left[\boldsymbol{B}, \; \boldsymbol{AB}, \; \boldsymbol{A}^2\boldsymbol{B}, \; \cdots, \; \boldsymbol{A}^{N-1}\boldsymbol{B}\right], \qquad (3.152)$$

is full-rank.

Proof. In the demonstration of the second reachability condition it was proved that $\operatorname{rank}(\boldsymbol{C}_1) = \operatorname{rank}(\boldsymbol{C})$. This also applies here and the criterion is proven. $\square$

Matrices $\boldsymbol{C}$ and $\boldsymbol{C}_1$ are denoted *discrete-time controllability matrices*. We note that this last criterion resembles the controllability criterion for continuous systems based on the rank of (3.82). We recall that our discrete-time models arise from discretizing continuous models using the Grünwald–Letnikov operator, and there exists a direct link between both controllability matrices.

3.9 Observability of Discrete-time LTI Commensurate-order Systems

Finally, the concept of observability is extended here to fractional-order discrete-time LTI systems of the form given by (3.114), (3.115) and (3.116). First the observability definition for discrete-time systems is presented.

Definition 3.16. The linear, discrete-time, fractional-order system modeled by (3.114), (3.115), and (3.116) is *observable* if and only if there exists a number N such that from the knowledge of the input sequence $\{u(0), u(1), \ldots, u(N-1)\}$ and the output sequence $\{y(0), y(1), \ldots, y(N-1)\}$ the initial state $\boldsymbol{x}_0 = \boldsymbol{x}(0)$ can be determined. $\square$

The next results were also originally developed in [37]. We also mention that an alternative approach to the observability problem, based on a generalized discrete-time observability Gramian, can be found in [30]. The observability condition obtained in [30] from this Gramian is more complex than that presented in [37], similarly to the comment made in the previous section about the controllability Gramian. However, [30] provides, besides its condition, an estimation of the initial state.

Observability Criterion 1. The system given by (3.114), (3.115), and (3.116) is observable (in N steps) if and only if matrix $\mathcal{O}$, defined as

$$
\mathcal{O} = \begin{bmatrix} \boldsymbol{C\Phi}(0) \\ \boldsymbol{C\Phi}(1) \\ \boldsymbol{C\Phi}(2) \\ \vdots \\ \boldsymbol{C\Phi}(N-1) \end{bmatrix}, \tag{3.153}
$$

is full-rank.

Proof. Substitute the solution of the system (3.114) and (3.115), given by (3.123), in the output equation (3.116). Then it is found that

$$
y(k) = \boldsymbol{C\Phi}(k)\boldsymbol{x}(0) + \boldsymbol{C}\sum_{i=0}^{k-1}\boldsymbol{\Phi}(k-1-i)\boldsymbol{B}T^{\alpha}u(i) + \boldsymbol{D}u(k), \quad k \geqslant 1. \tag{3.154}
$$

By denoting

$$
\hat{y}(k) = y(k) - \boldsymbol{D}u(k) - \boldsymbol{C}\sum_{i=0}^{k-1}\boldsymbol{\Phi}(k-1-i)\boldsymbol{B}T^{\alpha}u(i) \quad k \geqslant 1,
$$
$$
\hat{y}(0) = y(0) - \boldsymbol{D}u(0), \tag{3.155}
$$

(3.154) can be expressed as

$$
\hat{y}(k) = \boldsymbol{C\Phi}(k)\boldsymbol{x}(0), \quad k \geqslant 0. \tag{3.156}
$$

It is noted that $\hat{y}(k)$ can be easily calculated from (3.155) if $u(i), 0 \leqslant i \leqslant k$ and $y(k)$ were known.

If input $u(k)$ and output $y(k)$ were known for $0 \leqslant k \leqslant N-1$, and hence $\hat{y}(k)$, N equations would be obtained from particularizing (3.156) for $0 \leqslant k \leqslant N-1$. These equations can be expressed in a compact matrix form as

$$
\begin{bmatrix} \hat{y}(0) \\ \hat{y}(1) \\ \hat{y}(2) \\ \vdots \\ \hat{y}(N-1) \end{bmatrix} = \begin{bmatrix} \boldsymbol{C\Phi}(0) \\ \boldsymbol{C\Phi}(1) \\ \boldsymbol{C\Phi}(2) \\ \vdots \\ \boldsymbol{C\Phi}(N-1) \end{bmatrix} \boldsymbol{x}(0). \tag{3.157}
$$

This forms a set of N equations with n unknowns, which are the components of the initial state $\boldsymbol{x}(0)$. This set of equations has a solution if and only if matrix (3.153) is full-rank, and the criterion is proven. $\qquad\square$

A first consequence of this criterion is that $N \geqslant n$ must be verified.

Moreover the before criterion can be rewritten in an easier form as follows.

Observability Criterion 2. The system given by (3.114), (3.115), and (3.116) is observable (in N steps) if and only if matrix $\mathcal{O}_1$, defined as

$$
\mathcal{O}_1 = \begin{bmatrix} C \\ CA \\ CA^2 \\ \vdots \\ CA^{N-1} \end{bmatrix}, \tag{3.158}
$$

is full-rank.

Proof. Left multiplying (3.122) by C, substituting $k+1$ by k, splitting the first term on the right hand side of this expression, and then rearranging terms, it is shown that

$$
C\Phi(k) = T^\alpha CA\Phi(k-1) - \sum_{i=1}^{k} (-1)^i \binom{\alpha}{i} C\Phi(k-i), \quad k \geqslant 1, \tag{3.159}
$$

which shows that $C\Phi(k)$ can be expressed as the sum of $T^\alpha CA\Phi(k-1)$ and a linear combination of past terms $C\Phi(k-i)$, $1 \leqslant i \leqslant k$. Use (3.159) to define $C\Phi(k-i)$ on the left hand side of the equation. Then, substituting this in the first term on the right hand side, and taking into account the property (which is not proven here) that $A\Phi(i) = \Phi(i)A$, $i \geqslant 0$, yields

$$
C\Phi(k) = C(T^\alpha A)^2 \Phi(k-2) + \sum_{i=0}^{k-1} \hat{\nu}_i C\Phi(i), \tag{3.160}
$$

where $\hat{\nu}_i$ are scalars whose particular values are not of interest. And repeating this process recursively until $\Phi(0)$ appears in the first term of the right hand side of the expression, it is shown that

$$
C\Phi(k) = C(T^\alpha A)^k \Phi(0) + \sum_{i=0}^{k-1} \nu_i C\Phi(i), \tag{3.161}
$$

which shows that $C\Phi(k)$ can be expressed as the sum of $T^{k\alpha}CA^k$ and a linear combination of past terms $C\Phi(i)$, $0 \leqslant i < k$ (again values of scalars ν_i are not of interest), for $k \geqslant 1$.

Then row vector $C\Phi(k)$ of matrix $\mathcal{O}$ can be expressed as a linear combination of CA^k and previous row vectors $C\Phi(i)$, $0 \leqslant i < k$.

Matrix $\mathcal{O}$ would be full-rank if it had n row vectors linearly independent. According to the previous result (3.161), $C\Phi(k)$ would be linearly independent of the previous row vectors $C\Phi(i)$, $0 \leqslant i < k$ (rows located over in matrix $\mathcal{O}$), if and only if CA^k were linearly independent of these previous row vectors (or equivalently, linearly independent of previous CA^i). Then, taking into account that $C\boldsymbol{\Phi}(0) = C$, matrix $\mathcal{O}_1$ defined in (3.158) has the same rank as $\mathcal{O}$, and the criterion is proven.   □

Matrices $\mathcal{O}$ and $\mathcal{O}_1$ are denoted *discrete-time observability matrices*. This last criterion also resembles the observability criterion for continuous systems based on (3.100).

3.10 Summary

This chapter has developed models of commensurate-order LTI systems, both in the continuous-time and the discrete-time cases. The dynamic responses of these fractional-order models have been presented. Analysis techniques have also been presented for studying the stability, the controllability, and the observability of these systems, again in both the continuous-time and discrete-time cases. In this chapter the approach of studying discrete-time fractional-order models derived as exhibiting dynamics equivalent to continuous-time fractional-order models has been adopted.

Techniques shown in this chapter can be regarded as generalizations of the well known techniques for state-space integer-order models. In this sense, though the presented techniques have been developed for the case of SISO systems, these can easily be extended to the case of multiple inputs-multiple outputs systems. However the extension of these techniques to fractional-order systems of non-commensurate order is not simple, and remains as an open research area.

This chapter has focused only on modeling and analysis issues. The fundamentals of fractional-order control will be presented in the next chapter. The reader should note that control design techniques based on state-space representations as well as the design of observers for fractional-order systems are still areas of active research, and will not be developed in this book, nor in its parts devoted to controllers design. However the interested reader may find some results in the following citations.

For continuous-time fractional-order systems, we mention the design of PD^μ and PI^λ controllers in the state-space developed in [38], and the design of an observer and the observer-based controller presented in [39]. We also mention some robustness results obtained using state-space techniques in issues like stability [40] or controllability [41], and with sliding-mode control techniques [42].

For discrete-time fractional-order systems, we mention the design of observers [34], fractional-order Kalman filters [43], the application of these in control [44], and the adaptive control based on fractional-order systems identification, Kalman filters, and state feedback [45].

Chapter 4
Fundamentals of Fractional-order Control

This chapter gives a historical review of fractional-order control. The main basis of the application of fractional calculus to control is given. In order for the reader to understand the effects of the generalized control actions (derivative and integral ones), a section is devoted to this topic.

4.1 Why Fractional-order Control: Historical Review

Maybe the first sign of the potential of FOC, though without using the term "fractional," emerged with Bode [46, 47]. A key problem in the design of a feedback amplifier was to devise a feedback loop so that the performance of the closed-loop is invariant to changes in the amplifier gain. Bode presented an elegant solution to this robust design problem, which he called the *ideal cutoff characteristic*, nowadays known as *Bode's ideal loop transfer function*, whose Nyquist plot is a straight line through the origin giving a phase margin invariant to gain changes. Clearly, this ideal system is, from our point of view, a fractional-order integrator with transfer function $G(s) = (\omega_{cg}/s)^{\alpha}$, known as *Bode's ideal transfer function*, where ω_{cg} is the gain crossover frequency and the constant phase margin is $\varphi_m = \pi - \alpha\pi/2$, as we illustrated in Section 2.3.4. This frequency characteristic is very interesting in terms of robustness of the system to parameter changes or uncertainties, and several design methods have made use of it. In fact, the fractional-order integrator can be used as an alternative reference system for control [48].

This first step towards the application of fractional calculus in control led to the adaptation of the FC concepts to frequency-based methods. The frequency response and the transient response of the non-integer-order integral (in fact Bode's loop ideal transfer function) and its application to control systems was introduced by Manabe [15], and more recently in [49].

In what concerns automatic control, Oustaloup [50] studied the fractional-order algorithms for the control of dynamic systems and demonstrated the superior performance of the CRONE (Commande Robuste d'Ordre Non Entier, meaning Non-integer-order Robust Control) method over the PID controller. There are three generations of CRONE controllers, and [51] concentrates on the third generation that we will introduce in detail in Chapter 9. Podlubny [52] proposed a generalization of the PID controller, namely the $PI^\lambda D^\mu$ controller, involving an integrator of order λ and a differentiator of order μ. He also demonstrated the better response of this type of controller, in comparison with the classical PID controller, when used for the control of fractional-order systems. A frequency domain approach by using fractional-order PID controllers was also studied in [53].

Further research activities to define new effective tuning techniques for non-integer-order controllers used an extension of classical control theory. In this respect, in [54, 55] the extension of derivative and integration orders from integer to non-integer numbers provides a more flexible tuning strategy and therefore an easier way of achieving control requirements with respect to classical controllers. In [56] an optimal fractional-order PID controller based on specified gain and phase margins with a minimum integral squared error (ISE) criterion is designed. Other work [57, 58] takes advantage of the fractional orders introduced in the control action in order to design a more effective controller to be used in real-life models. The tuning of integer-order PID controllers is addressed in [59–61] by minimizing an objective function that reflects how far the behavior of the PID is from that of some desired fractional-order transfer function (FOTF), and in [62] with a somewhat similar strategy. The use of a new control strategy to control first-order systems with long time delay is also pursued in [63, 64]. Another approach is achieved in [65], where tuning and auto-tuning rules for fractional-order PID controllers are given. An interesting robustness constraint is considered in this work, forcing the phase of the open-loop system to be flat at the gain crossover frequency.

Fractional calculus also extends to other kinds of control strategies different from PID ones. For $\mathcal{H}_2$ and $\mathcal{H}_\infty$ controllers, for instance, [66] discusses the computation of the $\mathcal{H}_2$ norm of a fractional-order SISO system (without applying the result to the synthesis of controllers), and [67] suggests the tuning of $\mathcal{H}_\infty$ controllers for fractional-order SISO systems by numerical minimization.

Applications of fractional calculus in control are numerous. In [68] the control of viscoelastic damped structures is presented. Control applications to a flexible transmission [69, 70], an active suspension [71], a buck converter [72, 73], and a hydraulic actuator [74] are also found in the literature. The fractional-order control of a flexible manipulator is the objective in [75], rigid robots are treated in [76,77], and the fractional-order control of a thermal system in [78–81].

Through this range of design techniques and applications, though quite far from aiming at completeness, it is clear that FOC has become an important research topic. The generalization to non-integer-orders of traditional controllers or control schemes translates into more tuning parameters and more adjustable time and frequency responses of the control system, allowing the fulfilment of robust performance.

4.2 Generalized Fractional-order Control Actions

Starting from the block diagram of Figure 4.1, the effects of the basic control actions of type Ks^μ for $\mu \in [-1, 1]$ will be examined in this section. The basic control actions traditionally considered will be particular cases of this general case, in which:

- $\mu = 0$: proportional action,
- $\mu = -1$: integral action,
- $\mu = 1$: derivative action.

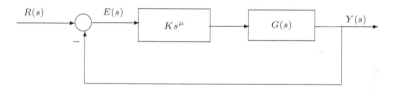

Figure 4.1 Block diagram of a closed-loop system with fractional-order control actions

4.2.1 Integral Action

As is known, the main effects of the integral actions are those that make the system slower, decrease its relative stability, and eliminate the steady-state error for inputs for which the system had a finite error.

These effects can be observed in the different domains. In the time domain, the effects over the transient response consist of the decrease of the rise time and the increase of the settling time and the overshoot. In the complex plane, the effects of the integral action consist of a displacement of the root locus of the system towards the right half-plane. Finally, in the frequency domain, these effects consist of an increment of $-20\,\mathrm{dB/dec}$ in the slopes of the magnitude curves and a decrement of $\pi/2\,\mathrm{rad}$ in the phase plots.

In the case of a fractional-order integral, that is, $\mu \in (-1,0)$, the selection of the value of μ needs consideration of the effects mentioned above.

In the time domain, the effects of the control action can be studied considering the effects of this action over a squared error signal.

If the error signal has the form

$$e(t) = \sum_{k=0}^{N} (-1)^k u_0(t - kT), \qquad k = 0, 1, 2, \cdots, N, \tag{4.1}$$

where $u_0(t)$ is the unit step, its Laplace transform is

$$E(s) = \sum_{k=0}^{N} (-1)^k \frac{e^{-kTs}}{s}. \tag{4.2}$$

So, the control action, as shown in the block diagram of Figure 4.1, will be given as

$$
\begin{aligned}
u(t) &= \mathscr{L}^{-1}\left\{U(s)\right\} = \mathscr{L}^{-1}\left\{ K \sum_{k=0}^{N} (-1)^k \frac{e^{-kTs}}{s^{1-\mu}} \right\} \\
&= K \sum_{k=0}^{N} \frac{(-1)^k}{\Gamma(1-\mu)} (t - kT)^{-\mu} u_0(t - kT).
\end{aligned}
\tag{4.3}
$$

Figure 4.2 shows the function $u(t)$ for the values $\mu = 0, -0.2, -0.5, -1$; $T = 30$; $N = 4$. As can be observed, the effects of the control action over the error signal vary between the effects of a proportional action ($\mu = 0$, square signal) and an integral action ($\mu = -1$, straight lines curve). For intermediate values of μ, the control action increases for a constant error, which results in the elimination of the steady-state error (see Table 2.1), and decreases when the error is zero, resulting in a more stable system.

In the complex plane, the root locus of the system with the control action is governed by

$$1 + Ks^{\mu}G(s) = 0, \tag{4.4}$$

or by the following equivalent conditions for the magnitude and phase:

$$|K| = \frac{1}{|s^{\mu}|\,|G(s)|}, \tag{4.5}$$

$$\arg\left[s^{\mu}G(s)\right] = (2n+1)\,\pi, \qquad l = 0, \pm 1, \pm 2, \cdots. \tag{4.6}$$

Taking into account that

$$s = |s|\,e^{j\phi} \Longrightarrow s^{\mu} = |s|^{\mu}\,e^{j\mu\phi}, \tag{4.7}$$

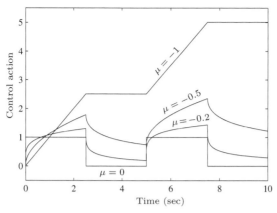

Figure 4.2 Integral control action for a square error signal and $\mu = 0, -0.2, -0.5, -1$

the conditions of magnitude and phase can be expressed by

$$|K| = \frac{1}{|s|^{\mu} |G(s)|}, \tag{4.8}$$

$$\arg\left[s^{\mu} G(s)\right] = \arg\left[G(s)\right] + \mu\phi = (2n + 1)\,\pi, \quad l = 0, \pm 1, \pm 2, \cdots. \tag{4.9}$$

The selection of the value of $\mu \in (-1, 0)$ affects the displacement of the root locus towards the right half-plane and the values of K that make that the magnitude condition is reached.

In the frequency domain, the magnitude curve is given by

$$20 \log \left|s^{\mu} G(s)\right|_{s=j\omega} = 20 \log \left|G(j\omega)\right| + 20\mu \log \omega, \tag{4.10}$$

and the phase plot by

$$\arg\left[s^{\mu} G(s)\right]_{s=j\omega} = \arg\left[G(s)\right] + \mu\frac{\pi}{2}. \tag{4.11}$$

By varying the value of μ between -1 and 0, it is possible:

- To introduce a constant increment in the slopes of the magnitude curve that varies between $-20\,\mathrm{dB/dec}$ and $0\,\mathrm{dB/dec}$.
- To introduce a constant delay in the phase plot that varies between $-\pi/2\,\mathrm{rad}$ and $0\,\mathrm{rad}$.

4.2.2 Derivative Action

It is known that the derivative action increases the stability of the system and tends to emphasize the effects of noise at high frequencies. In the time domain, a decrease in the overshoot and the settling time is observed. In the complex plane, the derivative action produces a displacement of the root locus of the system towards the left half-plane. In the frequency domain, this action produces a constant phase lead of $\pi/2$ rad and an increase of $20\,\mathrm{dB/dec}$ in the slopes of the magnitude curves.

Following a procedure similar to that for the integral action, it is easy to prove that all these effects can be weighted by the selection of the order of the derivative action, that is, $\mu \in (0,1)$.

In the time domain, the effects of the derivative control action can be studied considering the effects of this action over a trapezoidal error signal given by

$$e(t) = t\,u_0(t) - t(t-T)\,u_0(t-T) - t(t-2T)\,u_0(t-2T) + t(t-3T)\,u_0(t-3T),$$

whose Laplace transfer function is

$$E(s) = \frac{1}{s^2} - \frac{e^{-Ts}}{s^2} - \frac{e^{-2Ts}}{s^2} + \frac{e^{-3Ts}}{s^2}. \tag{4.12}$$

Therefore, and according to Figure 4.1, the control action will be given by the equation

$$
\begin{aligned}
u(t) = \mathscr{L}^{-1}\left\{U(s)\right\} = \mathscr{L}^{-1}\left\{K\left(\frac{1}{s^{2-\mu}} - \frac{e^{-Ts}}{s^{2-\mu}} - \frac{e^{-2Ts}}{s^{2-\mu}} + \frac{e^{-3Ts}}{s^{2-\mu}}\right)\right\} \\
= \frac{K}{\Gamma(2-\mu)}\left\{t^{1-\mu}u_0(t) - (t-T)^{1-\mu}\,u_0(t-T)\right. \\
\left. - (t-2T)^{1-\mu}\,u_0(t-2T) + (t-T)^{1-\mu}\,u_0(t-3T)\right\}.
\end{aligned}
\tag{4.13}
$$

The effects of the control action over the error signal are shown in Figure 4.3 and vary between the effects of a proportional action ($\mu = 0$, trapezoidal signal) and a derivative action ($\mu = 1$, square signal). For intermediate values of μ, the control action corresponds to intermediate curves. It must be noted that the derivative action is not zero for a constant error and the growth of the control signal is more damped when a variation in the error signal occurs, which implies a better attenuation of high-frequency noise signals.

In the frequency domain, the magnitude curve is given by (4.10) and the phase plot by (4.11). As can be observed, by varying the value of μ between 0 and 1, it is possible:

- to introduce a constant increment in the slopes of the magnitude curve that varies between $0\,\mathrm{dB/dec}$ and $20\,\mathrm{dB/dec}$,
- to introduce a constant delay in the phase plot that varies between $0\,\mathrm{rad}$ and $\pi/2\,\mathrm{rad}$.

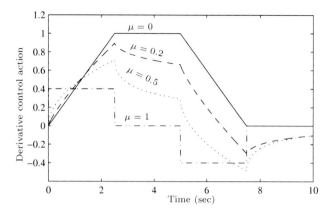

Figure 4.3 Derivative control action for a trapezoidal error signal and $\mu = 0, 0.2, 0.5, 1$

4.3 Generalized PID Controller

This section introduces a more generalized structure for the classical integer-order PID controller, keeping the simplicity of its formulation and making use of the generalized derivative and integral control actions described above. In order to show the characteristics and possibilities of application of the so-called fractional-order PID controller, a comparison with the standard PID will be given in the frequency domain.

4.3.1 Classical PID Controller

The classical PID controller can be considered as a particular form of lead-lag compensation in the frequency domain. Its transfer function can be expressed as

$$C(s) = \frac{U(s)}{E(s)} = K_p + \frac{K_i}{s} + K_d s, \tag{4.14}$$

or

$$C(s) = k \frac{(s/\omega_c)^2 + 2\delta_c s/\omega_c + 1}{s}, \tag{4.15}$$

with $\omega_c = \sqrt{K_i/K_d}$, $\delta = K_p/(2\sqrt{K_i K_p})$, $k = K_i$. Another form can be

$$C(s) = k \frac{(s+a)(s+b)}{s}. \tag{4.16}$$

Therefore, the contributions of the controller depend on one of:

- gains K_p, K_i, K_d;
- gain k and parameters ω_c, δ_c;
- gain k and location of zeros a and b.

In the frequency response of the controller, the selection of these gains or parameters is equivalent to the selection of the position, smoothness, and minimum value of the magnitude curve, and the slope of the phase plot of the controller at the frequency of this minimum. However, at high and low frequencies the values of the slopes in the magnitude curve and the values of the contributions in phase are fixed. This is illustrated in Figure 4.4 for $K_p = K_i = K_d = 1$ and Figure 4.5 for $K_p = 1$, $K_i = 0.5$, $K_d = 1$. Comparing these two figures, it is observed that both the value and position of the magnitude minima and the inflection point of the phase plot are modified by the value of K_i, whilst the slopes of the magnitude curves and the asymptotic values of the phase plots remain the same.

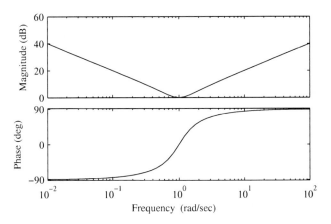

Figure 4.4 Frequency response of the classical PID controller with $K_p = K_i = K_d = 1$

4.3.2 Fractional-order PID Controller

The integro-differential equation defining the control action of a fractional-order PID controller is given by

$$u(t) = K_p e(t) + K_i \mathscr{D}^{-\lambda} e(t) + K_d \mathscr{D}^\mu e(t). \tag{4.17}$$

Applying Laplace transform to this equation with null initial conditions, the transfer function of the controller can be expressed by

$$C_f(s) = K_p + \frac{K_i}{s^\lambda} + K_d s^\mu = k \frac{(s/\omega_f)^{\lambda+\mu} + s\delta_f s^\lambda/\omega_f + 1}{s^\lambda}. \tag{4.18}$$

Figure 4.6 shows the frequency response of this controller for $k = 1$, $\omega_f = 1$, $\delta_f = 1$, and $\lambda = \mu = 0.5$.

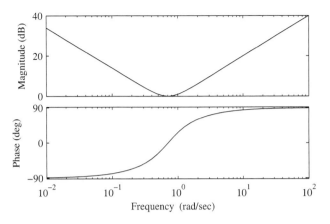

Figure 4.5 Frequency response of the classical PID controller with $K_p = 1$, $K_i = 0.5$, $K_d = 1$

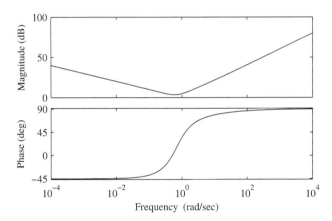

Figure 4.6 Frequency response of the fractional-order PID controller with $k = 1$, $\omega_f = 1$, $\delta_f = 1$, and $\lambda = \mu = 0.5$

As can be observed, this fractional-order controller allows us to select both the slope of the magnitude curve and the phase contributions at both high and low frequencies.

In a graphical way, the control possibilities using a fractional-order PID controller are shown in Figure 4.7, extending the four control points of the classical PID to the range of control points of the quarter-plane defined by selecting the values of λ and μ.

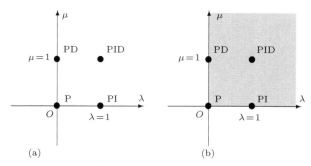

Figure 4.7 Fractional-order PID vs classical PID: from points to plane: (a) integer-order and (b) fractional-order

4.4 Summary

In this chapter, the fundamentals of fractional-order control have been introduced. After a review of the evolution of fractional-order control strategies in control systems, a study of the effects of the fractional-order in the basic control actions, derivative and integral, has been given. Finally, the generalized PID controller is discussed in the frequency domain.

This chapter aims to be an introduction to fractional-order control, taking the very well known PID controller as the starting point. In the following chapters of the book, other control strategies apart from PID control will be discussed in detail.

Part II
Fractional-order PID-Type Controllers

Chapter 5
Fractional-order Proportional Integral Controller Tuning for First-order Plus Delay Time Plants

PID (proportional integral derivative) controllers are the most popular controllers used in industry because of their simplicity, performance robustness, and the availability of many effective and simple tuning methods based on minimum plant model knowledge [82]. A survey has shown that 90% of control loops are of PI or PID structures [83,84]. In control engineering, a dynamic field of research and practice, better performance is constantly demanded; therefore, developing better and simpler control algorithms is a continuing objective.

In the past decade there has been an increase in research efforts related to fractional calculus [3,85,86] and its applications to control theory [58,87–89]. Clearly, for closed-loop control systems, there are four situations: (1) integer-order (IO) plant with IO controller; (2) IO plant with fractional-order (FO) controller; (3) FO plant with IO controller, and (4) FO plant with FO controller. In control practice, the fractional-order controller is more common, because the plant model may have already been obtained as an integer-order model in the classical sense. From an engineering point of view, improving or optimizing performance is the major concern [89]. Hence, our objective is to apply the fractional-order control (FOC) to enhance the (integer-order) dynamic system control performance [58,89]. Pioneering works in applying fractional calculus in dynamic systems and controls and other recent developments can be found in [15,58,90–95].

From Chapters 5 to 7 we will explain a clear and direct methodology for the design of fractional-order PI, fractional-order PD, and fractional-order PID controllers to be used in different types of dynamic systems, giving simulation and experimental results to illustrate the application and effectiveness of these tuning rules.

Let us begin with fractional-order control of first-order plus dead-time (FOPDT) plants which are widely seen in some process industrial environments. Specifically, we will focus on fractional-order proportional integral controllers, PI$^\lambda$. In this chapter, we will show how to develop a set of practical tuning rules for PI$^\lambda$ control of FOPDT plants. The tuning is optimum in the sense that the load disturbance rejection is optimized yet with a constraint on the maximum or peak sensitivity. The so-called MIGO (M_s constrained integral gain optimization) based controller tuning method is generalized to handle the PI$^\lambda$ case, called F-MIGO, given the fractional order α. The F-MIGO method is then used to develop tuning rules for the FOPDT class of dynamic systems. The final tuning rules developed only apply the relative dead time τ of the FOPDT model to determine the best fractional order α and, at the same time, the best PI$^\lambda$ gains. Extensive simulation results are included to illustrate the simple yet practical nature of these new tuning rules, whose development procedure is not only valid for FOPDT but also applicable for other general classes of plants, as illustrated at the end of this chapter.

5.1 Introduction

Much research has been devoted to developing tuning methods for PI$^\lambda$/PI$^\lambda$D$^\mu$ controllers [60,88,96–100]. The method applied in this chapter was motivated from the MIGO design method developed in [101,102], in which the motivation was to improve upon the Ziegler–Nichols tuning rules to overcome two major drawbacks: (1) very little process information was taken into account as the rules were based on the two characterization parameters of the system dynamics based on the step response data, and (2) the quarter amplitude damping design method exhibited very poor robustness. To overcome these drawbacks the authors of [101, 102] choose a new criterion for developing a tuning method for the PI controllers based on robust loop shaping. Here, in this chapter, we first create a generalized MIGO method. It aims to obtain the gains of the PI$^\lambda$ controller for any given fractional order α. As in [101, 102], the design focuses on the maximization of the integral gain with a constraint on maximum sensitivity M_s. The method assumes that the model of the system is available. To develop tuning rules, a test batch of FOPDT systems is chosen and F-MIGO is applied to scan them for different values of fractional order in the range $[0.1 : 0.1 : 1.9]$. The best fractional-order controller is then selected for each system based on the ISE criterion. The new tuning rules are then obtained by establishing relations between the process dynamics and the controller parameters. The final tuning rules developed only apply the relative dead time τ of the FOPDT model to determine the best fractional order α and, at the same time, the best PI$^\lambda$ gains.

The rest of the chapter is organized as follows. Section 5.2 describes the F-MIGO method. Section 5.3 shows the development of the tuning rules. Section 5.4 presents some simulation results to validate the design method, concluding with an overview of the chapter in Section 5.5.

5.2 F-MIGO: Fractional M_s Constrained Integral Gain Optimization Method

The most eminent and historically important work in the history of PID controller tuning is by Ziegler and Nichols [82,103]. The rules given by Ziegler and Nichols were simple, did not require the process transfer function, and were based only on the S-shaped step response data. The rules were effective and gave the designer a good start. A lot of research henceforth has gone into obtaining tuning rules for PID controllers based on different criteria like robust loop shaping, robustness to load disturbances, and robustness to parameter variations [103, 104]. Among these, one of the tuning rules worth mentioning is the MIGO design method developed by K. J. Åström, H. Panagopoulos, and T. Hägglund [101, 102]. Their work was based on improving the Ziegler–Nichols tuning method. The main idea was to come up with simple rules satisfying a very important industry design requirement which is robustness to load disturbance. These are optimization type rules and attempt to find the controller parameters with the objective of optimizing the load disturbance with a constraint on the maximum load disturbance-to-output sensitivity M_s [101, 102].

The most important assumption of this method is that the transfer function of the system has already been given. The system should be linear, and its transfer function must be analytical with finite poles and exhibit an essential singularity at infinity [101].

The PI$^\lambda$ controller can be described in time domain as

$$u(t) = K_p(sp(t) - y(t)) + K_i \mathscr{D}_t^{-\alpha}(sp(t) - y(t)), \qquad (5.1)$$

where $u(t)$ is the control signal, $sp(t)$ the set-point signal, and $y(t)$ the process output. The controller parameters are the proportional gain K_p, the integral gain K_i, and the non-integer-order of the integrator α. The $\mathscr{D}_t^\alpha x$ is the fractional operator as defined in [27]. The frequency domain description of the PI$^\lambda$ is given by

$$C(s) = K_p + \frac{K_i}{s^\alpha}. \qquad (5.2)$$

The primary design aim of this method is the load disturbance rejection. Load disturbances are typically low frequency signals and their attenuation is a very important characteristic of a controller. It is shown [103] that by maximizing the integral gain K_i the effect of load disturbance at output will be minimum. Some of the frequently used criteria are

$$\text{IAE} = \int_0^\infty |e(t)|\,\mathrm{d}t, \quad \text{IE} = \int_0^\infty e(t)\mathrm{d}t, \quad \text{ISE} = \int_0^\infty e^2(t)\mathrm{d}t. \qquad (5.3)$$

The load disturbance is defined by the integrated absolute error (IAE) due to a unit step load disturbance at the output. When the integral of the error (IE) is used, it has been proved [103] that IE $= 1/K_i$. Thus, under special circumstances when the system is well damped and the error is positive, then IE $=$ IAE. Hence, maximizing K_i will minimize the load disturbance. A system can be well damped by constraints imposed on the sensitivity functions. Here, we choose to use ISE. Of course, we can use IAE, but for fair comparison reasons to previous published results where ISE is used, we use it again for tuning rule development.

The loop transfer function is given by $L(s) = C(s)G(s)$, where $C(s)$ is the controller transfer function and $G(s)$ is the plant transfer function. We then define

$$S(s) = \frac{1}{1 + C(s)G(s)}, \quad T(s) = \frac{C(s)G(s)}{1 + C(s)G(s)}. \qquad (5.4)$$

$T(s)$ is called the complementary sensitivity function and it determines the suppression of load disturbances and good set-point tracking. $S(s)$ is called the sensitivity function and it determines the robustness to measurement noise and unmodeled system dynamics. It can be easily observed that $S(s) + T(s) = 1$, hence their sum is always one and both cannot be made zero at the same time. It has been observed in the real world that load disturbance signals and the reference signal are generally in the low frequency range and the measurement noise generally occupies a higher frequency band. So, to ensure good reference tracking and rejection of load disturbance at lower frequencies, $S(s) \approx 0$,, which implies $T(s) \approx 1$. At higher frequencies we need to ensure that the noise due to the measurement methods used is rejected, hence $T(s) \approx 0$, which implies $S(s) \approx 1$. Clearly, there is a design trade off between $S(s)$ and $T(s)$ in their frequency domain behaviors. The peak values of the sensitivity functions are denoted by M_s and M_p, respectively, which are given by

$$M_s = \max_{0 < \omega < \infty} |S(\mathrm{j}\omega)|, \quad M_p = \max_{0 < \omega < \infty} |T(\mathrm{j}\omega)|. \qquad (5.5)$$

The quantity M_s is also the inverse of the shortest distance to the Nyquist plot of the loop transfer function, $L(s) = C(s)G(s)$, from the critical point $(-1, \mathrm{j}0)$ in the complex plane. A circle drawn with center at -1 and radius $1/M_s$ is called the M_s circle. Therefore, by imposing a constraint on the value of M_s, we must ensure that the Nyquist plot of $L(s)$ lies outside the M_s circle. The typical values of M_s are in the range 1.3–2.0. The quantity M_p is the value of the resonance peak of the closed-loop system, being typically in the range 1.0–1.5. The M_p circles can also be drawn with center at $-M_\mathrm{p}^2/(M_\mathrm{p}^2-1)$ and radius $M_\mathrm{p}/(M_\mathrm{p}^2-1)$. Similarly, if we impose a constraint on M_p, we must ensure that the Nyquist plot of $L(s)$ lies outside the M_p circle.

It has been shown in [101, 102] that choosing M_s as the design parameter is useful, since decreasing or increasing its value causes significant changes in the step response of the system. However, it is also important that the value of M_p is not very high. Hence, this problem is overcome by choosing the design parameter to be a circle such that it encloses both the M_s and M_p circles. This circle has its center at C and radius R given by

$$C = \frac{M_\mathrm{s} - M_\mathrm{s}M_\mathrm{p} - 2M_\mathrm{s}M_\mathrm{p}^2 + M_\mathrm{p}^2 - 1}{2M_\mathrm{s}\left(M_\mathrm{p}^2 - 1\right)}, \tag{5.6}$$

$$R = \frac{M_\mathrm{s} + M_\mathrm{p} - 1}{2M_\mathrm{s}\left(M_\mathrm{p}^2 - 1\right)}. \tag{5.7}$$

The optimization problem can be stated as follows:

"Maximize K_i to obtain the controller parameters such that the closed-loop system is stable and the Nyquist plot of the loop transfer function lies outside the circle with center at $s = -C$ and radius R" [101].

Let us now define a function $f(K_\mathrm{p}, K_\mathrm{i}, \omega, \alpha)$ as

$$f(K_\mathrm{p}, K_\mathrm{i}, \omega, \alpha) = |1 + C(\mathrm{j}\omega)G(\mathrm{j}\omega)|^2. \tag{5.8}$$

Then, the sensitivity constraint can be expressed mathematically as

$$f(K_\mathrm{p}, K_\mathrm{i}, \omega, \alpha) \geqslant R^2. \tag{5.9}$$

Therefore, the optimization problem is the maximization of K_i subjected to the sensitivity constraint (5.9). Some important substitutions have to be made in (5.9) before we go any further with the analysis of the optimization problem:

- The PI^α controller transfer function is defined as

$$C(\mathrm{j}\omega) = K_\mathrm{p} + \frac{K_\mathrm{i}}{(\mathrm{j}\omega)^\alpha}, \tag{5.10}$$

where

$$(\mathrm{j}\omega)^\alpha = \mathrm{e}^{\mathrm{j}\pi\alpha/2}\omega^\alpha = \omega^\alpha \cos\gamma + \mathrm{j}\omega^\alpha \sin\gamma, \quad \gamma = \frac{\pi\alpha}{2}. \tag{5.11}$$

- The system transfer function can be expressed as a complex number in the frequency domain:

$$G(\mathrm{j}\omega) = a(\omega) + jb(\omega) = r(\omega)\mathrm{e}^{\mathrm{j}\phi(\omega)}, \tag{5.12}$$

where

$$a(\omega)=r(\omega)\cos\phi(\omega),\ b(\omega)=r(\omega)\sin\phi(\omega),\ r^2(\omega)=a^2(\omega)+b^2(\omega). \quad (5.13)$$

We now substitute (5.10), (5.11), (5.12) in the sensitivity constraint (5.9) to obtain a simplified optimization problem:

$$f=C^2+r^2K_\mathrm{p}^2+2CaK_\mathrm{p}+\frac{K_\mathrm{i}^2 r^2}{\omega^{2\alpha}}+\frac{2r^2 K_\mathrm{p} K_\mathrm{i}\cos\gamma}{\omega^\alpha}+\frac{2CK_\mathrm{i}\left(a\cos\gamma+b\sin\gamma\right)}{\omega^\alpha}\geqslant R^2.$$
$$(5.14)$$

5.2.1 Geometric Interpretation of Optimization Problems

From (5.14), we can see that it represents the general equation of an ellipse. Hence, at a given value of ω the sensitivity constraint represents the exterior of an ellipse. In [101], it has been assumed that positive values of K_i will ensure a stable closed-loop system. However, the difference here is that we have to plot the ellipses at varying values of α also. In Figure 5.1 we observe how the axis of the ellipse rotates with different values of α. In [101] it has been mentioned that the ellipses generate an envelope. Within the solution range, the envelope spans the positive and negative K_p axis and the positive K_i axis. The envelopes have two branches and only the lowest branch corresponds to the stable solution [101]. Therefore, the maximum value of K_i occurring at the lowest branch of the envelope will be the point of optimization. This is indicated in the graphs of Figure 5.1 with an arrow pointing at the position of the maximum value of K_i. Another observation we can make from Figure 5.1 is that a solution cannot be found at all fractional orders. We observe that in the other graphs it is not easy to find such a region and hence we can conclude that a solution satisfying the optimization constraint may not exist at these fractional-order controllers for this particular system.

The geometric illustration of the optimization problem is easy to under-stand. However, it is time consuming to find the envelopes at each order, since different systems may show different characteristics. Therefore, a reliable numerical approach is needed, which will be discussed in the next section.

5.2.2 Numerical Solution of Optimization Problems

Figure 5.1 represents the solution field, and the optimization condition implies that we find the maximum K_i on the envelope. However, it is not easy to generate the envelopes for each system, and efficient numerical methods have to be derived to give us a more effective solution. The envelopes tend to show the following characteristics:

- Some envelopes will have a continuous derivative at the maximum, as seen in Figure 5.1.
- Some maxima can occur at the corners.

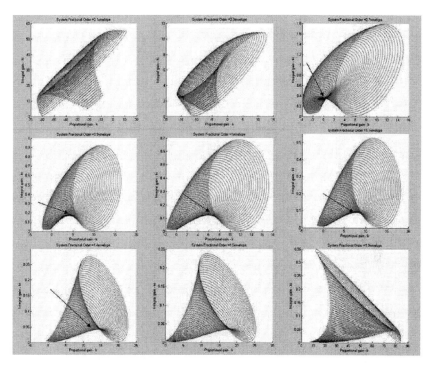

Figure 5.1 Geometric illustration of the envelopes generated by the optimization problem at different values of α for a typical system. The *arrow* mark indicates the frequency at which the optimum is achieved

The systems to be considered are of the first type and the numerical method will be developed based on that assumption. It is important to observe here that we will consider the fractional order to be a constant in the subsequent derivations, as we are trying to generalize the MIGO method.

The envelope can be described mathematically by the following equations:

$$f(K_p, K_i, \omega, \alpha) = R^2, \qquad \frac{\partial f}{\partial \omega}(K_p, K_i, \omega, \alpha) = 0. \tag{5.15}$$

The optimization condition, as explained before, implies finding the maximum K_i on the envelope defined by (5.15). Considering the case where the maximum occurs at the point where the envelope has a continuous derivative, we can observe that

$$df = \frac{\partial f}{\partial K_p} dK_p + \frac{\partial f}{\partial K_i} dK_i + \frac{\partial f}{\partial \omega} d\omega = 0. \tag{5.16}$$

Again, it is important to emphasize the fact that the fractional order α is treated as a constant in the algorithm. In (5.16), we also observe the following:

1. From (5.15) we have $\partial f / \partial \omega = 0$.
2. For the local maximum condition $dK_i = 0$.
3. We impose that, for random variations of dK_p, $\partial f / \partial K_p = 0$.

Hence, with the above-mentioned conditions, the mathematical definition of the optimization problem for the simplest scenario of maximum K_i occurring at the point of continuous derivative is given by

$$f(K_p, K_i, \omega, \alpha) = R^2, \quad \frac{\partial f}{\partial \omega}(K_p, K_i, \omega, \alpha) = 0, \quad \frac{\partial f}{\partial K_p}(K_p, K_i, \omega, \alpha) = 0. \tag{5.17}$$

So, now we have reduced the optimization problem to solving a set of algebraical equations. Some simplification methods will now be applied to the above set of equations to obtain a simple algebraical equation which can be solved using the Newton–Raphson technique [101]. The scenario for the corner case has not been investigated here as the first scenario is the most commonly encountered, but it can be assumed that it will follow the same methodology adopted in [101, 102].

Focusing on our problem, we have to solve a set of nonlinear equations of three variables and with the assumption that α is a constant. Some simple substitutions will give rise to a simple and efficient algorithm, as will be shown next.

First of all, from (5.14) and (5.16), it is found that

$$f = C^2 + r^2 K_p^2 + 2CaK_p + \frac{K_i^2 r^2}{\omega^{2\alpha}} + \frac{2r^2 K_p K_i \cos\gamma}{\omega^\alpha}$$
$$+ \frac{2CK_i \left(a\cos\gamma + b\sin\gamma \right)}{\omega^\alpha} = R^2, \tag{5.18}$$

$$\frac{\partial f}{\partial \omega} = 2K_p^2 rr' + 2CK_p a' + K_i^2 \left(\frac{r^2}{\omega^{2\alpha}} \right)' + 2K_p K_i \cos(\gamma) \left(\frac{r^2}{\omega^\alpha} \right)'$$
$$+ 2CK_i \left(\frac{a}{\omega^\alpha} \right)' \cos\gamma + 2CK_i \left(\frac{b}{\omega^\alpha} \right)' \sin\gamma = 0,$$

$$\frac{\partial f}{\partial K_p} = 2K_p r^2 + 2Ca + \frac{2r^2 K_i \cos\gamma}{\omega^\alpha} = 0. \tag{5.19}$$

In the above equations, the prime means differentiation with respect to ω. Solving (5.18) and (5.19), we can derive equations for K_p and K_i:

$$K_i = -\frac{R\omega^\alpha}{r\sin\gamma} - \frac{Cb\omega^\alpha}{r^2\sin\gamma},$$
(5.20)

$$K_p = R\frac{\cos\gamma}{\sin\gamma} + \frac{Cb\cos\gamma}{r^2\sin\gamma} - \frac{Ca}{r^2}.$$
(5.21)

Equations (5.20) and (5.21) represent the solution for finding the gains of the controller. However, they are dependent on the value of ω. Hence, we need to find a solution for ω. Substituting the equations for K_p and K_i into (5.19) for the final simplification of the optimization problem, the following equation is obtained:

$$\frac{\partial f}{\partial \omega} = \frac{2R^2}{r}r' + \frac{4RCb}{r^2}r' - \frac{2\alpha R^2}{\omega} - \frac{2\alpha RCb}{r\omega} - \frac{2RC}{r}b'.$$

This can be further simplified as in [101] to get a simple algebraic equation as shown below:

$$h(\omega) = \frac{\partial f}{\partial \omega} = 2R\left(\left[C\frac{b}{r} + R\right]\left[\frac{r'}{r} - \frac{\alpha}{\omega}\right] - C\left(\frac{b}{r}\right)'\right).$$
(5.22)

At this point, the optimization problem reduces to (5.22). Solving this equation will give the frequency ω_o at which K_i is maximized, and then we can compute K_p and K_i given by (5.20) and (5.21). However, another condition has to be validated to ensure that the solution is indeed the maximum, *i.e.*,

$$\frac{d^2 f}{d\omega^2} > 0.$$
(5.23)

The Newton–Raphson technique is used to solve (5.22). To ensure that the method converges quickly the initial conditions must be chosen suitably. This algorithm has been designed to eliminate that problem. Instead of applying the principle used in [101] for finding the initial solution, this algorithm allows the user to choose a range of initial solutions and (5.22) is then computed at each of these starting solutions. The final result is stored in an Excel file from which the user can choose the appropriate solution. The idea here is that, since the FOPDT system chosen for developing the tuning rules has one local maximum, starting from any initial solution within the system bandwidth will ensure that the optimal ω is reached.

Let us now summarize the F-MIGO algorithm:

1. Choose any stable system.
2. Choose a range of initial solutions to apply to (5.22). Example: Initial $\omega = [0.1, 0.3, 0.5, 1, 3]$.

3. Choose the fractional order α at which you want to find a controller for the system. Example: $\alpha = 1.1$.

4. Choose the values of the design parameters M_s and M_p. Example: $M_s = 1.4$ and $M_p = 2$.

5. Using Newton–Raphson technique, solve (5.22).

6. Evaluate the condition for maxima (5.23).

7. If Step (vi) is true, then check the values of K_p and K_i. Both should be positive. Else, go back to Step (iii) and change the order. For some fractional orders a stable solution does not exist.

8. If Step (vii) is true, then evaluate the values of M_s and M_p for the new loop transfer function.

9. If the values of M_s and M_p are satisfactory, then check if the system is stable.

10. If Step (ix) is true, then the solution is good and the Excel file will be created. Else, go to Step (iv).

11. If all the steps are true, then the solution is good.

12. Repeat the procedure for the next fractional order.

5.3 Development of the Tuning Rules

5.3.1 Introduction to the Method Used

The motivation for setting the tuning rules comes from the method in [105, 106]. Tuning rules for PI$^\lambda$ must be able to provide the user with the value of the fractional order and gains of the controller. The F-MIGO design algorithm gives us the flexibility of finding the controller gains at arbitrary α. Hence, the development of tuning rules follows a simple algorithm:

1. Choose a suitable test batch.

2. Using the F-MIGO algorithm, design fractional-order controllers for each system in the range [0.1:0.1:1.9].

3. Among the controllers, choose the best based on the ISE criterion [106].

4. Correlate the best controller parameters (α^*, K^*, K_i^*) with the relative dead time τ.

5.3.2 Test Batch

The first obvious choice for the test batch was the set of systems chosen in [105]. However, each of these systems can be approximated by an FOPDT model, whose structure is given by

$$G(s) = k \frac{e^{-Ls}}{Ts + 1},$$
(5.24)

where k is the process gain, which is assumed to be unity since all the systems are normalized; L and T are the delay and time constant of the system, respectively. The FOPDT models are characterized by a very important parameter called the relative dead time of the system, defined as

$$\tau = \frac{L}{L+T}. \qquad (5.25)$$

Parameter τ ranges between 0 and 1. Systems in which $L \gg T$ are called "delay dominant" and systems in which $T \gg L$ are called "lag dominant".

In the test batch parameters L and τ are in the range $L = (20, 10, 1)$ and $\tau = (0.99, 0.9, 0.8, 0.7, 0.6, 0.5, 0.4, 0.3, 0.2, 0.1, 0.05, 0.01, 0.009)$. The value of the time constant T can be derived from (5.25). We will use the terminology "Sys(xy)" to denote a system. Parameters x and y are the indexes pointing at the values of L and τ, respectively. That is, parameter x can take the values $(1, 2, 3)$, since there are 3 values considered for parameter L; and y can take the values $(1, 2, \cdots, 13)$, since there are 13 possible values for τ. Besides, we will denote a system as Type1, Type2, or Type3 according to the values $L = 20$, $L = 10$, and $L = 1$, respectively. For example, Sys15 refers to the system with $L = 20$ and $\tau = 0.6$, and it is a Type1 system $(L = 20)$.

5.3.3 F-MIGO Applied to the Test Batch

Figure 5.2 explains the steps followed in order to choose the best fractional order (α^*) controller for Type1 systems. The first step is to find the controller gains at all fractional orders in the range $[0.1{:}0.1{:}1.9]$. The orders which give a feasible solution are stored in an Excel file. "Feasible" here implies that all the conditions stated for the F-MIGO algorithm are satisfied. The first table in Figure 5.2 shows the solutions obtained for each Type1 system. The second one is the table of Type1 systems vs the fractional orders which give a good solution. Note that the empty areas imply that a solution could not be found. The closed-loop step response gives us the value of ISE for each controller. PI^λ controllers are simulated using Oustaloup's recursive approximation [51]. Details of the approximation is given in Section 12.1.2. This is summarized in the second table. The minimum value of ISE has been marked and it corresponds to the best fractional-order controller for that particular system. The third table summarizes the list of the best fractional order and gain values for the Type1 systems. This procedure is repeated for Type2 and Type3 systems to find the best PI^λ for the test batch.

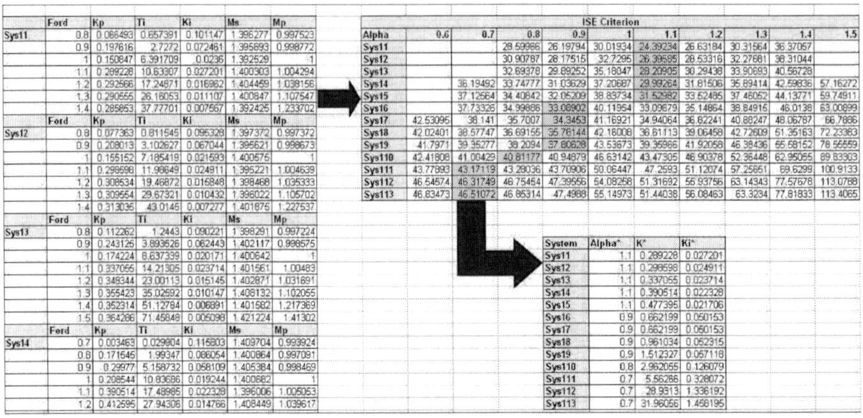

Left tables (Ford / Kp / Ti / Ki / Ms / Mp):

	Ford	Kp	Ti	Ki	Ms	Mp
Sys11	0.8	0.066493	0.657391	0.101147	1.396277	0.997523
	0.9	0.197616	2.7272	0.072461	1.395893	0.998772
	1	0.150647	6.391709	0.0236	1.392529	1
	1.1	0.209228	10.63907	0.027201	1.400303	1.004294
	1.2	0.292866	17.24871	0.016962	1.404459	1.038156
	1.3	0.290656	26.16053	0.011107	1.400847	1.107547
	1.4	0.285853	37.77701	0.007667	1.392426	1.233702
Sys12	0.8	0.077363	0.811545	0.095328	1.397372	0.997372
	0.9	0.208013	3.102627	0.067044	1.395621	0.998673
	1	0.155152	7.185419	0.021593	1.400675	1
	1.1	0.298698	11.99649	0.024911	1.395221	1.004639
	1.2	0.308534	19.46872	0.016848	1.398468	1.035333
	1.3	0.309554	29.67321	0.010432	1.396022	1.105702
	1.4	0.313036	43.0145	0.007277	1.401876	1.227537
Sys13	0.8	0.112262	1.2443	0.090221	1.398291	0.997224
	0.9	0.243125	3.893526	0.062443	1.402117	0.998575
	1	0.174224	8.637339	0.020171	1.400642	1
	1.1	0.337055	14.21305	0.023714	1.401561	1.00483
	1.2	0.345144	23.00113	0.015145	1.402871	1.031891
	1.3	0.395423	35.02592	0.010147	1.408132	1.102055
	1.4	0.352314	51.12784	0.006891	1.401582	1.217369
	1.5	0.364286	71.45648	0.006098	1.421224	1.41302
Sys14	0.7	0.003463	0.029804	0.115803	1.409704	0.993924
	0.8	0.171645	1.99347	0.086054	1.400864	0.997081
	0.9	0.29977	5.158732	0.058109	1.405384	0.998469
	1	0.298544	10.83686	0.019244	1.400682	1
	1.1	0.390514	17.48985	0.022328	1.396006	1.005053
	1.2	0.412595	27.94306	0.014766	1.408449	1.039617

Middle table — ISE Criterion:

	0.6	0.7	0.8	0.9	1	1.1	1.2	1.3	1.4	1.5
Sys11				29.59986	26.19794	30.01934	24.39234	26.63184	30.31564	36.37057
Sys12				30.90787	28.17515	32.7295	26.39685	26.53316	32.27681	38.31044
Sys13				32.89378	29.89252	35.18047	28.20905	30.29408	33.90693	40.56728
Sys14			36.19492	33.74777	37.20687	29.99264	31.61506	36.89414	42.59836	57.16272
Sys15			37.12564	34.40842	32.05209	38.89734	33.62485	37.48052	44.13771	59.74911
Sys16		37.73326	34.99686	33.09902	40.11954	33.09679	35.14864	38.84916	46.0138	63.00899
Sys17	42.53095	38.141	35.7007	34.3453	41.16921	34.94064	36.82241	40.88247	48.06787	66.7886
Sys18	42.02401	38.57747	36.69155	35.76144	42.16008	36.61113	39.06458	42.72609	51.35163	72.23363
Sys19	41.7971	39.35277	38.2094	37.60628	43.53673	39.35966	41.92058	46.38436	55.58152	78.56559
Sys110	42.41808	41.00429	40.81177	40.94879	46.63142	43.47305	46.90378	52.36448	62.95065	89.83303
Sys111	43.77993	43.17119	43.29036	43.70906	50.06447	47.2593	51.12074	57.25851	69.6299	100.9133
Sys112	46.54574	46.31749	46.75464	47.39556	54.08258	51.31682	56.93756	63.14343	77.57678	113.0789
Sys113	46.83473	46.51072	46.85314	47.4988	55.14973	51.44038	56.08463	63.3234	77.81833	113.4065

Right lower table:

System	Alpha*	K*	Ki*
Sys11	1.1	0.209226	0.027201
Sys12	1.1	0.298698	0.024911
Sys13	1.1	0.337055	0.023714
Sys14	1.1	0.390514	0.022328
Sys15	1.1	0.477395	0.021706
Sys16	0.9	0.662199	0.050153
Sys17	0.9	0.662199	0.050153
Sys18	0.9	0.961034	0.052315
Sys19	0.9	1.512327	0.057118
Sys110	0.8	2.962055	0.126079
Sys111	0.7	5.56266	0.328072
Sys112	0.7	28.9313	1.336192
Sys113	0.7	31.96056	1.456195

Figure 5.2 Flow for the selection of the best fractional-order controller for a given system

5.3.4 Tuning Tables

Figure 5.3 illustrates the relations between the best PI$^\lambda$ controller parameters and the process parameters.

Figure 5.3 (a) gives the relation between the best fractional order and τ. As shown in [105], the controller parameters are first normalized. The proportional gains K_p^* are multiplied by their respective process gain k, which in this case are all unity, and the integral gains $T_i^* = K_p^*/K_i^*$ are divided by their respective process time constant T, and plotted vs τ, as seen in Figure 5.3 (b,c), respectively.

5.3.4.1 Parameter α^* vs τ

Figure 5.3 (a) reveals some interesting observations. The fractional order depends on the value of τ and is almost invariant to the value of L. The ambiguous region between $[0.4 < \tau < 0.6]$ implies that the best fractional order is close to unity, indicating that a fractional-order controller may be unnecessary for these systems. This region has been approximated by a straight line. Delay dominant systems need an order higher than 1 and lag dominant systems can be controlled efficiently with a lower order controller. This relation can be approximated by

$$\alpha = \begin{cases} 1.1, & \text{if } \tau \geqslant 0.6 \\ 1.0, & \text{if } 0.4 \leqslant \tau < 0.6 \\ 0.9, & \text{if } 0.1 \leqslant \tau < 0.4 \\ 0.7, & \text{if } \tau < 0.1. \end{cases} \tag{5.26}$$

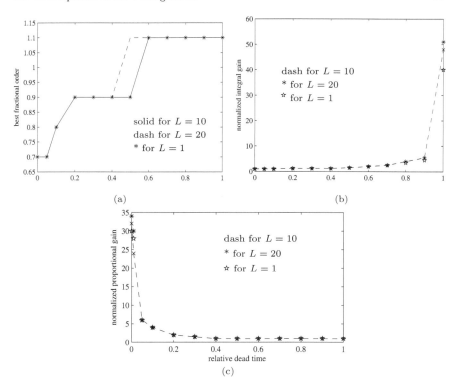

Figure 5.3 Normalized FOC parameters vs the relative dead time τ: (a) α^* vs τ, (b) kK_p^* vs τ, and (c) T_i^*T vs τ

5.3.4.2 Normalized Controller Gains vs τ

The Curve Fitting Toolbox of MATLAB has been used to find the tuning rules in Figure 5.3 (b,c). It was observed that the process of data fitting may not reproduce the exact results of the analytical tuning and hence a region of $\pm 15\%$ around the tuning rules should be considered. The interesting observations from the graph again reveal the dependency on the value of τ. For lag dominant systems the proportional gains are higher, whilst the value drops considerably for delay dominant systems. A similar but opposite trend is observed for the integral gain. These rules are summarized as follows:

$$K_p^* = \frac{1}{k}\left(\frac{0.2978}{\tau + .000307}\right), \quad T_i^* = T\left(\frac{0.8578}{\tau^2 - 3.402\tau + 2.405}\right). \tag{5.27}$$

5.3.5 Summary of Tuning Rules

The tuning rules reveal a very good dependency of the controller parameters
on the relative dead time of the system. Thus, given a system transfer function
or its step response, the tuning rules can be summarized as follows:

1. Find the FOPDT model of the system and define the values k, L, T.
2. Find the relative dead time τ of the system.
3. From the value of τ, calculate the fractional order α from (5.26).
4. Find the controller gains from (5.27). Allow a 15% range around the
gain values.

5.4 Simulation Results

5.4.1 Validity of the F-MIGO Method

For the validation of the method we will consider the systems chosen in [101]
and will obtain the parameters of the PI controller when $\alpha = 1$. Table 5.1
summarizes the same results as shown in Table 1 in [101]. Observe that for
$\alpha = 1$ the results obtained match with those of the MIGO method. Hence,
this method can be used to design an M_s constrained PI$^\lambda$ controller at any
given value of α.

Table 5.1 Validation of the F-MIGO method for the systems $G_1(s)$–$G_6(s)$ as in [101]

Process	α	M_s	K_p	K_i	M_p	ω_o	Process	α	M_s	K_p	K_i	M_p	ω_o
$G_1(s)$	1.0	1.4	0.633	1.945	1.00	0.737	$G_2(s)$	1.0	1.4	1.930	0.744	1.10	3.378
	1.6		0.861	1.868	1.04	0.787		1.6		2.741	0.671	1.27	3.820
	1.8		1.053	1.816	1.24	0.827		1.8		3.469	0.622	1.46	4.180
	2.0		1.218	1.779	1.45	0.860		2.0		4.112	0.587	1.66	4.470
$G_3(s)$	1.0	1.4	0.156	5.862	1.00	0.096	$G_4(s)$	1.0	1.4	0.167	13.64	1.40	0.293
	1.6		0.201	5.667	1.00	0.098		1.6		0.231	10.43	1.50	0.343
	1.8		0.233	5.480	1.02	0.100		1.8		0.285	8.967	1.62	0.377
	2.0		0.258	5.352	1.17	0.102		2.0		0.332	7.960	1.77	0.407
$G_5(s)$	1.0	1.4	0.177	1.758	1.00	0.385	$G_6(s)$	1.0	1.4	0.313	0.373	1.04	1.985
	1.6		0.223	1.685	1.00	0.397		1.6		0.386	0.343	1.15	2.038
	1.8		0.264	1.627	1.04	0.407		1.8		0.440	0.325	1.26	2.078
	2.0		0.292	1.583	1.20	0.415		2.0		0.482	0.312	1.37	2.108

5.4.2 Three Types of FOPDT Systems

This section explores the advantages of applying the tuning rules given in (5.26) and (5.27). The tuned PI^λ is then compared with the Ziegler–Nichols (ZN), modified ZN (MZN) and AMIGO [105] design methods. In this section, six processes have been considered for comparison. The processes are listed below and Table 5.2 lists the parameters obtained after approximating them by an FOPDT model:

$$G_1(s) = \frac{1}{0.05s+1}e^{-s}, \quad G_2(s) = \frac{1}{(s+1)^3}e^{-15s},$$

$$G_3(s) = \frac{1}{(s+1)^4}, \quad G_4(s) = \frac{9}{(s+1)(s^2+2s+9)},$$

$$G_5(s) = \frac{1}{(1+s)(1+0.2s)(1+0.04s)(1+0.008s)}, \quad G_6(s) = \frac{1}{(s+1)(0.2s+1)}.$$

These six systems have been considered so that we have two delay dominant systems $(L \gg T)$, two balanced lag and delay systems $(L \approx T)$, and two lag dominant systems $(L \ll T)$.

Table 5.2 FOPDT parameters for systems $G_1(s) - G_6(s)$

System	k	L	T	τ	Type	System	k	L	T	τ	Type
$G_1(s)$	1	1	0.09	0.92	Delay dominant	$G_2(s)$	1	16.23	1.76	0.9	Delay dominant
$G_3(s)$	1	1.42	2.90	0.33	Balanced	$G_4(s)$	1	0.59	0.745	0.44	Balanced
$G_5(s)$	1	0.1436	2.65	0.051	Lag dominant	$G_6(s)$	1	0.105	1.11	0.09	Lag dominant

5.4.2.1 Delay Dominated Systems

Table 5.3 summarizes the results obtained for delay dominant systems for the different tuning strategies. Figure 5.4 (a,b) shows the step response and load disturbance response for the two delay dominant systems. It is observed that the F-MIGO controlled systems show a fairly better response in comparison with the sluggish integer-order counterparts, implying that systems with large dead time need a little more than just an integrator to improve their closed-loop control performance.

5.4.2.2 Balanced Lag and Delay Systems

Table 5.3 summarizes the results obtained for the balanced systems for the different tuning strategies. Figure 5.4 (c,d) shows the step response and load disturbance response for these two systems. Systems whose relative dead time is in the range $0.3 < \tau < 0.6$ can be considered as balanced systems. It has

Table 5.3 Controller parameters for systems $G_1(s)$ – $G_6(s)$

		$G_1(s)$				$G_2(s)$			
Method	α	K_p	K_i	M_s	ISE	K_p	K_i	M_s	ISE
F-MIGO	1.1	0.32	0.53	1.4	1.32	0.33	0.032	1.4	20.8
ZN	1.0	0.41	0.24	1.7	2.00	0.42	0.014	1.7	32.3
MZN	1.0	0.44	0.46	1.9	1.33	0.44	0.028	1.9	21.7
AMIGO	1.0	0.16	0.42	1.4	1.63	0.17	0.026	1.4	26.5
		$G_3(s)$				$G_4(s)$			
Method	α	k_p	k_i	M_s	ISE	k_p	k_i	M_s	ISE
F-MIGO	0.9	0.90	0.51	1.4	2.44	0.67	1.16	1.4	0.96
ZN	1.0	1.55	0.40	2.0	2.10	1.06	0.70	1.9	0.89
MZN	1.0	1.71	0.39	2.1	2.08	1.18	0.70	2.0	0.88
AMIGO	1.0	0.71	0.34	1.5	2.75	0.50	0.77	1.5	1.09
		$G_5(s)$				$G_6(s)$			
Method	α	k_p	k_i	M_s	ISE	k_p	k_i	M_s	ISE
F-MIGO	0.7	5.76	5.66	1.40	0.29	3.43	7.64	1.41	.20
ZN	1.0	11.85	26.3	2.41	0.31	6.90	21.3	2.32	.21
MZN	1.0	13.28	12.99	2.18	0.24	7.73	10.5	2.14	.17
AMIGO	1.0	5.31	7.80	1.42	0.35	3.1	7.72	1.43	.25

been observed that, for these systems, the fractional order tends to be close to 1. Besides, from the responses, it can be observed that the best controller cannot be decided for these systems. This leads us to believe that a fractional order may be unnecessary in the case of balanced systems.

5.4.2.3 Lag Dominated Systems

Table 5.3 summarizes the results obtained for lag dominant systems for the different tuning strategies. Figure 5.4 (e,f) shows the step response and load disturbance response for the two lag dominant systems. From the responses, it is very clear that the F-MIGO controlled system performance is very good in comparison with the integer-order counterparts, which show overshoot and oscillatory response. Even though the AMIGO method is comparable, it shows a slightly larger overshoot compared with the F-MIGO controller. This leads to the conclusion that systems with very small dead time may not need a full integrator to give a good closed-loop response.

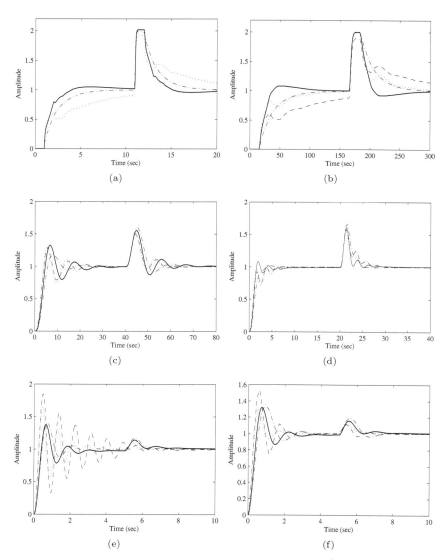

Figure 5.4 Step response and load disturbance response for PI^λ controllers designed using the tuning rules for FOPDT systems (*thick solid line*), ZN (*dashed line*), modified ZN (*dotted line*), AMIGO (*dashed dotted line*): (a) $G_1(s)$, (b) $G_2(s)$, (c) $G_3(s)$, (d) $G_4(s)$, (e) $G_5(s)$, and (f) $G_6(s)$. The controller gains have been listed in Table 5.3

5.4.3 Special Systems

This section deals with those systems which show complex dynamics [101]. They cannot be typically approximated by a simple FOPDT model. The F-MIGO algorithm is applied to them and the fractional order is scanned in the range $0.1 \leqslant \alpha \leqslant 1.9$. The best controller is then picked based on the ISE criterion.

A pure time delay system is given by the transfer function

$$G(s) = e^{-s}. \tag{5.28}$$

This can be considered an extreme case of the systems with large values of relative dead time, *i.e.*, $\tau = 1$. Table 5.4 shows the values obtained via the F-MIGO method for the range of α considered. Systems with τ close to 1 seem to have a good solution only in the range $0.8 < \alpha < 1.4$, as shown in the table. Choosing the lowest value of ISE corresponds to the fractional order 1.1, which again reinforces what has already been discussed for delay dominant systems. Figure 5.5 shows the response for the obtained solutions.

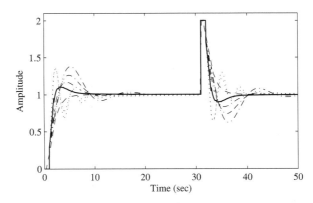

Figure 5.5 Pure time delay controlled with PI$^\lambda$ controllers, best at $\alpha = 1.1$, $k = 0.292$, $K_i = 0.73$ (*solid line*). The one with integer-order controller is given in *dashed lines*. The closed-loop shows oscillatory response for $\alpha \leqslant 1$, in *dotted lines*, and high overshoot for $\alpha > 1.1$, in *dashed dotted lines*

Table 5.4 Scan of fractional order for pure delay system

α	K_p	K_i	w_o	M_s	M_p	ISE	α	K_p	K_i	w_o	M_s	M_p	ISE
0.8	0.075	1.104	1.81	1.40	1.0	1.374	0.9	0.204	1.074	1.83	1.40	1.0	1.278
1.0	0.157	0.472	1.73	1.40	1.0	1.485	**1.1**	0.292	0.729	1.79	1.41	1.0	**1.209**
1.2	0.291	0.606	1.77	1.41	1.0	1.330	1.3	0.291	0.540	1.75	1.41	1.1	1.513
1.4	0.288	0.500	1.73	1.41	1.2	1.8143							

A pure integrator with time delay is given by

$$G(s) = \frac{e^{-s}}{s}. \tag{5.29}$$

This system can be considered as an extreme case of an FOPDT model with a very small value of the relative dead time, *i.e.*, $\tau = 0$, which implies that $T \approx \infty$. Following a similar approach, we scan the system at different values of α and pick the best fractional-order PI controller based on the least ISE value. Figure 5.6 and Table 5.5 give the summary of the results. It has been observed for lag dominating systems that solutions can be obtained in the range $0.4 < \alpha < 1.5$. The best fractional order in this case is 0.7.

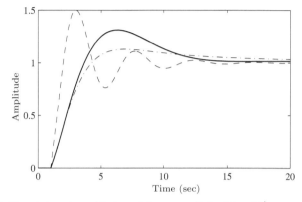

Figure 5.6 Pure integrator with time delay controlled with a PI$^\lambda$ controller: closed-loop response without controller (*dashed line*); best fractional order at $\alpha = 0.7$ with $K_p = 0.324$ and $K_i = 0.103$ (*solid line*); integer-order controller with $K_p = 0.365$ and $K_i = 0.042$ (*dotted dashed line*)

Table 5.5 Scan of fractional order for pure integrator with time delay system

α	K_p	K_i	w_o	M_s	M_p	ISE	α	K_p	K_i	w_o	M_s	M_p	ISE
0.4	0.141	0.173	0.564	1.40	1.2	2.448	0.5	0.209	0.147	0.587	1.40	1.2	2.379
0.6	0.262	0.123	0.599	1.40	1.2	2.359	**0.7**	0.324	0.103	0.606	1.41	1.1	**2.348**
0.8	0.351	0.087	0.606	1.40	1.1	2.375	0.9	0.282	0.078	0.612	1.40	1.1	2.407
1.0	0.365	0.042	0.544	1.40	1.1	2.793	1.1	0.368	0.057	0.599	1.40	1.1	2.645
1.2	0.375	0.051	0.593	1.40	1.2	2.874	1.3	0.373	0.048	0.588	1.40	1.2	3.245
1.4	0.377	0.044	0.578	1.40	1.2	4.027	1.5	0.157	0.043	0.567	1.40	1.3	5.863

5.5 Summary

The results presented in this chapter can be divided into two main parts. In the first part, a generalized MIGO design method for obtaining the parameters of the PI$^\lambda$ controller has been derived based on the principles presented in [101], and it has been validated for the integer case $\alpha = 1$. Hence, for any given system, at any given α for the PI$^\lambda$ controller, a solution can be given if it satisfies the design constraints.

The second part uses this generalized method to scan a set of FOPDT systems for the best fractional order based on the ISE constraint. From the best fractional-order controller obtained, a relation has been established between the controller parameters and the relative dead time τ of the FOPDT systems. This relation has been found to be highly dependent on the value of τ. Tuning rules have been then obtained from these relations for the FOPDT systems. Hence, given the step response of a system, if its FOPDT model can be found, then a fractional-order controller can be suggested for better control.

A comparison has been carried out with the existing popular tuning methods for integer-order PI controllers. From these comparisons, the following conclusions can be drawn. Given the FOPDT model of a system, if the relative dead time is very small, then it has been observed that a fractional-order PI controller of order $\alpha \approx 0.7$ is found to outperform the integer-order counterparts. For systems with a balanced lag and delay values, an advantage of using a fractional-order controller cannot be established. For systems with relative dead time close to unity, it has been observed that the fractional order $\alpha = 1.1$ speeds up the response compared with the sluggish integer-order counterparts, though with the disadvantage of having slightly higher overshoot.

Therefore, in conclusion, a relation between the need for a fractional-order controller and the relative dead time of the system has been established. The need for a little more than an integrator, in some cases, and the lack of a complete integer-order controller, in some others, has been justified.

Chapter 6
Fractional-order Proportional Derivative Controller Tuning for Motion Systems

In Chapter 5, we focused on fractional-order proportional integral controller tuning. The plants to be controlled are assumed to be FOPDT plants. In this chapter, we focus on fractional-order proportional derivative controllers, PD$^\mu$, for another class of plants that are very common in motion control applications. A new tuning method for PD$^\mu$ controllers is proposed to ensure that specifications of gain crossover frequency and phase margin are fulfilled. Furthermore, the derivative of the phase in Bode plot of the open-loop system with respect to the frequency is forced to be zero at the given gain crossover frequency so that the closed-loop system is robust to gain variations. The design method proposed is practical and simple to apply. Simulation and experimental results show that the closed-loop system can achieve favorable dynamic performance and robustness.

6.1 Introduction

As introduced in the previous chapter, the application of fractional-order controllers is currently increasing [2, 52, 107–109]. A better control performance can be achieved with this type of controller due to the introduction of the fractional orders, as demonstrated in [52], where a PI$^\lambda$D$^\mu$ controller was proposed and compared with the classical PID controller. In general, there is not a systematic way for setting the fractional-order parameters, specially in complex cases [82, 89, 110, 111]. However, we may be able to get practical and simple FOC parameter tuning methods for certain specific plants.

In this chapter, a design method for fractional-order PD$^\mu$ controller for a typical second-order plant is discussed. The PD$^\mu$ controller has the following form of transfer function:

$$C(s) = K_{\mathrm{p}}(1 + K_{\mathrm{d}}s^\mu), \tag{6.1}$$

where $\mu \in (0,1]$. Clearly, this is a specific form of the most common PI$^\lambda$D$^\mu$ controller which involves an integrator of order λ ($\lambda = 0$, in this chapter) and a differentiator of order μ. The typical second-order plant discussed here is

$$G(s) = \frac{1}{s(\tau s + 1)}, \tag{6.2}$$

which can approximately model a DC motor position servo system. Note that the plant gain is normalized to 1 without loss of generality, since the proportional factor in the transfer function (6.2) can be introduced through the controller gain K_p.

The rest of the chapter is organized as follows. Sections 6.2 and 6.3 describe the design method proposed for the PD$^\mu$ controller. Section 6.4 shows simulation examples of application, and Section 6.5 presents the experimental results obtained when controlling a dynamometer platform. Finally, chapter summary is given in Section 6.6.

6.2 Fractional-order PD Controller Design for a Class of Second-order Plants

Let us restrict our attention to a class of second-order plants $G(s)$ described by (6.2). The transfer function of the fractional-order PD$^\mu$ controller has the form of (6.1). The phase and magnitude of the plant in the frequency domain are given by

$$\arg[G(\mathrm{j}\omega)] = -\tan^{-1}(\omega\tau) - \frac{\pi}{2}, \tag{6.3}$$

$$|G(\mathrm{j}\omega)| = \frac{1}{\omega\sqrt{1 + (\omega\tau)^2}}. \tag{6.4}$$

The fractional-order PD$^\mu$ controller described by (6.1) can be written as

$$C(\mathrm{j}\omega) = K_\mathrm{p}\left[1 + K_\mathrm{d}(\mathrm{j}\omega)^\mu\right] = K_\mathrm{p}\left(1 + K_\mathrm{d}\omega^\mu \cos\frac{\mu\pi}{2} + \mathrm{j}K_\mathrm{d}\omega^\mu \sin\frac{\mu\pi}{2}\right). \tag{6.5}$$

In the frequency domain,

$$\arg[C(\mathrm{j}\omega)] = \tan^{-1}\frac{\sin\frac{(1-\mu)\pi}{2} + K_\mathrm{d}\omega^\mu}{\cos\frac{(1-\mu)\pi}{2}} - \frac{(1-\mu)\pi}{2}, \tag{6.6}$$

$$|C(\mathrm{j}\omega)| = K_\mathrm{p}\sqrt{\left(1 + K_\mathrm{d}\omega^\mu \cos\frac{\mu\pi}{2}\right)^2 + \left(K_\mathrm{d}\omega^\mu \sin\frac{\mu\pi}{2}\right)^2}. \tag{6.7}$$

The open-loop transfer function $G(s)$ is

$$L(s) = C(s)G(s). \tag{6.8}$$

From (6.3) and (6.6), the phase of $G(s)$ is

$$\arg[G(\mathrm{j}\omega)] = \tan^{-1} \frac{\sin \dfrac{(1-\mu)\pi}{2} + K_{\mathrm{d}}\omega^{\mu}}{\cos \dfrac{(1-\mu)\pi}{2}} + \frac{\mu\pi}{2} - \pi - \tan^{-1}(\omega\tau). \quad (6.9)$$

Here, three interesting specifications to be met by the fractional-order PD^{μ} controller are proposed. From the basic definition of gain crossover frequency and phase margin:

1. Phase margin specification

$$\arg[G(\mathrm{j}\omega_{\mathrm{cg}})] = \arg[C(\mathrm{j}\omega_{\mathrm{cg}})G(\mathrm{j}\omega_{\mathrm{cg}})] = -\pi + \varphi_m,$$

2. Robustness to variation in the gain of the plant

$$\left. \frac{\mathrm{d}(\arg(C(\mathrm{j}\omega)G(\mathrm{j}\omega)))}{\mathrm{d}\omega} \right|_{\omega=\omega_{\mathrm{cg}}} = 0,$$

that is, the derivative of the phase of the open-loop system with respect to the frequency is forced to be zero at the gain crossover frequency so that the closed-loop system is robust to gain variations, and therefore the overshoots of the response are almost invariant.

3. Gain crossover frequency specification

$$|G(\mathrm{j}\omega_{\mathrm{cg}})|_{\mathrm{dB}} = |C(\mathrm{j}\omega_{\mathrm{cg}})G(\mathrm{j}\omega_{\mathrm{cg}})|_{\mathrm{dB}} = 0.$$

6.2.1 Conventional PD Controllers ($\mu = 1$)

From (6.9) and according to specification 2,

$$\left. \frac{\mathrm{d}(\arg(G(\mathrm{j}\omega)))}{\mathrm{d}\omega} \right|_{\omega=\omega_{\mathrm{cg}}} = \frac{K_{\mathrm{d}}}{1 + (K_{\mathrm{d}}\omega_{\mathrm{cg}})^2} - \frac{\tau}{1 + (\tau\omega_{\mathrm{cg}})^2} = 0,$$

we obtain that

$$K_{\mathrm{d}} = \frac{1}{\tau\omega_{\mathrm{cg}}^2}, \quad \arg[G(\mathrm{j}\omega)] = \tan^{-1}\left(\frac{1}{\tau\omega_{\mathrm{cg}}}\right) - \frac{\pi}{2} - \tan^{-1}(\omega_{\mathrm{cg}}\tau).$$

That means that specifications 1 and 2 cannot be satisfied simultaneously for traditional PD controllers.

6.2.2 Fractional-order PD$^\mu$ Controllers

According to specification 1, the phase of $G(s)$ can be expressed as

$$\arg[G(\mathrm{j}\omega)]\Big|_{\omega=\omega_{\mathrm{cg}}} = \tan^{-1}\frac{\sin\dfrac{(1-\mu)\pi}{2}+K_{\mathrm{d}}\omega_{\mathrm{cg}}^{\mu}}{\cos\dfrac{(1-\mu)\pi}{2}}+\frac{\mu\pi}{2}-\pi-\tan^{-1}(\omega_{\mathrm{cg}}\tau)$$

$$= -\pi + \varphi_{\mathrm{m}}. \tag{6.10}$$

From (6.10), the relation between K_{d} and μ can be established as follows:

$$K_{\mathrm{d}} = \frac{1}{\omega_{\mathrm{cg}}^{\mu}}\tan\left[\varphi_{\mathrm{m}}+\tan^{-1}(\omega_{\mathrm{cg}}\tau)-\frac{\mu\pi}{2}+\pi\right]\cos\frac{(1-\mu)\pi}{2}-\frac{1}{\omega_{\mathrm{cg}}^{\mu}}\sin\frac{(1-\mu)\pi}{2}. \tag{6.11}$$

According to specification 2 about the robustness to gain variations in the plant,

$$\frac{\mathrm{d}(\mathrm{Arg}(G(\mathrm{j}\omega)))}{\mathrm{d}\omega}\Bigg|_{\omega=\omega_{\mathrm{cg}}}$$

$$= \frac{\mu K_{\mathrm{d}}\omega_{\mathrm{cg}}^{\mu-1}\cos\dfrac{(1-\mu)\pi}{2}}{\cos^2\dfrac{(1-\mu)\pi}{2}+\left(\sin\dfrac{(1-\mu)\pi}{2}+K_{\mathrm{d}}\omega_{\mathrm{cg}}^{\mu}\right)^2}-\frac{\tau}{1+(\tau\omega_{\mathrm{cg}})^2}=0. \tag{6.12}$$

From (6.12), we can establish the following equation for K_{d}:

$$A\omega_{\mathrm{cg}}^{2\mu}K_{\mathrm{d}}^2 + BK_{\mathrm{d}} + A = 0, \tag{6.13}$$

that is

$$K_{\mathrm{d}} = \frac{-B\pm\sqrt{B^2-4A^2\omega_{\mathrm{cg}}^{2\mu}}}{2A\omega_{\mathrm{cg}}^{2\mu}}, \tag{6.14}$$

where

$$A = \frac{\tau}{1+(\omega_{\mathrm{cg}}\tau)^2}, \quad B = 2A\omega_{\mathrm{cg}}^{\mu}\sin\frac{(1-\mu)\pi}{2}-\mu\omega_{\mathrm{cg}}^{\mu-1}\cos\frac{(1-\mu)\pi}{2}.$$

According to specification 3, the equation established for K_{p} is

$$|G(\mathrm{j}\omega_{\mathrm{cg}})| = |C(\mathrm{j}\omega_{\mathrm{cg}})||G(\mathrm{j}\omega_{\mathrm{cg}})|$$

$$= \frac{K_{\mathrm{p}}\sqrt{\left(1+K_{\mathrm{d}}\omega_{\mathrm{cg}}^{\mu}\cos\dfrac{\mu\pi}{2}\right)^2+\left(K_{\mathrm{d}}\omega_{\mathrm{cg}}^{\mu}\sin\dfrac{\mu\pi}{2}\right)^2}}{\omega_{\mathrm{cg}}\sqrt{1+(\omega_{\mathrm{cg}}\tau)^2}}=1. \tag{6.15}$$

Clearly, we can solve (6.11), (6.14), and (6.15) to get μ, K_{d}, and K_{p}.

6.3 Design Procedure

It can be observed from (6.11) and (6.14) that μ and K_d can be obtained jointly. Fortunately, a graphical method can be used as a practical and simple way to get these parameters. The procedure to tune the fractional-order PD^μ controller is as follows:

- Given ω_{cg}, the gain crossover frequency.
- Given φ_m, the desired phase margin.
- Plot curve 1 in Figure 6.1, corresponding to K_d with respect to μ, according to (6.11).
- Plot curve 2 corresponding to K_d with respect to μ, according to (6.14).
- Obtain μ and K_d from the intersection point between Curves 1 and 2.
- Calculate K_p from (6.15).

Remark 6.1. Design specifications should not be chosen too aggressive because of the constraint in (6.14) and the intersection point between the two curves. □

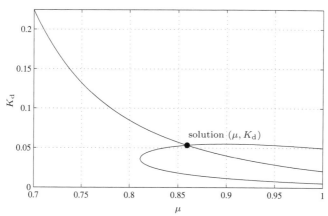

Figure 6.1 K_d vs μ

6.4 Simulation Example

The fractional-order PD^μ controller design method above is illustrated via a numerical simulation. In the simulation, the plant parameter τ in (6.2) is 0.05 sec. The specifications of interest are set as $\omega_{cg} = 60\,\mathrm{rad/sec}$, $\varphi_m = 70°$, and the robustness to gain variations in the plant is required.

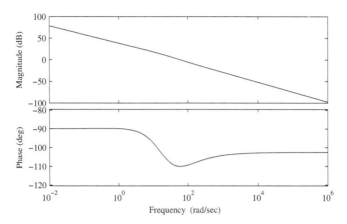

Figure 6.2 Bode plot with fractional-order PD$^\mu$ controller

According to (6.11) and (6.14), two curves are plotted easily in Figure 6.1. From the intersection point of these curves, it is found that $\mu = 0.86$ and $K_\mathrm{d} = 0.053$. Then, K_p is calculated from (6.15) easily, resulting in $K_\mathrm{p} = 84.89$.

Actually, the PD$^\mu$ fractional-order controller itself is an infinite-dimensional linear filter due to the fractional-order differentiator μ. A band-limit implementation is important in practice. Finite dimensional approximation of the FOC should be utilized in a proper range of frequency of practical interest. The approximation method used here is the Oustaloup recursive algorithm [112]. Assuming that the frequency range selected is $(\omega_\mathrm{b}, \omega_\mathrm{h})$, the approximate finite-dimensional transfer function of a continuous filter for s^γ with the Oustaloup algorithm can be obtained. Details on Oustaloup approximation and implementation is given in Section 12.1.2.

In our simulation, for the approximation of fractional-order differentiator s^μ, the frequency range of practical interest is set from 0.0001 Hz to 10,000 Hz. The Bode plots of the resulting open-loop system are shown in Figure 6.2. As can be seen, the gain crossover frequency specification, $\omega_\mathrm{cg} = 60\,\mathrm{rad/sec}$, and phase margin specification, $\varphi_\mathrm{m} = 70°$, are satisfied. The phase is forced to be flat at ω_cg.

In order to compare the integer-order controller and our designed fractional-order controller fairly, two simulation cases are presented.

6.4.1 Step Response Comparison with Varying K_p

It is well known that a proportional controller is adopted commonly for the typical second-order plant discussed in this chapter, and that the Integral of Time and Absolute Error (ITAE) optimum proportional parameter is $K = 1/(2\tau)$ [113]. Therefore, the proportional parameter is set to 10 in this example if the commonly used proportional controller is used.

In Figure 6.3, applying the ITAE optimal proportional controller, the unit step responses are plotted with the open-loop plant gain varying from 8 to 12 ($\pm 20\%$ variations from the desired value 10). In Figure 6.4, applying the fractional-order PD^{μ} controller, the unit step responses are plotted with gains varying from 67.9 to 101.8 ($\pm 20\%$ variations from desired value 84.89).

It can be seen from Figure 6.3 and 6.4 that the fractional-order PD^{μ} controller designed by the proposed method in this chapter is effective. With the fractional-order controller, faster responses are achieved and besides, the overshoots of the step responses remain almost constant under gain variations, i.e., an iso-damping property is exhibited. This means that the system is more robust to gain changes.

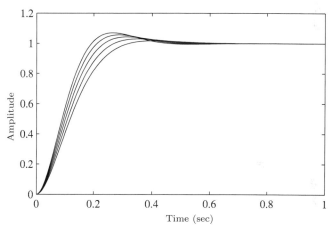

Figure 6.3 Step responses with the ITAE optimum proportional controller, with $K_{\mathrm{p}} = 8 \sim 12$

6.4.2 Ramp Response Comparison with Varying K_{p}

In this simulation case, we use the unit ramp input response to compare the PI controller with the fractional-order PD^{μ} controller. The ITAE optimum parameters of the PI controller are designed as follows [113]:

$$C_{\mathrm{PI}}(s) = K_{\mathrm{p}}(1 + K_{\mathrm{i}}/s), \qquad (6.16)$$

where $K_{\mathrm{p}} = 21.2245$ and $K_{\mathrm{i}} = 74.6356$.

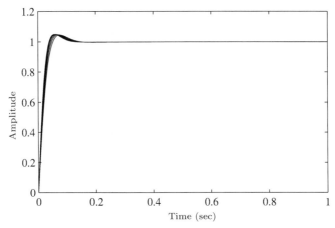

Figure 6.4 Step responses with the PD$^\mu$ controller, with K_p variation of 80 %~120 %

In Figure 6.5, the ITAE optimum PI controller is applied and the unit ramp responses are plotted with open-loop gain varying from 16.9796 to 25.4694 ($\pm$20% variations from the desired value 21.2245). In Figure 6.6, the fractional-order PD$^\mu$ controller is applied and the unit ramp responses are plotted with gains varying from 67.8 to 101.8 ($\pm$20% variations from the desired value 84.89).

From Figures 6.5 and 6.6, it is obvious that, with the designed PD$^\mu$ controller using the proposed tuning rule, the overshoots are almost constant under gain variations and are much lower than those with the ITAE optimal PI controller.

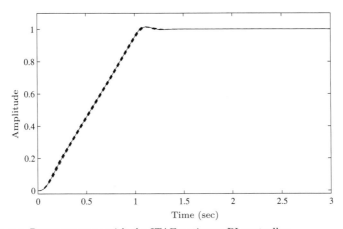

Figure 6.5 Ramp response with the ITAE optimum PI controller

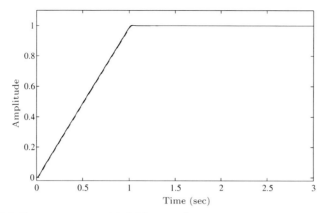

Figure 6.6 Ramp responses with PD$^\mu$ controller

6.5 Experiments

6.5.1 Introduction to the Experimental Platform

A fractional horsepower dynamometer has been developed as a general purpose experiment platform [114]. The architecture of the dynamometer control system is shown in Figure 6.7. The dynamometer includes the DC motor to be tested, a hysteresis brake for applying torque load to the motor, a load cell to provide force feedback, an optical encoder for position feedback, and a tachometer for velocity feedback. The dynamometer has been modified to connect to a Quanser MultiQ4 terminal board in order to control the system through MATLAB/Simulink Real-Time Workshop (RTW) based software. This terminal board connects with the Quanser MultiQ4 data acquisition card. Then, we use the MATLAB/Simulink^{TM} environment, which uses the WinCon application from Quanser, to communicate with the data acquisition card. Thus complex nonlinear control schemes can be tested since the programmable load can be used to emulate nonlinear relations. This enables rapid prototyping and experiment capabilities to many nonlinear models of arbitrary form.

Without loss of generality, consider the servo control system modeled by

$$\dot{x}(t) = v(t), \tag{6.17}$$

$$\dot{v}(t) = K_{\mathrm{u}}u(t) - K_{\mathrm{b}}v, \tag{6.18}$$

where x is the position state, v is the velocity, and u is the control input; K_{u} and K_{b} are positive coefficients.

Through a simple system identification process, the DC motor can be approximately modeled by a transfer function $1.52/(0.4s + 1)$.

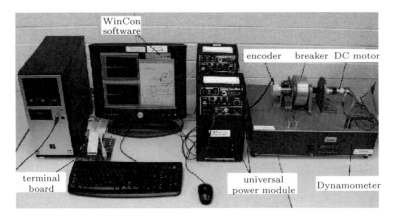

Figure 6.7 The dynamometer setup

6.5.2 Experimental Model Simulation

Since we have already experimentally modeled the dynamometer with a transfer function $1.52/(0.4s + 1)$, we can test the simulation effect first in Simulink. Then, the simulation results can be compared with the real-time experiments on the dynamometer. This way, the verification of our proposed method is more effective.

For the fractional-order PD$^\mu$ controller, the gain crossover frequency is set as $\omega_{cg} = 10$ rad/sec. Correspondingly, for the approximation of the fractional-order differentiator s^μ, the frequency range of practical interest is set from 1Hz to 100Hz, and the desired phase margin is set as $\varphi_m = 70°$. Moreover, the robustness to gain variations is required. According to the numerical method in Section 6.3, we can obtain that $\mu = 0.844$, $K_d = 0.368$ and $K_p = 13.860$.

First, the unit step responses are tested to compare the ITAE optimum P controller with the PD$^\mu$ controller. The ITAE optimum P parameter is $K_p = 1/(2\tau) = 1.25$ [113]. In Figure 6.8, applying the ITAE optimal P controller, the unit step responses are plotted with the open-loop gain varying from 1 to 1.5 ($\pm 20\%$ variations from the desired value 1.25). In Figure 6.9, applying the fractional-order PD$^\mu$ controller, the unit step responses are plotted with gains varying from 11.088 to 16.632 ($\pm 20\%$ variations from the desired value 13.86).

From Figures 6.8 and 6.9, it is obvious that, with the designed PD$^\mu$ controller, faster unit step responses are achieved, and the overshoots remain almost constant under gain variations and are much lower than that with the optimal P controller, demonstrating that the controlled system using the PD$^\mu$ controller is more robust to gain changes in the loop.

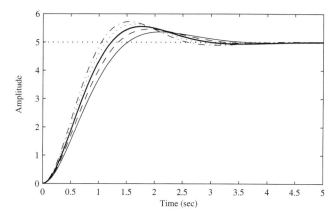

Figure 6.8 Dynamometer simulation model. Step position responses with the ITAE optimal proportional controller, with *thick solid line* for $K_{\mathrm{p}} = 1.25$, *thin solid line* for $K_{\mathrm{p}} = 1$, and *dashed, dotted*, and *dashed dotted lines* for $K_{\mathrm{p}} = 1.125, 1.375$ and 1.5, respectively

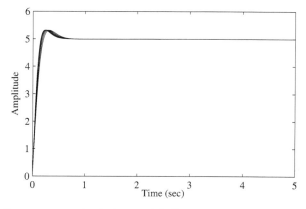

Figure 6.9 Dynamometer simulation model. Step position responses with the PD$^{\mu}$ controller

Next, the unit ramp responses are tested to compare the ITAE optimum PI controller with the PD$^{\mu}$ controller. The ITAE optimum PI parameters are designed as $K_{\mathrm{p}} = 2.6531$ and $K_{\mathrm{i}} = 1.1662$ [113]. In Figure 6.10, applying the ITAE optimal PI controller, the unit ramp responses are plotted with the open-loop gain varying from 1.8531 to 3.1837 ($\pm 20\%$ variations from the desired value 2.6531). In Figure 6.9, applying the fractional-order PD$^{\mu}$ controller, the unit ramp responses are plotted with gains varying from 11.088 to 16.632 ($\pm 20\%$ variations from the desired value 13.86).

It can be seen from Figures 6.10 and 6.11 that the fractional-order PD$^{\mu}$ controller designed by the proposed tuning method is more effective.

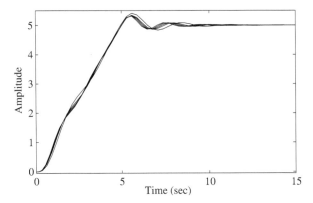

Figure 6.10 Dynamometer simulation model. Ramp position responses with the ITAE optimal PI controller

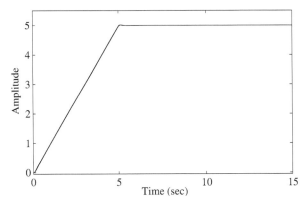

Figure 6.11 Dynamometer simulation model. Ramp position responses with the PD$^\mu$ controller

6.5.3 Experiments on the Dynamometer

Substituting the DC motor simulation model for the real dynamometer platform, the proposed PD$^\mu$ controller is tested in a hardware-in-the-loop manner.

Figures 6.12 and 6.13 show the unit step position responses for the ITAE optimal P controller and the unit ramp position responses for the ITAE optimal PI controller, respectively. Figures 6.14 and 6.15 present the unit step and ramp position responses for the fractional-order PD$^\mu$ controller designed by the proposed tuning method in this chapter, respectively. It is obvious that, with the designed PD$^\mu$ controller, faster responses are achieved and the overshoots remain almost constant under gain variations. The overshoots are much lower than those with the ITAE optimal P or PI controllers.

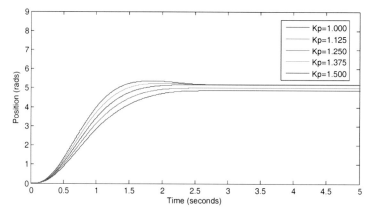

Figure 6.12 Dynamometer real-time experiment. Step position responses with the ITAE optimal proportional controller

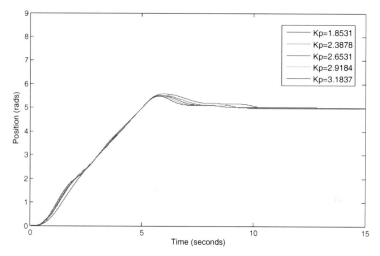

Figure 6.13 Dynamometer real-time experiment. Ramp position responses with the ITAE optimal PI controller

6.6 Summary

In this chapter, we have presented a new tuning method for fractional-order proportional and derivative controllers PD$^\mu$ for a class of second-order plants. The PD$^\mu$ controller is tuned to ensure that the given gain crossover frequency and the phase margin are achieved, and also to guarantee the robustness of the system to gain variations. The tuning method proposed, aiming at typical second-order plants, is practical and simple. Simulation and experimental results show that the closed-loop system can achieve favorable dynamic performance and robustness.

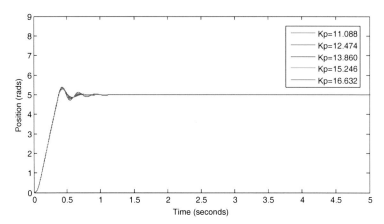

Figure 6.14 Dynamometer real-time experiment. Step position responses with the PD$^{\mu}$ controller

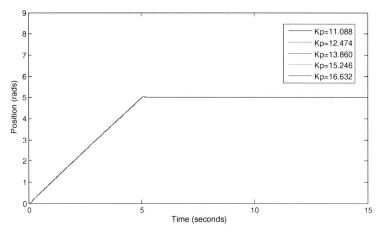

Figure 6.15 Dynamometer real-time experiment. Ramp position response with the PD$^{\mu}$ controller

Chapter 7
Fractional-order Proportional Integral Derivative Controllers

This chapter deals with the design of fractional-order $PI^\lambda D^\mu$ controllers, in which the orders of the integral and derivative parts, λ and μ respectively, are fractional. The purpose is to take advantage of the introduction of these two parameters and fulfil additional specifications of design, ensuring a robust performance of the controlled system with respect to gain variations and noise. A method for tuning the $PI^\lambda D^\mu$ controller is proposed to fulfil five different design specifications. Experimental results show that the requirements are fully met for the platform to be controlled.

7.1 Introduction

It is important to realize that there is a very wide range of control problems and, consequently, also a need for a wide range of design techniques. There are already many tuning methods available, but a replacement of the Ziegler–Nichols method is long overdue. On the research side, it appears that the development of design methods for integer-order PID control is approaching the point of diminishing returns. There are few difficult problems that remain to be solved.

Therefore, this chapter proposes the application of fractional-order $PI^\lambda D^\mu$ controllers as an alternative to solve some of the control problems that can arise when dealing with industrial applications, as will be commented on later. On the one hand, a new method for the design of fractional-order controllers is proposed, and more concretely for the tuning of a generalized $PI^\lambda D^\mu$ controller of the form

$$C(s) = K_p + \frac{K_i}{s^\lambda} + K_d s^\mu, \tag{7.1}$$

where λ and μ are the fractional orders of the integral and derivative parts of the controller, respectively. Since this kind of controller has five parameters to tune $(K_\mathrm{p}, K_\mathrm{d}, K_\mathrm{i}, \lambda, \mu)$, up to five design specifications for the controlled system can be met, that is, two more than in the case of a conventional PID controller, where $\lambda = 1$ and $\mu = 1$. It is essential to study which specifications are more interesting as far as performance and robustness are concerned, since it is the aim to obtain a controlled system robust to uncertainties of the plant model, load disturbances, and high-frequency noise. All these constraints will be taken into account in the tuning technique in order to take advantage of the introduction of the fractional orders.

The chapter is organized as follows. The tuning method proposed for fractional-order PI$^\lambda$D$^\mu$ controllers is described in Section 7.2. Section 7.3 shows the experimental results obtained when controlling an experimental platform with the controller designed. Finally, some relevant concluding remarks are presented in Section 7.4.

7.2 Design Specifications and Tuning Problem

As stated in the introduction, our objective is to design a fractional-order controller so that the system fulfils different specifications regarding robustness to plant uncertainties, load disturbances, and high-frequency noise. For that reason, specifications related to phase margin, sensitivity functions, and robustness constraints are going to be considered in this design method, due to their important features regarding performance, stability, and robustness. Of course, other kinds of specifications can be met, depending on the particular requirements of the system. Therefore, the design problem is formulated as follows:

- **Phase margin φ_m and gain crossover frequency ω_cg specifications.** Gain and phase margins have always served as important measures of robustness. It is known that the phase margin is related to the damping ratio of the system and therefore can also serve as a performance measure [115]. As seen in the previous chapter, the equations that define the phase margin and the gain crossover frequency are

$$|C(\mathrm{j}\omega_\mathrm{cg})G(\mathrm{j}\omega_\mathrm{cg})|_\mathrm{dB} = 0\,\mathrm{dB}, \quad \arg(C(\mathrm{j}\omega_\mathrm{cg})G(\mathrm{j}\omega_\mathrm{cg})) = -\pi + \varphi_\mathrm{m}. \quad (7.2)$$

- **Robustness to variations in the gain of the plant.** The next constraint can be considered in this case [116]:

$$\left. \frac{\mathrm{d}\arg(F(s))}{\mathrm{d}\omega} \right|_{\omega=\omega_\mathrm{cg}} = 0. \quad (7.3)$$

This condition has already been considered in the previous chapter, and forces the phase of the open-loop system $F(s) = C(s)G(s)$ to be flat at ω_{cg} and so, to be almost constant within an interval around ω_{cg}. It means that the system is more robust to gain changes and the overshoot of the response is almost constant within a gain range, also known as *iso-damping property* of the time response. It must be noted that the interval of gains for which the system is robust is not fixed with this condition. That is, the user cannot force the system to be robust for a particular gain range. This range depends on the frequency range around ω_{cg} for which the phase of the open-loop system keeps flat. This frequency range will be longer or shorter, depending on the resulting controller and the frequency characteristics of the plant.

- **High-frequency noise rejection.** A constraint on the complementary sensitivity function $T(j\omega)$ can be established:

$$\left| T(j\omega) = \frac{C(j\omega)G(j\omega)}{1 + C(j\omega)G(j\omega)} \right|_{dB} \leqslant A\,dB,$$
(7.4)

$$\forall \omega \geqslant \omega_t\,rad/sec \; \Rightarrow \; |T(j\omega_t)|_{dB} = A\,dB,$$

with A the desired noise attenuation for frequencies $\omega \geqslant \omega_t\,rad/sec$.

- **To ensure a good output disturbance rejection.** A constraint on the sensitivity function $S(j\omega)$ can be defined:

$$\left| S(j\omega) = \frac{1}{1 + C(j\omega)G(j\omega)} \right|_{dB} \leqslant B\,dB,$$
(7.5)

$$\forall \omega \leqslant \omega_s\,rad/sec \Rightarrow |S(j\omega_s)|_{dB} = B\,dB,$$

with B the desired value of the sensitivity function for frequencies $\omega \leqslant \omega_s\,rad/sec$ (desired frequency range).

- **Steady-state error cancelation.** Properly implemented, a fractional-order integrator of order $k + \alpha, k \in \mathbb{N}, 0 < \alpha < 1$, is, for steady-state error cancelation, as efficient as an integer-order integrator of order $k + 1$ [92], as demonstrated in Section 2.3.3.3, and the steady-state error is shown in Table 2.1.

  However, though the Final-Value Theorem states that the fractional-order system exhibits null steady-state error if $\alpha > 0$, the fact of being $\alpha < 1$ makes the output converge to its final-value more slowly than in the case of an integer-order controller. Furthermore, the fractional effect has to be band-limited when it is implemented. Therefore, the fractional-order integrator must be implemented as $1/s^\alpha = s^{1-\alpha}/s$, ensuring this way the effect of an integer-order integrator $1/s$ at very low frequencies.

Similarly to the fractional-order integrator, the fractional-order differentiator, s^μ, also has to be band-limited when implemented, ensuring in this way a finite control effort and noise rejection at high frequencies.

Using the fractional-order PI$^\lambda$D$^\mu$ controller of (7.1), up to five of these design specifications can be fulfilled, since it has five parameters to tune. For fractional-order controllers such as a PI$^\lambda$ or a PD$^\mu$, three design specifications could be met (one for each parameter). Therefore, for the general case of a PI$^\lambda$D$^\mu$ controller the design problem is based on solving the system of five nonlinear equations (given by the corresponding design specifications) and five unknown parameters K_p, K_d, K_i, λ, μ.

However, the complexity of this set of nonlinear equations is very significant, specially when fractional orders of the Laplace variable s are introduced, and finding out the solution is not trivial. In fact, a nonlinear optimization problem must be solved, in which the best solution of a constrained nonlinear equation has to be found.

Global optimization is the task of finding the absolutely best set of admissible conditions to achieve an objective under given constraints, assuming that both are formulated in mathematical terms. Some large-scale global optimization problems have been solved by current methods, and a number of software packages are available that reliably solve most global optimization problems in small (and sometimes larger) dimensions. However, finding the global minimum, if one exists, can be a difficult problem (very dependent on the initial conditions). Superficially, global optimization is a stronger version of local optimization, whose great usefulness in practice is undisputed. Instead of searching for a locally feasible point, one wants the globally best point in the feasible region. However, in many practical applications, finding the globally best point, though desirable, is not essential, since any sufficiently good feasible point is useful and usually an improvement over what is available without optimization (this particular case). Besides, sometimes, depending on the optimization problem, there is no guarantee that the optimization functions will return a global minimum, unless the global minimum is the only minimum and the function to minimize is continuous [117]. Taking all this into account, and considering that the set of functions to minimize in this case are continuous and can only present one minimum in the feasible region, any of the optimization methods available could be effective, *a priori*. For this reason, and taking into account that MATLAB is a very appropriate tool for the analysis and design of control systems, MATLAB optimization toolbox has been used to reach out the best solution with the minimum error. The function used for this purpose is called `fmincon()`, which finds the constrained minimum of a function of several variables. It solves problems of the form $\min_x f(x)$ subject to $C(x) \leqslant 0$,

$C_{eq}(\boldsymbol{x}) = 0$, $\boldsymbol{x}_m \leqslant \boldsymbol{x} \leqslant \boldsymbol{x}_M$, where $f(\boldsymbol{x})$ is the function to minimize; $C(\boldsymbol{x})$ and $C_{eq}(\boldsymbol{x})$ represent the nonlinear inequalities and equalities, respectively (nonlinear constraints); $\boldsymbol{x}$ is the minimum looked for; $\boldsymbol{x}_m$ and $\boldsymbol{x}_M$ define a set of lower and upper bounds on the design variables $\boldsymbol{x}$.

In this particular case, the specification in (7.2) is taken as the main function to minimize, and the rest of specifications (7.2) $\sim$ (7.5) are taken as constraints for the minimization, all of them subjected to the optimization parameters defined within the function fmincon. The success of this design method depends mainly on the initial conditions considered for the parameters of the controller.

The tuning method proposed here is illustrated next with the results obtained from an experimental platform consisting of a liquid level system.

7.3 Experimental Results

The experimental platform "Basic Process Rig 38-100 Feedback Unit" has been used to test the fractional-order controllers designed by the optimization tuning method proposed. The platform consists of a low pressure flowing water circuit which is bench mounted and completely self contained. The water circuit is arranged in front of a vertical panel, as can be seen in Figure 7.1.

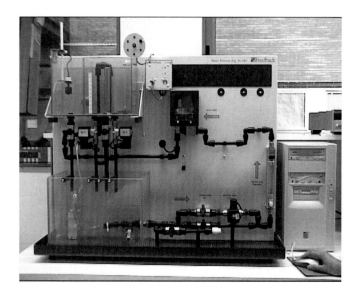

Figure 7.1 Photo of the Basic Process Rig 38-100 Feedback Unit

For the characterization of the plant and implementation of the controller, a data acquisition board PCL-818H, by PC-LabCard, has been used, running on MATLAB 5.3 and using its real time toolbox "Real-Time Windows Target." A computer Pentium II, 350MHz, 64M RAM, supports the data acquisition board and the program in C programming language (from MAT-LAB) corresponding to the controller.

After the characterization of the system, the resulting transfer function is

$$G(s) = \frac{k}{\tau s + 1}\mathrm{e}^{-Ls} = \frac{3.13}{433.33s + 1}\mathrm{e}^{-50s}, \tag{7.6}$$

that is, the liquid level system is modeled by a first-order transfer function with time delay $L = 50$ sec, gain $k = 3.13$ and time constant $\tau = 433.33$ sec. The design specifications required for the system are:

- gain crossover frequency, $\omega_{cg} = 0.008$ rad/sec;
- phase margin, $\varphi_m = 60°$;
- robustness to variations in the gain of the plant must be fulfilled;
- sensitivity function: $|S(\mathrm{j}\omega)|_{dB} \leqslant -20$ dB, $\forall\, \omega \leqslant \omega_s = 0.001$ rad/sec;
- noise rejection: $|T(\mathrm{j}\omega)|_{dB} \leqslant -20$ dB, $\forall\, \omega \geqslant \omega_t = 10$ rad/sec.

Applying the optimization method described previously, the fractional-order PI$^\lambda$D$^\mu$ controller obtained to control the system is

$$C(s) = 0.6152 + \frac{0.0100}{s^{0.8968}} + 4.3867s^{0.4773}. \tag{7.7}$$

In this particular case the fractional-order integral and derivative parts have been implemented by the Oustaloup continuous approximation of the fractional-order integrator ([16, 51], Chapter 12), choosing a frequency band from 0.001 rad/sec to 100 rad/sec and an order of the approximation equal to 5 (number of poles and zeros). Once the continuous-time fractional-order controller is obtained, it is discretized by using the Tustin rule with a sampling period $T = 1$ sec and a prewarp frequency ω_{cg} [118].

The Bode plots of the open-loop system $F(s) = C(s)G(s)$ are shown in Figure 7.2. As can be observed, the specifications of gain crossover frequency and phase margin are met. Besides, the phase of the system is forced to be flat at ω_{cg} and so, to be almost constant within an interval around ω_{cg}. It means that the system is more robust to gain changes and the overshoot of the response is almost constant within this interval, as can be seen in Figure 7.3, where a step input of 0.47 has been applied to the closed-loop system in simulation. Variations in the gain of the plant have been considered from 2.75 to 3.75. The magnitudes of the functions $S(s)$ and $T(s)$ for the nominal plant are shown in Figures 7.4 and 7.5, respectively, fulfilling the specifications.

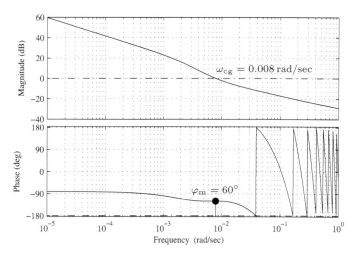

Figure 7.2 Bode plots of the open-loop system $F(s) = C(s)G(s)$

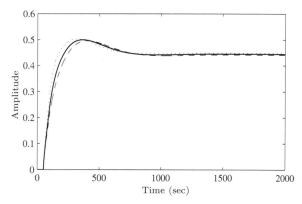

Figure 7.3 Simulation step responses of the controlled system with controller $C(s)$, with *solid line* for $K = 3.13$, *dash line* for $K = 2.75$, and *dotted line* for $K = 3.75$

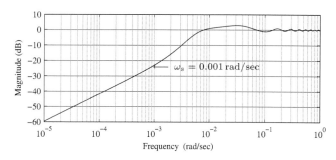

Figure 7.4 Magnitude of $S(j\omega)$

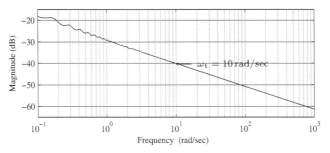

Figure 7.5 Magnitude of $T(j\omega)$

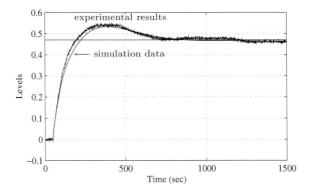

Figure 7.6 Comparison between simulated and experimental levels for $k = 3.13$ with controller $C(s)$

The experimental results obtained when controlling the liquid level plant in real time are shown next. Figure 7.6 shows the comparison between simulated and experimental levels for the nominal gain $k = 3.13$. In Figure 7.7 the experimental responses for different gains (set via software) are scoped, fulfilling the robustness constraint to gain changes (within the variation range selected). Figure 7.8 shows the experimental control laws obtained for each value of gain. As far as the control laws are concerned, only a slight variation in the peak value of the signal is produced when the gain changes, which is an important feature as far as the actuator saturation is concerned. In this case, the peak value is very far from the saturation value of 10 V for the servo valve.

From these results, the potential of the fractional-order PI$^\lambda$D$^\mu$ controller in practical industrial settings, regarding performance and robustness aspects, is clear. However, the design method proposed here involves complex equations relating the specifications of design and, sometimes, it may be difficult to find a solution to the problem.

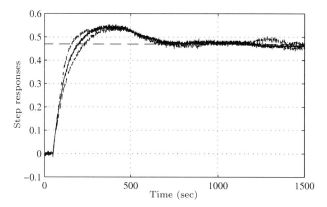

Figure 7.7 Experimental step responses of the controlled system with controller $C(s)$ with *solid line* for $K = 3.13$, *dash line* for $K = 2.75$, and *dash dotted line* for $K = 3.75$

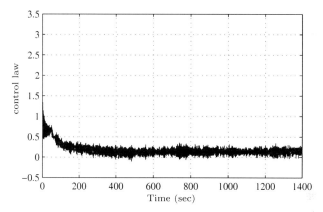

Figure 7.8 Experimental control laws of the controlled system with controller $C(s)$, very similar for $K = 3.13$, $K = 2.75$, and $K = 3.75$

7.4 Summary

In this chapter, a synthesis method for fractional-order $PI^\lambda D^\mu$ controllers has been developed to fulfil five different design specifications for the closed-loop system, that is, two more specifications than in the case of a conventional PID controller. An optimization method to tune the controller has been used for that purpose, based on a nonlinear function minimization subject to some given nonlinear constraints. Experimental results show that the requirements are totally satisfied for the platform to be controlled. Thus, advantage has been taken of the fractional orders λ and μ to fulfil additional specifications of design, ensuring a robust performance of the controlled system to gain changes and noise.

Part III of the book deals with tuning methods for fractional-order lead-lag compensators. These methods will avoid the nonlinear minimization problem presented here, giving very simple relations between the parameters of the controller to fulfil the specifications and preserving the robustness characteristics regarding performance, gain variations, and noise.

Part III
Fractional-order Lead-lag Compensators

Chapter 8
Tuning of Fractional-order Lead-lag Compensators

This part of the book will concentrate on the tuning and auto-tuning of fractional-order lead-lag compensators, that is, a generalization of the classical lead-lag compensator. The similarities between this structure and the fractional-order $PI^\lambda D^\mu$ controller in the frequency domain will be discussed. The design method proposed here will avoid the nonlinear minimization and initial conditions problems presented in Part II of the book.

In this chapter, a design method for fractional-order lead-lag compensators (FOLLC) is presented. Simple relations among the parameters of the fractional-order controller are obtained and specifications of steady-state error constant K_{ss}, phase margin φ_m, and gain crossover frequency ω_{cg} are fulfilled, following a robustness criterion based on the flatness of the phase curve of the compensator. The tuning method proposed will be taken as a first step for a later generalization of these lead-lag compensators to the fractional-order $PI^\lambda D^\mu$ controllers, as will be explained in Chapter 9.

8.1 Introduction

The transfer function of an FOLLC is given by

$$C(s) = K_c \left(\frac{s + 1/\lambda}{s + 1/(x\lambda)} \right)^\alpha = K_c x^\alpha \left(\frac{\lambda s + 1}{x\lambda s + 1} \right)^\alpha, \quad 0 < x < 1, \quad (8.1)$$

where α is the fractional order of the controller, $1/\lambda = \omega_{zero}$ is the zero frequency, and $1/(x\lambda) = \omega_{pole}$ is the pole frequency (when $\alpha > 0$). As can be observed, this compensator corresponds to a fractional-order lead compensator when $\alpha > 0$ and $0 < x < 1$, and to a fractional-order lag compensator when $\alpha < 0$ and $0 < x < 1$. The condition $0 < x < 1$ is maintained in both cases. Assuming that the lead compensator behaves similarly to a fractional-order PD^μ controller and the lag compensator similarly to a fractional-order PI^λ controller, the first step for a latter generalization of these structures to the fractional-order $PI^\lambda D^\mu$ controller would be overcome.

This transmittance corresponds to a frequency-bounded fractional-order differentiator/integrator which is at the very origin of the CRONE control [16, 119]. An infinite-dimensional state-space representation for this kind of controller has been studied in [120]. It has also been used on the modeling and the feedback control laws for the stability of viscoelastic control systems [121].

The frequency characteristics of this fractional-order compensator when $\alpha > 0$ (lead compensator) are shown in Figure 8.1. For values of $\alpha < 0$ (lag compensator) the slope of the magnitude curve is negative and the compensator introduces a phase lag. As can be seen in the figure, the value of x sets the distance between the fractional zero ω_{zero} and pole ω_{pole} and the value of λ sets their position in the frequency axis. These two values depend on the value of α. It is observed that, for a fixed pair (x, λ), the higher the absolute value of α, the higher the slope of the magnitude of $C(s)$ and the higher the maximum phase ϕ_m that the compensator can give.

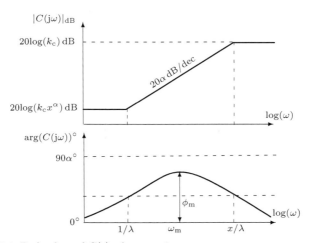

Figure 8.1 Bode plots of $C(s)$ when $\alpha > 0$

As in the case of an integer-order lead-lag compensator, the frequency ω_m is the geometric mean of the corner frequencies ω_{zero} and ω_{pole}, and its equation is given by $\omega_m = \sqrt{x}/\lambda$. At this frequency the characteristics of the compensator $C(s)$ are

$$\left|\frac{C(s)}{K_c x^\alpha}\right|_{\omega=\omega_m} = |C'(s)|_{\omega=\omega_m} = \left(\sqrt{\frac{(\lambda\omega_m)^2 + 1}{(x\lambda\omega_m)^2 + 1}}\right)^\alpha = \left(\frac{1}{\sqrt{x}}\right)^\alpha, \qquad (8.2)$$

$$\arg\left(C'(s)\right)_{\omega=\omega_m} = \phi_m = \alpha \sin^{-1}\left(\frac{1-x}{1+x}\right). \qquad (8.3)$$

The contribution of parameter α is remarkable. The lower the value of α, the longer the distance between the zero and pole and *vice versa*, so that the contribution of phase at a certain frequency stands still. This fact makes the controller more flexible and allows considerations of robustness in the design, as will be explained next.

The rest of the chapter is organized as follows. Section 8.2 describes the design method proposed for the FOLLC. Section 8.3 shows a simulation example of application. Finally, some conclusions are drawn in Section 8.4.

8.2 The Design Method

In this section, an analytical design method is proposed for this type of controller, based on the lead-lag regions defined for the compensator in the complex plane, depending on the value of α. This method allows a flexible and direct selection of the parameters of the fractional structure through the knowledge of the plant and the specifications of error constant, K_{ss}, gain crossover frequency ω_{cg} and phase margin φ_m, also following a robustness criterion based on the flatness of the phase curve of the compensator.

First of all, the value of the compensator gain $k' = K_c x^\alpha$ can be set in order to fulfil an error constant specification for the controlled system. For a general plant model of the form

$$G(s) = \frac{k \prod_i (\tau_i s + 1)}{s^n \prod_j (\tau_j s + 1)}, \tag{8.4}$$

and an error constant K_{ss}, the next equation is obtained:

$$
\begin{aligned}
K_{ss} &= \lim_{s \to 0} s^n C(s) G(s) \\
&= \lim_{s \to 0} s^n k' \left(\frac{\lambda s + 1}{x \lambda s + 1} \right)^\alpha \frac{k \prod_i (\tau_i s + 1)}{s^n \prod_j (\tau_j s + 1)} = \lim_{s \to 0} \frac{s^n k' k}{s^n} = k' k.
\end{aligned} \tag{8.5}
$$

That is, $k' = K_c x^\alpha = K_{ss}/k$.

For a specified phase margin φ_m and gain crossover frequency ω_{cg}, the next relation for the open-loop system is given in the complex plane:

$$
\begin{aligned}
G(j\omega_{cg}) k' \left(\frac{j\lambda\omega_{cg} + 1}{jx\lambda\omega_{cg} + 1} \right)^\alpha &= e^{j(\pi + \varphi_m)} \\
\Rightarrow \left(\frac{j\lambda\omega_{cg} + 1}{jx\lambda\omega_{cg} + 1} \right)^\alpha &= \frac{e^{j(\pi + \varphi_m)}}{k' G(j\omega_{cg})} = a_1 + jb_1 \\
\Rightarrow \left(\frac{j\lambda\omega_{cg} + 1}{jx\lambda\omega_{cg} + 1} \right) &= (a_1 + jb_1)^{1/\alpha} = a + jb,
\end{aligned} \tag{8.6}
$$

where $G(s)$ is the plant to control and (a_1, b_1) is called the "design point."

Doing some calculations, the equations for x and λ will by given by

$$x = \frac{a-1}{a(a-1)+b^2}, \quad \lambda = \frac{a(a-1)+b^2}{b\omega_{cg}}. \tag{8.7}$$

Studying the conditions for a and b to find a solution, it can be concluded that a lead compensator is obtained when $a > 1$ and $b > 0$, and a lag compensator when $\dfrac{1 - \sqrt{1 - 4b^2}}{2} < a < \dfrac{1 + \sqrt{1 - 4b^2}}{2}$ and $-1/2 < b < 0$. Figure 8.2 shows these lead and lag regions in the complex plane for this integer compensator.

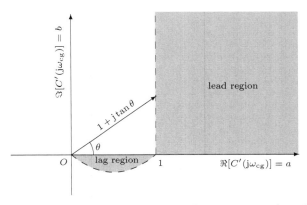

Figure 8.2 Lead and lag regions for the integer-order compensator ($\alpha = 1$)

Let us focus first on the lead compensation. It is clear that for the conventional lead compensator ($\alpha = 1$) the vector $a + jb = a_1 + jb_1$ is perfectly known through the knowledge of the plant and the specifications of phase margin and gain crossover frequency required for the system, as can be seen in (8.6). Knowing the pair (a, b), the values of x and λ are directly obtained by (8.7), and the compensator is therefore designed.

As shown in Figure 8.2, the vector $1 + j \tan \theta$ defines the borderline of the lead region. Using the polar form of this vector

$$\sqrt{1 + \tan^2 \theta}\, e^{j\theta} = \frac{1}{\cos \theta}\, e^{j\theta}, \tag{8.8}$$

and expressing the vector $(a_1 + jb_1)^{1/\alpha}$ in its polar form

$$\left(\sqrt{a_1^2 + b_1^2}\right)^{1/\alpha} e^{j\tan^{-1}(b_1/a_1)/\alpha} = \rho^{1/\alpha} e^{j\delta/\alpha}, \tag{8.9}$$

where $\rho = \left(\sqrt{a_1^2 + b_1^2}\right)$ and $\delta = \tan^{-1}(b_1/a_1)$, the next relations are established from (8.6):

$$\delta = \theta\alpha, \quad \rho^{1/\alpha} = \frac{1}{\cos \theta} \Rightarrow 1 = \rho \left[\cos \left(\frac{\delta}{\alpha}\right)\right]^{\alpha}. \tag{8.10}$$

Then, solving numerically the function $1 = \rho\left[\cos\left(\delta/\alpha\right)\right]^{\alpha}$, the lead compensation regions in the complex plane for different positive values of α are obtained, as shown in Figure 8.3.

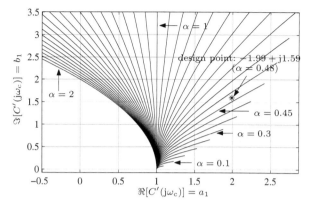

Figure 8.3 Lead regions for the fractional-order compensator $(0 \leqslant \alpha \leqslant 2)$

The zone to the right of each curve is the lead region, and any design point in this zone can be fulfilled with a fractional-order compensator having a value of α equal to or bigger than the one defining the curve which passes through the design point $\alpha_{\min}$. For instance, for the design point in Figure 8.3 the value of $\alpha_{\min}$ is 0.48.

By choosing the minimum value $\alpha_{\min}$, the distance between the zero and the pole of the compensator will be the maximum possible (minimum value of parameter x). In this case, the phase curve of the compensator is the flattest possible and variations in a frequency range centered at ω_{cg} will not produce a significant phase change as in other cases, improving the robustness of the system.

Figure 8.4 shows the pairs (x, λ) obtained for each value of α in the range $\alpha_{\min} \leqslant \alpha \leqslant 2$, with $\alpha_{\min} = 0.48$ (compensation of the design point in Figure 8.3). It is observed that the minimum value of x is obtained for $\alpha_{\min}$ (maximum robustness).

Therefore, through the curves in Figures 8.3 and 8.4, the selection of the parameters of the compensator is flexible and direct.

This method proposed for the fractional-order lead compensator [122, 123] can be used for the design of a fractional-order lag compensator with some corrections that are explained next.

First of all, to determine whether a lead or lag compensator is required to fulfil the specification of phase margin, a simple computation has to be done:

1. if $\pi + \arg(G(j\omega_c)) < \varphi_m \rightarrow$ Lead Compensator;

2. if $\pi + \arg(G(j\omega_c)) > \varphi_m \rightarrow$ Lag Compensator.

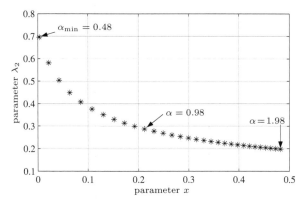

Figure 8.4 Pairs (x, λ) for $\alpha_{\min} \leqslant \alpha \leqslant 2$

In case a phase lag (φ_{lag}) is required for the system, the compensator will be designed as a lead one giving a phase $\varphi_{\text{lead}} = -\varphi_{\text{lag}}$, and then the sign of α is changed. Therefore, in order to keep the specification of phase margin, the phase φ_{lag} of the compensator will now be given by

$$\varphi_{\text{lag}} = -\pi + 2[\pi + \arg(G(\mathrm{j}\omega_{\text{c}}))] - \varphi_{\text{m}} - \arg(G(\mathrm{j}\omega_{\text{c}})). \tag{8.11}$$

Let us remember that in the case of a lead compensation the phase of the compensator is given by $\varphi_{\text{lead}} = -\pi + \varphi_{\text{m}} - \arg(G(\mathrm{j}\omega_{\text{c}}))$.

Besides, it has to be taken into account that the fact of changing the sign of α for the lag compensation also changes the magnitude of the compensator designed (makes it the inverse). So, in order to keep the gain invariable (fulfilling the specification of gain crossover frequency), the lag compensator will be multiplied by a gain $K_{\text{lag}} = 1/(k' |G(\mathrm{j}\omega_{\text{c}})|)^2$. Therefore, the fractional-order lag compensator will be given by

$$C_{\text{lag}}(s) = K_{\text{lag}}k' \left(\frac{\lambda s + 1}{x\lambda s + 1} \right)^{-\alpha}, \tag{8.12}$$

with α a positive real number.

8.3 Simulation Results

In this section, the tuning method proposed is illustrated by a simulation example. The plant to control is given by

$$G(s) = \frac{2}{s(0.5s + 1)}.$$

The gain crossover frequency will be $\omega_{cg} = 10\,\text{rad/sec}$. At this frequency, the plant has a magnitude of $-28.1188\,\text{dB}$ and a phase of $-169.65°$. The velocity error constant will be set to $K_v = 20$, and the phase margin to $\varphi_m = 50°$. As can be observed, a lead compensator is needed in this case. Using the method proposed in Section 8.2, the resulting compensator is

$$C(s) = 10 \left(\frac{0.6404s + 1}{0.0032s + 1} \right)^{0.5},$$

with $k' = 10$, $x = 0.0050$, $\lambda = 0.6404$, and $\alpha = 0.5$. The Bode plots of this compensator are shown in Figure 8.5.

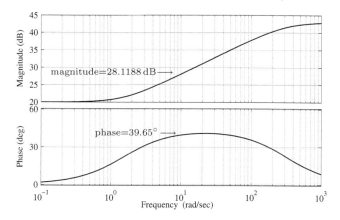

Figure 8.5 Bode plots of controller $C(s)$

At the gain crossover frequency $\omega_{cg} = 10\,\text{rad/sec}$, the compensator has a magnitude of $28.1188\,\text{dB}$ and a phase of $39.65°$. At that frequency the magnitude of the open-loop system $F(s)$ is $0\,\text{dB}$, and the phase margin obtained is $50°$, as shown in Figure 8.6. Therefore, the specifications are correctly fulfilled.

For the sake of implementation, this compensator has been approximated by using a frequency identification method. This way, an integer-order transfer function is obtained that fits the frequency response of the FOLLC in a frequency range of interest. A detailed explanation for the implementation of fractional-order controllers is given in Chapter 12. The resulting step response of the closed-loop system using this controller approximation is shown in Figure 8.7.

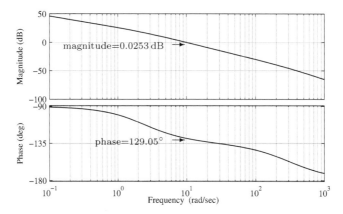

Figure 8.6 Bode plots of the open-loop system with controller $C(s)$

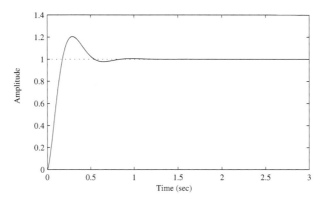

Figure 8.7 Step response of the system with controller $C(s)$

8.4 Summary

In this chapter we have presented a tuning method for FOLLC. Simple relations among the parameters of this fractional-order controller are obtained and specifications of steady-state error constant K_{ss}, phase margin φ_m, and gain crossover frequency ω_{cg} are fulfilled, following a robustness criterion based on the flatness of the phase curve of the compensator. Simulation results are given to show the goodness of the method. For a better understanding of the potential of this tuning technique and its robustness feature, please see Chapters 9 and 17 for a real mechatronic application case.

Chapter 9
Auto-tuning of Fractional-order Lead-lag Compensators

In this chapter, an auto-tuning method for the FOLLC described in the previous chapter will be discussed. We will study to what extent the auto-tuning method proposed here can be used for the tuning of the $PI^\lambda D^\mu$ controller, avoiding in this way the nonlinear minimization and initial conditions problems presented in Part II of the book.

9.1 Introduction

Many process control problems can be adequately and routinely solved by conventional PID-control strategies. The reason why the PID controller is so widely accepted is its simple structure, which has proven to be appropriate for many commonly met control problems such as set-point regulation/tracking and disturbance attenuation. However, although tuning guidelines are available, the tuning process can still be time consuming with the result that many control loops are often poorly tuned and full potential of the control system is not achieved. These methods require a fair amount of *a priori* knowledge such as, for instance, sampling period, dead time, model order, and desired time response. This knowledge may either be given by a skilled engineer or may be acquired automatically by some kind of experimentation. The second alternative, commonly known as auto-tuning, is preferable, not only to simplify the task of the operator, but also for the sake of robustness.

There is a wide variety of auto-tuning methods for integer controllers. Some of them aim in some way at tuning the robustness of the controlled system [124], for example, forcing the phase of the open-loop system to be flat around the gain crossover frequency so that the system is robust to gain variations [62, 116]. However, the complexity of the equations relating the parameters of the controller increases when some kinds of robustness constraints are required for the controlled system. The implementation of these types of auto-tuning methods for industrial purposes will be really complex since, in general, industrial devices, such as a PLC, cannot solve sets of complex nonlinear equations.

For that reason, an auto-tuning method for fractional-order $PI^\lambda D^\mu$ controllers based on the relay test is proposed, that allows the fulfilment of robustness constraints for the controlled system by simple relations among the parameters of the controller, simplifying the later implementation process.

The final aim is to find out a method to auto-tune a fractional-order $PI^\lambda D^\mu$ controller formulated as

$$C(s) = K_c x^\mu \left(\frac{\lambda_1 s + 1}{s}\right)^\lambda \left(\frac{\lambda_2 s + 1}{x\lambda_2 s + 1}\right)^\mu. \tag{9.1}$$

As can be observed, this controller has two different parts given by the following equations:

$$PI^\lambda(s) = \left(\frac{\lambda_1 s + 1}{s}\right)^\lambda, \tag{9.2}$$

$$PD^\mu(s) = K_c x^\mu \left(\frac{\lambda_2 s + 1}{x\lambda_2 s + 1}\right)^\mu. \tag{9.3}$$

Equation 9.2 corresponds to a fractional-order PI^λ controller and (9.3) to a fractional-order lead compensator that can be identified as a PD^μ controller plus a noise filter. In this method, the fractional-order PI^λ controller will be used to cancel the slope of the phase of the plant at the gain crossover frequency ω_{cg}. This way, a flat phase around the frequency of interest is ensured. Once the slope is canceled, the PD^μ controller will be designed to fulfil the design specifications of gain crossover frequency ω_{cg} and phase margin φ_m, following a robustness criterion based on the flatness of the phase curve of this compensator, as explained in Section 8.2 This way, the resulting phase of the open-loop system will be the flattest possible, ensuring the maximum robustness to plant gain variations.

From now on, we will refer to the lead part of the compensator as a PD^μ controller and to the lag part as a PI^λ one.

Let us first give some ideas about the relay test used for the auto-tuning problem.

9.2 Relay Test for Auto-tuning

The relay auto-tuning process has been widely used in industrial applications [125]. The choice of relay feedback to solve the design problem is justified by the possible integration of system identification and control into the same design strategy, giving birth to relay auto-tuning. In this work, a variation of the standard relay test is used, shown in Figure 9.1, where a delay θ_a is introduced after the relay function.

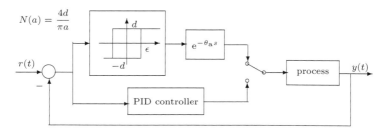

Figure 9.1 Relay auto-tuning scheme with delay

With this scheme, as explained in [116], the following relations are given:

$$\arg\left(G(j\omega_c)\right) = -\pi + \omega_c\theta_a, \tag{9.4}$$

$$|G(j\omega_c)| = \frac{\pi a}{4d} = \frac{1}{N(a)}, \tag{9.5}$$

where $G(j\omega_c)$ is the transfer function of the plant at the frequency ω_c, which is the frequency of the output signal y corresponding to the delay θ_a, d is the relay output, a is the amplitude of the output signal y in Figure 9.1, and $N(a)$ is the equivalent relay gain. This way, for each value of θ_a a different point on the Nyquist plot of the plant is obtained. Therefore, a point on the Nyquist plot of the plant at a particular desired frequency ω_c can be identified, e.g., at the gain crossover frequency required for the controlled system ($\omega_c = \omega_{cg}$).

The problem would be how to select the right value of θ_a which corresponds to a specific frequency ω_c. An iterative method can be used to solve this problem, as presented in [116]. The artificial time delay parameter can be updated using the simple interpolation/extrapolation scheme

$$\theta_n = \frac{\omega_c - \omega_{n-1}}{\omega_{n-1} - \omega_{n-2}}(\theta_{n-1} - \theta_{n-2}) + \theta_{n-1}, \tag{9.6}$$

where n represents the current iteration number. With the new θ_n, after the relay test, the corresponding frequency ω_n can be recorded and compared with the frequency ω_c so that the iteration can continue or stop. Two initial values of the delay (θ_{-1} and θ_0) and their corresponding frequencies (ω_{-1} and ω_0) are needed to start the iteration. Therefore, first of all, a value for θ_{-1} is selected and the relay test is carried out, obtaining an output signal with frequency ω_{-1}. Then, in a second iteration, another value is given for θ_0, obtaining an output signal with frequency ω_0. With these two pairs, $(\theta_{-1}, \omega_{-1})$ and (θ_0, ω_0), the next value of θ_n is automatically obtained by using the interpolation/extrapolation scheme of (9.6).

Let us now concentrate on the design of the fractional-order PI^λ controller.

9.3 Design of the Fractional-order Lag Part

The fractional-order PI^λ controller of (9.2) will be used to cancel the slope of the phase of the plant in order to obtain a flat phase around the frequency point ω_{cg}. The value of this slope is given by

$$v = \frac{\phi_u - \phi_{n-1}}{\omega_u - \omega_{n-1}} \sec, \tag{9.7}$$

where ω_{n-1} is the frequency $n-1$ experimented with the relay test and ϕ_{n-1} its corresponding plant phase, and ϕ_u the plant phase corresponding to the frequency of interest $\omega_u = \omega_{cg}$.

The phase of the fractional-order PI^λ controller is given by

$$\psi = \arg\left(PI^\lambda(s)\right) = \lambda\left(\arctan(\lambda_1\omega) - \frac{\pi}{2}\right). \tag{9.8}$$

In order to cancel the slope of the phase curve of the plant, v, the derivative of the phase of $PI^\lambda(s)$ at the frequency point ω_{cg} must be equal to $-v$, resulting in the equation

$$\psi' = \left.\frac{d\psi}{d\omega}\right|_{\omega=\omega_{cg}} = \lambda\frac{\lambda_1}{1 + (\lambda_1\omega_{cg})^2} = -v. \tag{9.9}$$

The parameters λ and λ_1 must be selected so that this equation is fulfiled. Studying the function (9.9) and differentiating with respect to parameter λ_1 in (9.10), it is found that it has a maximum at $\lambda_1 = 1/\omega_{cg}$ in (9.11), as can be observed in Figure 9.2:

$$\frac{d\psi'}{d\lambda_1} = \frac{\lambda\left[(\lambda_1\omega_{cg})^2 - 1\right]}{(1 + (\lambda_1\omega_{cg})^2)^2}, \tag{9.10}$$

$$\frac{d\psi'}{d\lambda_1} = 0 \Rightarrow (\lambda_1\omega_{cg})^2 - 1 = 0 \Rightarrow \lambda_1 = \frac{1}{\omega_{cg}}. \tag{9.11}$$

That is, choosing $\omega_{zero} = 1/\lambda_1 = \omega_{cg}$ the slope of the plant at the frequency ω_{cg} will be canceled with the maximum slope of the fractional-order controller. Once λ_1 is fixed, the value of λ is easily determined by

$$\lambda = \frac{-v\left(1 + (\lambda_1\omega_{cg})^2\right)}{\lambda_1}. \tag{9.12}$$

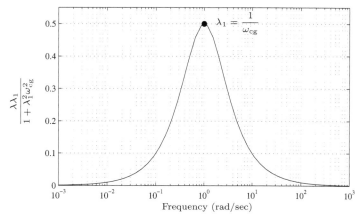

Figure 9.2 Derivative of the phase of the PI^λ controller at $\omega = \omega_{cg}$, for $\lambda = 1$ and $\omega_{cg} = 1$

It is observed that the value of λ will be minimum when $\lambda_1 = 1/\omega_{cg}$. Variations of the frequency ω_{zero} up or down the frequency ω_{cg} will produce higher values of the parameter λ. Therefore, selecting $\omega_{zero} = \omega_{cg}$ the phase lag of the resulting $PI^\lambda(s)$ controller will be the minimum one with minimum λ. This fact is very interesting from the robustness point of view. The less the phase lag of the controller $PI^\lambda(s)$, the less the phase lead of the controller $PD^\mu(s)$ at the frequency ω_{cg}, favoring the flatness of its phase curve.

Then, considering this robustness criterion, the value of λ_1 will be fixed to $1/\omega_{cg}$ and λ will be obtained by (9.12).

Remember that the real value of ω_{cg} to be used in the design is ω_u, which is that obtained with the relay test and very close to ω_{cg}.

9.4 Design of the Fractional-order Lead Part

Defining the system $G_{flat}(s) = G(s)PI^\lambda(s)$, now the controller $PD^\mu(s)$ will be designed so that the open-loop system $F(s) = G_{flat}(s)PD^\mu(s)$ satisfies the specifications of gain crossover frequency ω_{cg} and phase margin φ_m following a robustness criterion based on the flatness of the phase curve of this compensator [122, 123], as explained in Section 8.2.

For a specified phase margin φ_m and gain crossover frequency ω_{cg}, the following relations for the open-loop system can be given in the complex plane:

$$G_{flat}(j\omega_{cg})k'\left(\frac{j\lambda_2\omega_{cg}+1}{jx\lambda_2\omega_{cg}+1}\right)^\mu = e^{j(-\pi+\varphi_m)}$$

$$\Rightarrow C'(j\omega_{cg}) = \left(\frac{j\lambda_2\omega_{cg}+1}{jx\lambda_2\omega_{cg}+1}\right)^\mu = \frac{e^{j(-\pi+\varphi_m)}}{G_{flat}(j\omega_{cg})k'} = a_1 + jb_1 \qquad (9.13)$$

$$\Rightarrow \left(\frac{j\lambda_2\omega_{cg}+1}{jx\lambda_2\omega_{cg}+1}\right) = (a_1 + jb_1)^{1/\mu} = a + jb,$$

where $k' = K_c x^\mu = 1$ in this case, $G_{flat}(s)$ is the plant to be controlled, and (a_1, b_1) is the "design point."

As can be seen, we are in the case of the design of the FOLLC described in Section 8.2. Therefore, the parameters of the PD^μ controller will be tuned according to the explanations in that chapter. Please review it for a better understanding of this auto-tuning method.

Let us then sum up how the $PI^\lambda D^\mu$ controller is auto-tuned. The following steps can be solved by a simple computer, using a data acquisition system to control and monitor the real process (as explained in the section for experimental results). A PLC could also be used for the determination of the parameters of the controller, due to the simplicity of the equations involved in the auto-tuning method:

1. Once the specifications of design are given (ω_{cg} and φ_m), the relay test is applied to the plant and the resulting pairs (θ_n, ω_n) obtained from the n iterations of the test are saved and used for the calculation of the phase and magnitude of the plant at each frequency ω_n (following (9.4) and (9.5)). As explained previously, these values are used for obtaining the slope of the plant phase v (9.7). With the value of the slope, the parameters λ and λ_1 of the PI^λ controller are directly obtained by (9.9) and (9.11). Then the system $G_{flat}(j\omega_{cg})$ is determined.

2. Once the system $G_{flat}(j\omega_{cg})$ is defined, and according to (9.13), the parameters of the fractional-order compensator in (9.3) are obtained by simple calculations summarized next, following the robustness feature explained in Section 8.2.

3. Select a very small initial value of μ, for example $\mu = 0.05$. For this initial value, calculate the value of x and λ_2 using the relations in (9.13) and (8.7).

4. If the value of x obtained is negative, then the value of μ is increased a fixed step and Step 2 is repeated again. The smaller the fixed increase of μ, the more accurate the selection of parameter μ_{min}. Repeat Step 2 until the value of x obtained is positive.

5. Once a positive value of x is obtained, the value of μ must be recorded as $\mu_{\min}$. This value of x will be close to zero and will ensure the maximum flatness of the phase curve of the compensator (iso-damping constraint). The value of λ_2 corresponding to this value $\mu_{\min}$ is also recorded.

Therefore, all the parameters of the $\text{PI}^\lambda\text{D}^\mu$ controller have been obtained through this iterative process. Then the controller is implemented and starts to control the process through the switch illustrated in Figure 9.1, concluding the auto-tuning procedure.

9.5 Formulation of the Resulting Controller

Once the parameters of the fractional-order $\text{PI}^\lambda\text{D}^\mu$ controller of (9.1) are obtained by following the design method explained above, these parameters can be related to those of the standard $\text{PI}^\lambda\text{D}^\mu$ controller given by

$$C_{\text{std}}(s) = K_{\text{p}} \left(1 + \frac{1}{T_{\text{i}}s} \right)^\lambda \left(1 + \frac{T_{\text{d}}s}{1 + T_{\text{d}}s/N} \right)^\mu . \tag{9.14}$$

Carrying out some calculations in (9.14), the following transfer function is obtained:

$$C_{\text{std}}(s) = \frac{K_{\text{p}}}{(T_{\text{i}})^\lambda} \left(\frac{T_{\text{i}}s + 1}{s} \right)^\lambda \left(\frac{T_{\text{d}}(1 + 1/N)s + 1}{1 + T_{\text{d}}s/N} \right)^\mu . \tag{9.15}$$

Comparing (9.1) and (9.14), the relations obtained are $T_{\text{i}} = \lambda_1$, $K_{\text{p}} = k'(\lambda_1)^{-\lambda}$, $N = (1 - x)/x$ and $T_{\text{d}} = \lambda_2(1 - x)$.

9.6 Summary

In this chapter, an auto-tuning method for the fractional-order $\text{PI}^\lambda\text{D}^\mu$ as a generalization of the FOLLC using the relay test has been proposed. This method allows a flexible and direct selection of the parameters of the controller through the knowledge of the magnitude and phase of the plant at the frequency of interest, obtained with the relay test. Specifications of gain crossover frequency ω_{cg} and phase margin φ_{m} can be fulfilled with a robustness property based on the flatness of the phase curve of the open-loop system, guaranteeing the iso-damping property of the time response of the system to gain variations.

Chapter 17 shows a real mechatronic application case in which this auto-tuning technique is implemented experimentally.

Part IV
Other Fractional-order Control
Strategies

Chapter 10
Other Robust Control Techniques

10.1 CRONE: Commande Robuste d'Ordre Non Entier

In Chapter 4 we introduced the idea of robust control and basic controllers to fulfil this characteristic. In the frequency domain, we can say that Bode's ideal loop transfer function in Section 2.3.4 is the reference model to achieve a robust performance. The purpose is to obtain an open-loop characteristic similar to that of this reference model, ensuring in this way a constant phase margin around a frequency of interest and, therefore, a constant overshoot of the time responses to plant gain variations.

This idea was developed extensively by Oustaloup [50], who studied the fractional-order algorithms for the control of dynamic systems and demonstrated the superior performance of the CRONE (Commande Robuste d'Ordre Non Entier, meaning Non-integer-order Robust Control) method over the PID controller. There are three generations of CRONE controllers [51], and we will briefly review all of them in this chapter.

Some of the characteristics of the CRONE techniques are:

- Frequency domain based methodology using fractional-order differentiation as high-level design parameter (since 1975).
- Continuous-time or discrete-time control of disturbed SISO and MIMO systems.
- Use of the common unity-feedback configuration.
- Robustness of the stability degree with respect to parametric plant disturbances (no over-estimation).
- Avoiding over-estimation of plant disturbance leads to non-conservative robust control systems and to as good performance as possible.
- Control of minimum-phase or non-minimum-phase plants, unstable plants, or plants with mechanical bending modes, time-varying plants, and non-linear plants.

From now on, we will consider the control system in Figure 10.1, with controller $C(s)$ and plant $G(s)$.

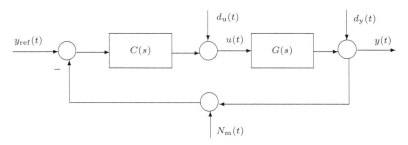

Figure 10.1 Control system

Let us concentrate on each of the CRONE generations.

10.1.1 First Generation CRONE Controller

The first generation CRONE controller is very suitable for gain-like plant disturbance models and for constant plant phase around a frequency of interest. Its transfer function is given by

$$C(s) = C_0 s^\alpha, \tag{10.1}$$

with α and $C_0 \in \mathbb{R}$.

The Bode plots of this controller are shown in Figure 10.2. The controller is defined within a frequency range (ω_b, ω_h) around the desired open-loop gain crossover frequency ω_{cg}. The Oustaloup recursive approximation can be used to implement this controller, as described in Section 12.1.2. However, any approximating formula may be used as long as it allows us to obtain a rational transfer function whose frequency response fits the frequency response of the original irrational-order transfer function in a desired frequency range (ω_b, ω_h).

This type of controller is useful when the plant to be controlled already has a constant phase, at least in a frequency range around the gain crossover frequency (asymptotic plant frequency response within this band). In that case, the loop will be robust to plant gain variations, since even though the gain crossover frequency may change, the plant phase margin will not, and neither will the controller phase.

However, due to control effort limitations, it is sometimes impossible to choose an open-loop gain crossover frequency within an asymptotic behavior

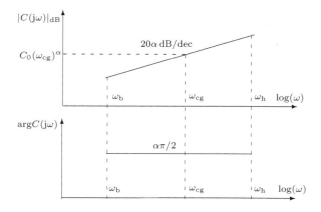

Figure 10.2 Bode plots of first generation CRONE controller

frequency band of the plant. Therefore, when the desired ω_{cg} is outside an asymptotic behavior band, the first generation CRONE controller cannot ensure the robustness of the closed-loop system stability margins.

Nevertheless, as Bode first stated, for the design of single-loop absolutely stable amplifiers whose gains vary, the robust controller is the one which yields an open-loop transfer function defined by a constant phase (and constant dB/dec gain slope) in a useful band. This leads to the second generation CRONE controller.

10.1.2 Second Generation CRONE Controller

As already stated, if a plant does not have a constant phase, first generation CRONE controllers will not be robust to plant gain variations. However, it is possible to devise a controller that ensure a constant open-loop phase [50]. When ω_{cg} is within a frequency band where the plant uncertainties are gain-like, the CRONE approach defines the open-loop transfer function, in the frequency band $(\omega_{\text{b}}, \omega_{\text{h}})$, by that of a fractional-order integrator:

$$F(s) = C(s)G(s) = \left(\frac{\omega_{\text{cg}}}{s}\right)^{\alpha}, \tag{10.2}$$

with $\alpha \in \mathbb{R}$ and $\alpha \in [1,2]$. It is in fact Bode's ideal loop transfer function defined in Section 2.3.4.

Figure 10.3 shows the open-loop Nichols chart for the second generation CRONE approach. The vertical straight line, which is the desired shape of the open-loop Nichols chart, is called *the template*.

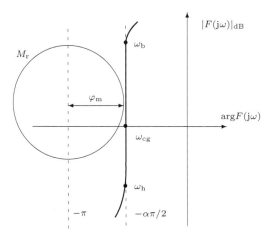

Figure 10.3 Open-loop Nichols chart for the second generation CRONE approach

At the time of plant disturbances, the vertical displacement of the vertical template ensures the robustness of:

- the phase margin, $\varphi_m = (2 - \alpha)\pi/2$;
- the gain margin, $M_g = \sin(\alpha\pi/2)$;
- the resonant peak, $M_r = 1/\sin(\alpha\pi/2)$;
- the damping ratio, which is related to the closed-loop poles, $\zeta = -\cos(\pi/\alpha)$.

Once the optimal open-loop Nichols chart is obtained, the fractional-order controller $C(s)$ is obtained from (10.2) as

$$C(s) = \frac{F(s)}{G(s)}. \tag{10.3}$$

There are two ways to synthesize the rational form of the controller. In its original formulation, the Oustaloup recursive approximation for the fractional-order integrator $F(s)$ was to be used (Section 12.1.2 and [50]). To manage the control effort level and the steady-state errors, the fractional-order open-loop transfer function has to be band-limited and is more complicated by including integral and low-pass effects:

$$F_R(s) = k' \left(\frac{\omega_b}{s} + 1\right)^{n_b} \left(\frac{\omega_h + s}{\omega_b + s}\right)^n \frac{1}{(s + \omega_h)^{n_F}}, \tag{10.4}$$

with $\omega_b, \omega_h, k' \in \mathbb{R}^+$, and $n_b, n_F \in \mathbb{N}^+$ [50]. Then, the controller will have the form

$$C(s) = \frac{F_R(s)}{G(s)}. \tag{10.5}$$

Another way to implement $C(s)$ consists of replacing it by a rational-order transfer function which has the same frequency response as $F(s)/G(s)$ in a frequency range of interest, using any frequency-based identification method as shown in Section 12.3.

10.1.3 Third Generation CRONE Controller

Both first and second generation CRONE controllers aim at being robust to plant gain variations. However, other types of model uncertainty, including pole and zero misplacement, are not covered. Third generation CRONE controllers [51] try to take these cases into account.

Before identifying the objective of this third generation, let us introduce two important sets of curves in the Nichols chart [50, 51, 126]: curves of constant closed-loop gain and curves of approximately constant damping ratio.

Let us consider an open-loop system $F(\mathrm{j}\omega)$ whose frequency response is given by

$$\Theta = |F(\mathrm{j}\omega)|, \tag{10.6}$$

$$\theta = \arg[F(\mathrm{j}\omega)]. \tag{10.7}$$

The closed-loop will be given by

$$L(\mathrm{j}\omega) = \frac{F(\mathrm{j}\omega)}{1 + F(\mathrm{j}\omega)}, \tag{10.8}$$

and therefore, after simple calculations,

$$|L(\mathrm{j}\omega)| = \left| \frac{\Theta e^{\mathrm{j}\theta}}{1 + \Theta e^{\mathrm{j}\theta}} \right| = \frac{\Theta}{\sqrt{1 + \Theta^2 + 2\Theta \cos\theta}}. \tag{10.9}$$

Figure 10.4 represents the curves of constant values of $|L(\mathrm{j}\omega)|$ in a Nichols chart. They have a periodicity of 2π rad in the phase axis and are symmetric with respect to all vertical straight lines given by $x = k\pi$, $k \in \mathbb{Z}$ [50].

In [50], Oustaloup shows that the damping ratio is approximately given by

$$\zeta = -\cos \frac{\pi^2}{2\arccos\left(\dfrac{1 + \Theta^2 + 2\Theta \cos\theta}{2\Theta} - 1 \right)}, \tag{10.10}$$

the exact relationship depending on the plant.

Figure 10.5 represents curves of constant values of ζ in a Nichols chart. They have a periodicity of 2π rad in the phase axis and are symmetric with respect to all vertical straight lines given by $x = k\pi$, $k \in \mathbb{Z}$ [50].

Now that these two concepts are explained, we can discuss the objective of a third generation CRONE controller [50,51], which is to ensure that the closed-loop gain will never get beyond a certain value, even if some parameters

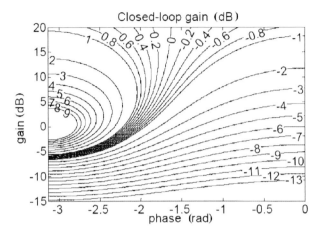

Figure 10.4 Nichols chart with curves of constant values of closed-loop gain

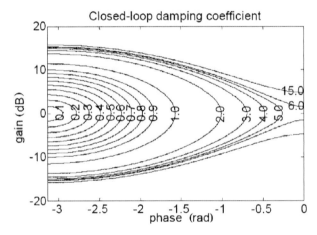

Figure 10.5 Nichols chart with curves of constant values of closed-loop damping coefficient

of the plant vary within a known range. Or the objective may be to ensure that the closed-loop damping ratio will never get below a certain value, also for a known range of variation of some plant parameters. In other words, the objective is to prevent the Nichols chart of the open-loop system from approaching the zones where the closed-loop gain is high and the damping ratio is low, that is, zones around points $(0\,\mathrm{dB}, (2k+1)\pi\,\mathrm{rad})$, $k \in \mathbb{Z}$.

Once the desired open-loop frequency behavior is known, the controller behavior is obtained similarly to the case of the second generation CRONE controller in (10.3). And the same considerations for the implementation of the controller can be taken into account (see [50, 51] and Chapter 12).

An example of application is that presented in Figure 10.6, which is a benchmark for digital robust control presented in [16, 127]. F_1, F_2, and F_3 correspond to three different configurations of a discrete-time plant model in the open-loop:

$$F_1(z^{-1}) = \frac{0.50666z^{-4} + 0.28261z^{-3}}{0.88642z^{-4} - 1.31608z^{-3} + 1.58939z^{-2} - 1.41833z^{-1} + 1}, \quad (10.11)$$

$$F_2(z^{-1}) = \frac{0.18123z^{-4} + 0.10276z^{-3}}{0.89413z^{-4} - 1.84083z^{-3} + 2.20265z^{-2} - 1.99185z^{-1} + 1}, \quad (10.12)$$

$$F_3(z^{-1}) = \frac{0.10407z^{-4} + 0.06408z^{-3}}{0.87129z^{-4} - 1.93353z^{-3} + 2.31962z^{-2} - 2.09679z^{-1} + 1}, \quad (10.13)$$

with sampling period $T = 0.05\,\mathrm{sec}$.

The Nichols charts of such models come dangerously close to points where the closed-loop gain is rather high, as can be observed in Figure 10.6. A controller is required for this plant that can deal with the three configurations ensuring that the closed-loop gain is never higher than 1 dB, which is fulfilled with a third generation CRONE controller as shown in Figure 10.7. The transfer function of this controller is

$$C(z^{-1}) = \frac{\begin{aligned}&-0.14614z^{-7} + 0.607639z^{-6} - 0.580113z^{-5} - 0.80347z^{-4}\\ &+1.60318z^{-3} - 0.2082z^{-2} - 0.87043z^{-1} + 0.41053\end{aligned}}{\begin{aligned}&-0.05328z^{-7} + 0.10323z^{-6} - 0.54187z^{-5} + 0.58878z^{-4}\\ &+1.32411z^{-3} - 1.68605z^{-2} - 0.73491z^{-1} + 1\end{aligned}}. \quad (10.14)$$

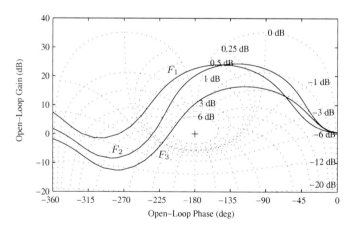

Figure 10.6 Performance in open-loop without controller

The step responses of the three plants under the same controller can be obtained as shown in Figure 10.8. It can be seen that the CRONE controller is robust for the three plants.

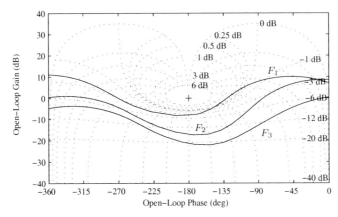

Figure 10.7 Performance in open-loop with a third generation CRONE controller

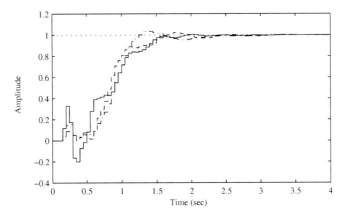

Figure 10.8 Step responses of the closed-loop system, with *solid lines*, *dashed lines*, and *dashed dotted lines*, respectively, for plants F_1, F_2, and F_3

10.2 QFT: Quantitative Feedback Theory

As is known, a wide range of research activities deal with the application of Quantitative Feedback Theory (QFT) for the design of different control structures. QFT consists of shaping the open-loop function to a set of restrictions (or *boundaries*) given by the design specifications and the uncertain model of the plant.

The optimal loop computation is a nonlinear and non-convex optimization problem for which it is difficult to find a satisfactory solution, since there is no optimization algorithm which guarantees a global optimum solution.

A possible approach is to simplify the equations in order to obtain a different optimization problem for which there exists a closed-form solution or an optimization algorithm which does guarantee a global optimum. Thus, there is a trade-off between the conservative simplification of the problem and the computational solvability. This is the approach in [128, 129]. Some authors have also researched the loop shaping problem in terms of particular rational structures, with a certain degree of freedom, which can be shaped to the particular problem to be solved [130, 131]. Another possibility is to use evolutionary algorithms, able to solve nonlinear and non-convex optimization problems [132, 133]. All these approaches, which use rational controllers and typically a low order controller, give a poor result, in general. Thus, very high order rational controllers have to be used to obtain a close to optimal solution. This is a main drawback of the above automatic loop shaping techniques, since the resulting optimization problem is considerably more complex as the number of parameters (directly related to the controller order) is increased.

A solution to this problem in the sense adopted in CRONE is the use of fractional-order controllers for minimizing the search space by using a minimum number of parameters, still having a good frequency performance. This idea has been used in [134, 135], where some fractional-order controllers, including CRONE and second-order fractional structures, have been considered.

In this section, we will briefly describe how to use QFT for the robust tuning of an FOLLC of the form in (8.1).

10.2.1 Some Preliminaries About QFT

The basic idea of QFT [136] is to define and take into account, along the control design process, the quantitative relation between the amount of uncertainty to deal with and the amount of control effort to use. Typically, the QFT control system configuration in Figure 10.9 considers two degrees of freedom: a controller $C(s)$ in closed-loop, which cares for the satisfaction of robustness specifications despite uncertainty; and a pre-controller $P(s)$ designed after $C(s)$, which allows one to achieve the desired frequency response once the uncertainty has been controlled.

For a given plant $G_p(s)$, its template $\mathcal{G}$ is defined as the set of possible plant frequency responses due to uncertainty. A *nominal* plant, $G_{p0} \in \mathcal{G}$, is chosen.

The design of the controller $C(s)$ is accomplished in the Nichols chart, in terms of the nominal open-loop transfer function, $F_0(s) = C(s)G_{p0}(s)$. A discrete set of design frequencies Ω is chosen. Given quantitative specifica-

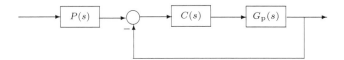

Figure 10.9 Two degrees of freedom control system configuration

tions on robust stability and robust performance for the closed-loop system, *boundaries* $\mathcal{B}_\omega$, $\omega \in \Omega$, are computed. $\mathcal{B}_\omega$ defines the allowed regions for $F_0(j\omega)$ in the Nichols chart, so that $\mathcal{B}_\omega$ not being violated by $F_0(j\omega)$ implies specification satisfaction by $F(j\omega) = C(j\omega)G_p(j\omega)$, $\forall G_p(s) \in \mathcal{G}$. The basic step in the design process, *loop shaping*, consists of the design of $F_0(j\omega)$ which satisfies boundaries and is reasonably close to optimum. QFT optimization criterion is the minimization of the high-frequency gain [137], *i.e.*, K_{hf} in

$$K_{hf} = \lim_{s \to \infty} s^{n_{pe}} F(s), \tag{10.15}$$

where n_{pe} is the pole-zero excess.

This definition of K_{hf} allows one to compare (only) transfer functions with equal n_{pe}. The universal high-frequency boundary (UHFB) is a special boundary. It is a robust stability boundary which should be satisfied by $F_0(j\omega)$, $\forall \omega \geq \omega_U$. It has been shown [47, 137, 138] that, to get close to the optimum, $F_0(j\omega)$ should be as close as possible to the critical point $(0\,\mathrm{dB}, -180°)$ in the vicinity of the gain crossover frequency, ω_{cg}, and so should get as close as possible to the lower right side of the UHFB. Frequency ω_{cg} is defined as the first frequency such that $F(j\omega_{cg}) = 0\,\mathrm{dB}$.

Loop shaping is traditionally carried out manually with the help of tools, such as MATLAB. This leads to a trial-error process, whose resulting quality is strongly determined by the designer's experience and intuition. There is no commercial tool for this purpose yet. An automatic loop shaping procedure is, therefore, a key issue which is still of great interest.

10.2.2 Case Study with a Fractional-order Lead-lag Compensator

Let us consider the system in (7.6), that is, the experimentally identified liquid level system in Chapter 7, with a nominal plant of

$$G(s) = \frac{k}{\tau s + 1} e^{-Ls} = \frac{3.13}{433.33s + 1} e^{-5s}. \tag{10.16}$$

In addition, parameters k and τ have an interval uncertainty given by $k \in [2, 9.8]$ and $\tau \in [380, 1200]$ sec. The value of the delay to be considered for the loop shaping process is $L = 5$ sec.

The (nominal) design specifications are given basically in terms of a gain crossover frequency ω_{cg}, a phase margin φ_{m}, a noise rejection specification given by $|T(\mathrm{j}\omega)|_{\mathrm{dB}} \leqslant A \, \mathrm{dB}$, $\forall \omega \geqslant \omega_{\mathrm{t}}$, and an output disturbance rejection specification over a frequency interval given by $|S(\mathrm{j}\omega)|_{\mathrm{dB}} \leqslant B \, \mathrm{dB}$, $\forall \omega \leqslant \omega_{\mathrm{s}}$.

It should be pointed out that the above specifications are given for a nominal closed-loop design (in particular for $k = 3.13$ and $\tau = 433.33$ sec). Since QFT can consider specifications for the whole set of plants $\mathcal{G}$, these specifications should be relaxed accordingly to make a fair comparison of QFT with other techniques. In particular, for a given value of the delay L, the design specifications are considered to be:

- (nominal) gain crossover frequency, ω_{cg};
- (worst case) phase margin, $\varphi_{\mathrm{m}} \geqslant \varphi_{\mathrm{wcm}}$, $\forall k \in [2, 9.8]$ and $\forall \tau \in [380, 1200]$ sec;
- noise rejection, $|T(\mathrm{j}\omega)|_{\mathrm{dB}} \leqslant A \, \mathrm{dB}$, $\forall \omega \geqslant \omega_{\mathrm{t}}$, $\forall k \in [2, 9.8]$ and $\forall \tau \in [380, 1200]$ sec;
- output disturbance rejection, $|S(\mathrm{j}\omega)|_{\mathrm{dB}} \leqslant B \, \mathrm{dB}$, $\forall \omega \leqslant \omega_{\mathrm{s}}$, $\forall k \in [2, 9.8]$ and $\forall \tau \in [380, 1200]$ sec.

Since these design specifications are somehow stronger that those based on nominal values, there is some degree of freedom in relaxing some of them in terms of the parameters φ_{wcm}, ω_{s}, and ω_{t}.

The controller considered is the FOLLC in (8.1). For a better behavior of the optimization algorithm used for loop shaping, the controller is finally formulated as [139]:

$$C(s) = k_c x^\mu \left(\frac{s + 1/\lambda_1}{s} \right)^\lambda \left(\frac{\lambda_2 s + 1}{x \lambda_2 s + 1} \right)^\mu . \tag{10.17}$$

In addition, it is convenient to have a fast roll-off at high frequencies. To have an additional parameter that enables this behavior, some extra structure is introduced in the controller. This is analogous to term C_2 of the TID controller in [107], and is given by

$$C_2(s) = \frac{1}{(1 + s/\omega_{\mathrm{h}})^{n_{\mathrm{h}}}} . \tag{10.18}$$

So, the final structure is

$$C(s)C_2(s) = k_c x^\mu \left(\frac{s + 1/\lambda_1}{s} \right)^\lambda \left(\frac{\lambda_2 s + 1}{x \lambda_2 s + 1} \right)^\mu \frac{1}{(1 + s/\omega_{\mathrm{h}})^{n_{\mathrm{h}}}} . \tag{10.19}$$

The controller will be implemented by identification of its global frequency response, as described in Section 12.3.

As stated at the beginning of this section, CRONE controllers have also been a case study for this QFT technique [134]. In that case, second and third generation CRONE controllers have been considered, with additional terms to shape low and high frequency responses. The consideration of these CRONE controllers implies slight modifications in the QFT algorithm with respect to that used for the controller in (10.19), due to the similarities between these structures.

10.2.3 Simulation Results

The design process is performed by minimizing K_{hf}, the high frequency gain of the nominal loop, subject to the restrictions given by specifications. The optimization problem is solved by means of an evolutionary algorithm, using the tool described in [134].

The design specifications are:

- $\omega_{\mathrm{cg}} = 0.01\,\mathrm{rad/sec}$;
- $\varphi_{\mathrm{m}} \geqslant \varphi_{\mathrm{wcm}} = 50°$, $\forall k \in [2, 9.8]$, and $\forall \tau \in [380, 1200]\,\mathrm{sec}$;
- $|T(\mathrm{j}\omega)_{\mathrm{dB}}| \leqslant -20\,\mathrm{dB}, \forall \omega \geqslant \omega_{\mathrm{t}} = 10\,\mathrm{rad/sec}$, $\forall k \in [2, 9.8]$, and $\forall \tau \in [380, 1200]\,\mathrm{sec}$;
- $|S(\mathrm{j}\omega)_{\mathrm{dB}}| \leqslant -20\,\mathrm{dB}, \forall \omega \leqslant \omega_s = 0.002\,\mathrm{rad/sec}$, $\forall k \in [2, 9.8]$, and $\forall \tau \in [380, 1200]\,\mathrm{sec}$.

The controller obtained is

$$C(s)C_2(s) = 1.8393 \left(\frac{s+0.011}{s}\right)^{0.96} \left(\frac{8.8 \times 10^{-5}s+1}{8.096 \times 10^{-5}s+1}\right)^{1.76} \frac{1}{(1+s/0.29)^2},$$
$$(10.20)$$

with $K_{\mathrm{hf}} = -60.7\,\mathrm{dB}$. Figures 10.10 and 10.11 show the frequency behavior of the resulting open-loop system. The corresponding step responses are given in Figures 10.12 and 10.13, for different gain values and different time constant values, respectively.

As can be seen, the system performs in a robust way, taking into account the large parametric uncertainties and the design specifications. From Figure 10.10 we can conclude that the fractional-order controller exhibits a rich frequency behavior, making the open-loop gain fit closely to the stability boundary. Note that the optimal design should fit this stability boundary as close as possible, especially in the lower right corner. In contrast to other tuning techniques that guarantee nominal closed-loop specifications, QFT

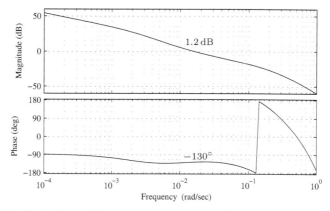

Figure 10.10 Bode plots of the nominal open-loop for $L = 5$ sec

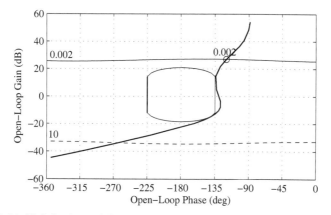

Figure 10.11 Nichols chart of the nominal open-loop for $L = 5$ sec

guaranties the satisfaction of robust stability and performance specifications for all the considered parametric uncertainties.

On the other hand, evolutionary algorithms have efficiently arrived at the given solution with a reasonable number of iterations, mainly due to the reduced number of parameters of the $PI^\lambda D^\mu$ structure.

10.3 Summary

In this chapter, a brief description of other robust control techniques has been given. The three generations of CRONE control have been introduced, presenting their main design goals in the frequency domain. The QFT technique applied to fractional-order controllers has also been described,

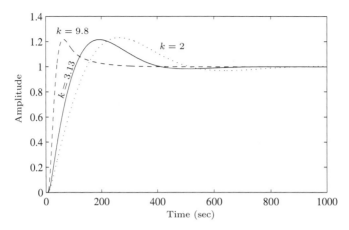

Figure 10.12 Step responses of the controlled system for $L = 5$ sec and different values of k

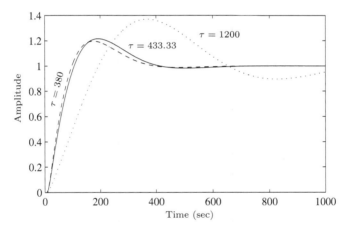

Figure 10.13 Step responses of the controlled system for $L = 5$ sec and different values of τ

including some simulation results. Since the structure of QFT controllers may sometimes be rather complicated, the implementation of the controller should also be taken into account. This matter is also presented in Chapter 12.

Chapter 11
Some Nonlinear Fractional-order Control Strategies

11.1 Sliding Mode Control

The purpose of this section is to present the consequences of combining fractional-order control (FOC) and sliding mode control (SMC) in two ways, taking into account the two components of the sliding mode design approach. So, FOC is introduced in SMC by using, on the one hand, fractional-order sliding surfaces or switching functions, and, on the other hand, control laws with fractional-order derivatives and integrals. For a clear illustration of the proposed control strategies, the well known double integrator is considered as the system to be controlled.

11.1.1 Introduction

Pioneering works in variable structure control systems (VSCS) were made in Russia in the 1960s, and the fundamental ideas were published in English in the mid-1970s. After that, VSCS concepts have been used in robust control, adaptive, and model reference systems, tracking, state observes and fault detection (see [140, 141] for additional information). Sliding mode control is a particular type of VSCS designed to drive and then constrain the system state to lie within a neighborhood of the decision rule or switching function, and it is well known for its robustness to disturbances and parameters variations. On the other hand, fractional-order control, that is, the use of fractional-order derivatives and integrals in the control laws, has been recognized as an alternative strategy for solving robust control problems (see [58, 126] for a survey on the topic). The purpose of this work is to study the consequences of introducing FOC in SMC in two ways, taking into account the two components of the sliding mode design approach, the first one involving the design of a switching function so that the sliding motion satisfies the design specifications, and the second one concerning the selection of a control law which will enforce the sliding mode; therefore existence and reachability conditions are satisfied. So, FOC is introduced in SMC by using, on the one hand, fractional-order sliding surfaces or switching functions, and, on the other hand, control laws with fractional-order derivatives and integrals. The first approach was introduced in [142] applied to a power electronic Buck converter with very interesting and successful experimental results. In this work we study in a more extended way this first approach and introduce the second one. Furthermore, considerations about existence and reachability conditions are made. For a clear illustration of the proposed control strategies, the well known double integrator is considered as the system to be controlled. This system, representing a one degree-of-freedom translational and rotational motion, can be considered as a linear approximation of some interesting mechanical and electromechanical systems, such as the low friction servo or the pendulum.

11.1.2 SMC of the Double Integrator

The main part of this section has been taken from [140]. Consider the double integrator given by

$$\ddot{y}(t) = u(t). \tag{11.1}$$

Initially consider the feedback proportional control law

$$u(t) = -ky(t), \qquad k \in \mathbb{R}^+. \tag{11.2}$$

So, substituting for the control action in (11.1), multiplying the resulting equation throughout by $\dot{y}$, and integrating the resulting equations, the following relationship between velocity and position is obtained:

$$\dot{y}^2 + ky^2 = c, \tag{11.3}$$

where c represents a strictly positive constant of integration resulting from the initial conditions. Equation 11.3 corresponds to an ellipse, and the plots of $\dot{y}$ against y (phase portrait) for different values of k would be ellipses with different relations between axes; that is, this control law is not appropriate since the variables do not move toward the origin.

Consider now the control law

$$u(t) = \begin{cases} -k_1 y(t), & y\dot{y} < 0, \\ -k_2 y(t), & \text{otherwise}, \end{cases} \tag{11.4}$$

where $0 < k_1 < 1 < k_2$, the controlled system can be considered as a VSCS, and an asymptotically stable motion is obtained, as can be observed in Figure 11.1.

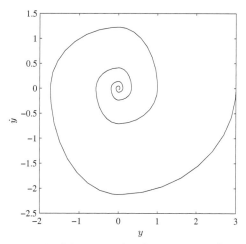

Figure 11.1 Phase portrait of the system (11.1) under control (11.4), and $k_1 = 0.5, k_2 = 1.5$

In Figure 11.2, the output and the control signals for this system are represented. In the latter, the discontinuous nature of the control signal can be clearly observed.

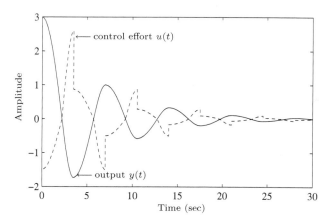

Figure 11.2 Output and control effort of the system (11.1) under the control (11.4), $k_1 = 0.5, k_2 = 1.5$

By defining the switching function

$$S(y, \dot{y}) = my + \dot{y}, \quad m \in \mathbb{R}^+,\qquad(11.5)$$

and using the variable structure control law given by

$$u(t) = \begin{cases} -1, & S(y, \dot{y}) > 0, \\ 1, & S(y, \dot{y}) < 0, \end{cases}\qquad(11.6)$$

or, alternatively,

$$u(t) = -\text{sgn}\left(S(t)\right),\tag{11.7}$$

high-frequency switching between the two different control structures will take place as the system trajectories repeatedly cross the line

$$\mathcal{L}_s = \{(y, \dot{y}) : S(y, \dot{y}) = 0\}.\tag{11.8}$$

The motion when confined to the line $\mathcal{L}_s$ satisfies the differential equation obtained from rearranging $S(y, \dot{y}) = 0$, namely

$$\dot{y}(t) = -my(t).\tag{11.9}$$

Such dynamic behavior is described as an *ideal sliding mode* or an *ideal sliding motion* and the line $\mathcal{L}_s$ is termed the *sliding surface*. During the sliding motion the system behaves as a reduced-order system which is apparently independent of the control. The control action, rather than prescribing the dynamic performance, ensures instead that the condition for sliding motion is satisfied, that is, that the reachability condition $S\dot{S} < 0$ is satisfied. So, in terms of design, the choice of the switching function, represented in this situation by the parameter m, governs the performance response; whilst the control law itself is designed to guarantee that the reachability condition is satisfied.

Suppose at time t_s the switching surface is reached and an ideal sliding motion takes place. It follows that the switching function satisfies $S(t) = 0$ for all $t > t_s$, which in turns implies $\dot{S} = 0$ for all $t \geqslant t_s$. So, a control law which maintains the motion on $\mathcal{L}_s$ is obtained as

$$\begin{aligned} \dot{S}(t) &= m\dot{y}(t) + \ddot{y}(t) = m\dot{y}(t) + u(t) = 0 \\ u(t) &= -m\dot{y}(t), \quad (t \geqslant t_s). \end{aligned}\tag{11.10}$$

This control law is referred as the *equivalent control* action, and may be thought of as the control signal which is applied "in average."

Figure 11.3 shows some significant results of applying (11.6) and (11.7) to (11.1) for different values of parameter m, with $y(0) = 3$, $\dot{y}(0) = 0$. The switched nature of the control signal can be observed. This high-frequency switching will produce wear and tear on the actuators, and should be minimized. So, different alternatives should be considered.

The two properties of an ideal sliding motion are *disturbance rejection* and *order reduction*. These properties are only obtained once sliding is induced. So, the time taken to induce sliding, t_s, should be minimized, and the region in which sliding takes place, $\Omega = \left\{(y, \dot{y}) : S\dot{S} < 0\right\}$, maximized.

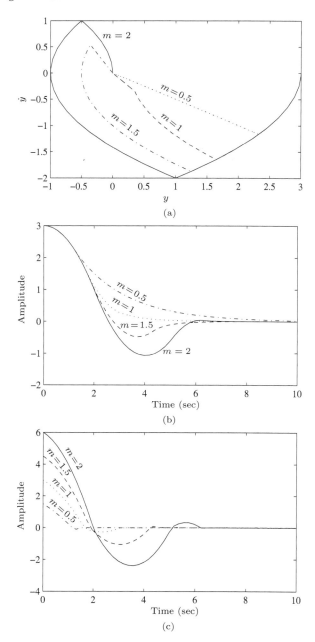

Figure 11.3 System (11.1) under control (11.7) and switching function (11.5) for different values of the parameter m: (a) phase plane portraits, (b) outputs of the system, and (c) switching functions

The main objectives of the design will be then:

1. Minimize t_s;
2. Maximize Ω;
3. Ensure that once the trajectories reach the sliding surface they are forced to remain there;
4. Reduce the amplitude of the high-frequency switching for limiting wear and tear on the actuators.

A control law candidate for this is

$$u(t) = - (m + \Phi)\, \dot{y}(t) - \Phi m y(t) - \rho \operatorname{sgn}(S(t)), \qquad (11.11)$$

where m, Φ, ρ are positive scalars. The parameter Φ corresponds to the rate at which the sliding surface is attained, the parameter m corresponds to the dynamics of the sliding motion, and the parameter ρ allows reduction of the amplitude of the high-frequency switching.

Figures 11.4 show the results for $m = 1$, $\Phi = 1$, $\rho = 0.3$, $y(0) = 3$, $\dot{y}(0) = 0$. It can be observed how the four objectives mentioned above are satisfied.

11.1.3 Fractional Sliding

11.1.3.1 Fractional Switching Function

There are several ways for introducing fractional orders in SMC. The first is by using a *fractional-order switching function*. By introducing the notation $\mathscr{D}^\lambda y \triangleq y^{(\lambda)}$, a generalization of (11.5) can be expressed of the form

$$S(y, y^{(\alpha)}) = my + y^{(1+\alpha)}, \quad 0 < \alpha < 1. \qquad (11.12)$$

By doing so the motion when confined to the line $\mathcal{L}_s$ satisfies the differential equation obtained from rearranging $S(y, y^{(\alpha)}) = 0$, namely

$$y^{(1+\alpha)} = -my. \qquad (11.13)$$

The characteristic equation is

$$s^{1+\alpha} + m = 0, \qquad (11.14)$$

corresponding to a fractional-order integrator in closed-loop. This dynamics exhibit an overshoot governed by the parameter α and a rate governed by the couple (m, α), namely in the frequency domain, a gain crossover frequency $\omega_c = m^{1/(1+\alpha)}$.

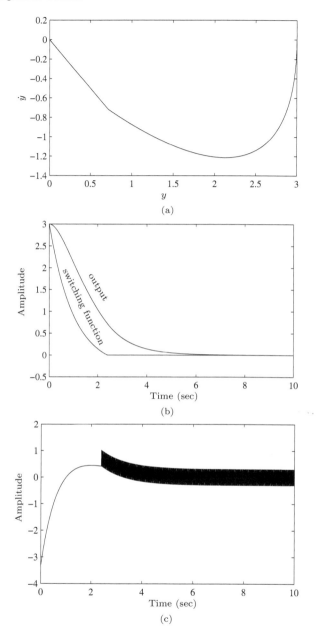

Figure 11.4 System under control (11.11) and switching function (11.5): (a) phase portrait, (b) output and switching surface, and (c) control signal

11.1.3.2 Fractional-order Control Law

The equivalent control corresponding to (11.10) can be obtained as follows:

$$S = my + y^{(1+\alpha)} \implies \dot{S} = m\dot{y} + y^{(2+\alpha)} = m\dot{y} + u^{(\alpha)} = 0$$
$$\implies u = -my^{(1-\alpha)}. \tag{11.15}$$

So, following the reasoning of the previous section, for the control action we propose to use

$$u(t) = -my^{(1-\alpha)} - \rho \mathscr{I}^{\beta} \mathrm{sgn}\left(S\left(t\right)\right), \quad 0 < \alpha, \beta < 1, \tag{11.16}$$

where the linear term is equal to the equivalent control, and $\mathscr{I}^{\beta}\left(\cdot\right)$ means the fractional-order integral. This last term can be seen as a low-pass filter or an attenuation selective in frequency, and allows reduction of the amplitude of the high-frequency switching.

11.1.3.3 Reachability Condition

Taking into account (11.12) and (11.16), for obtaining the reachability condition we have

$$\dot{S} = m\dot{y} + y^{(2+\alpha)} = m\dot{y} + u^{(\alpha)}$$
$$= m\dot{y} - m\dot{y} - \rho \mathscr{I}^{\beta-\alpha} \mathrm{sgn}\left(S\left(t\right)\right) = -\rho \mathscr{I}^{\beta-\alpha} \mathrm{sgn}\left(S\left(t\right).\right) \tag{11.17}$$

Taking into account that

$$\mathscr{I}^{\gamma} \mathrm{sgn}\left(x\right) = \begin{cases} \mathscr{I}^{\gamma}\left\{1\right\}, & x \geqslant 0, \\ \mathscr{I}^{\gamma}\left\{-1\right\}, & x < 0, \end{cases} \tag{11.18}$$

and using the Riemann–Liouville definition for the fractional-order derivative, it can be concluded that

$$\mathrm{sgn}\left[\mathscr{I}^{\gamma} \mathrm{sgn}\left(x\right)\right] = \mathrm{sgn}(x), \quad -1 < \gamma < 1. \tag{11.19}$$

So, for $\rho > 0$, and $-1 < \beta - \alpha < 1$, $S > 0$ implies $\dot{S} < 0$, and $S < 0$ implies $\dot{S} > 0$, and the reachability condition $S\dot{S} < 0$ is always fulfilled.

11.1.3.4 Simulation Results

By choosing $\alpha = 0.2, \beta = 0.4$, with $m = 1, \rho = 1.8, y(0) = 3, \dot{y}(0) = 0$, the results obtained are presented in Figure 11.5. In this figure it can be observed that both t_{s} and the amplitude of the high-frequency switching are reduced. So, similar results can be obtained by using the control actions (11.16). The main difference can be observed on the output, with overshoot for the first one due to the fact of having selected $\alpha > 0$. For the simulations, the fractional-order operators have been approximated by using the fifth-order Oustaloup's filter, with the frequency range of interest of $\left[10^{-2}, 10^2\right]$ rad/sec.

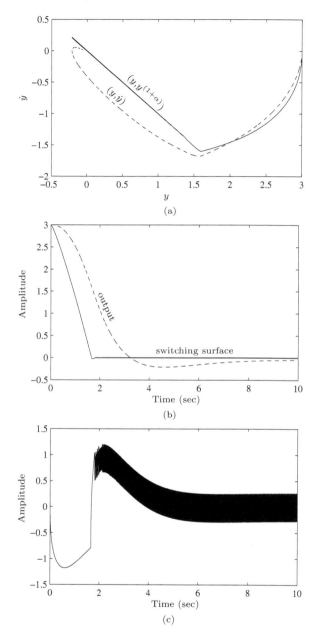

Figure 11.5 System (11.1) under control (11.19) and switching function (11.15): (a) phase portrait, (b) outputs and switching surfaces, and (c) control signals

11.2 Model Reference Adaptive Control

11.2.1 Introduction

This section investigates the use of fractional calculus in conventional model reference adaptive control (MRAC) systems. Two modifications to the conventional MRAC are presented, *i.e.*, the use of fractional-order parameter adjustment rule and the employment of fractional-order reference model. Through examples, benefits from the use of FOC are illustrated together with some remarks for further research.

This section is organized as follows: in Section 11.2.2, the MRAC is briefly reviewed . In Section 11.2.3 the use of fractional calculus into MRAC via fractional-order adjustment rule and fractional-order reference model is presented together with some illustrative simulation results. Section 11.2.4 presents the conclusions.

11.2.2 MRAC: A Brief Review

The model reference adaptive system is one of the main approaches to adaptive control, in which the desired performance is expressed in terms of a reference model (a model that describes the desired input-output properties of the closed-loop system) and the parameters of the controller are adjusted based on the error between the reference model output and the system output. These basic principles are illustrated in Figure 11.6, with two loops: an inner-loop which provides the ordinary control feedback, and an outer-loop which adjusts the parameters in the inner-loop.

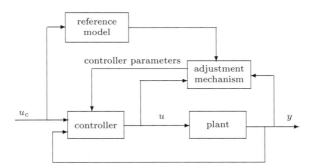

Figure 11.6 Basic principles of model reference adaptive system

The gradient approach to MRAC is based on the assumption that the parameters change more slowly than the other variables in the system. This assumption, which admits a quasi-stationary treatment, is essential for the computation of the sensitivity derivatives that are needed in the adaptation.

Let e denote the error between the system output, y, and the reference output, y_m. Let θ denote the parameters to be updated. By using the criterion

$$J(\theta) = \frac{1}{2}e^2, \qquad (11.20)$$

the adjustment rule for changing the parameters in the direction of the negative gradient of J is that

$$\frac{d\theta}{dt} = -\gamma \frac{\partial J}{\partial \theta} = -\gamma e \frac{\partial e}{\partial \theta}. \tag{11.21}$$

If it is assumed that the parameters change much more slowly than the other variables in the system, the derivative $\partial e/\partial\theta$, i.e., the sensitivity derivative of the system, can be evaluated under the assumption that θ is constant.

There are many variants about the MIT rules for the parameter adjustment. For example, the *sign-sign* algorithm is widely used in communication systems [143]; the PI-adjustment rule is used in [144]. Here, we will introduce a new variant of the MIT rules for the parameter adjustment by using fractional calculus. In addition, we shall extend the reference model to the fractional-order.

11.2.3 Using Fractional Calculus in MRAC Scheme

In this section, fractional calculus is introduced into MRAC scheme in two ways. One is the use of fractional-order derivatives for the MIT adjustment rules and the other one is the use of fractional-order reference models. The modified MRAC schemes are explained with some simulation illustrations.

11.2.3.1 Fractional-order Adjustment Rule

As can be observed in (11.21), the rate of change of the parameters depends solely on the adaptation gain, γ. Taking into account the properties of the fractional-order differential operator, it is possible to make the rate of change depend on both the adaptation gain, γ, and the derivative order, α, by using the adjustment rule

$$\frac{d^\alpha \theta}{d^\alpha t} = -\gamma \frac{\partial J}{\partial \theta} = -\gamma e \frac{\partial e}{\partial \theta}, \tag{11.22}$$

where α is a real number. In other words, the above parameter updating rule can be expressed as

$$\theta = -\gamma \mathscr{I}^\alpha \left[\frac{\partial J}{\partial \theta}\right] = -\gamma \mathscr{I}^\alpha \left[e \frac{\partial e}{\partial \theta}\right]; \quad \mathscr{I}^\alpha \equiv \mathscr{D}^{-\alpha}. \tag{11.23}$$

For example, consider the first-order SISO system to be controlled:

$$\frac{dy}{dt} + ay = bu, \tag{11.24}$$

where y is the output, u is the input, and the system parameters a and b are unknown constants or unknown slowly time-varying. Assume that the corresponding reference model is given by

$$\frac{dy_m}{dt} + a_m y_m = b_m u_c, \tag{11.25}$$

where u_c is the reference input signal for the reference model, y_m is the output of the reference model, and a_m and b_m are known constants. Perfect model-following can be achieved with the controller defined by

$$u(t) = \theta_1 u_c(t) - \theta_2 y(t), \tag{11.26}$$

where

$$\theta_1 = \frac{b_m}{b}; \quad \theta_2 = \frac{a_m - a}{b}, \tag{11.27}$$

From (14) and (16), assuming that $a + b\theta_1 \approx a_m$, and taking into account that b can be absorbed in γ, the equations for updating the controller parameters can be designed as [143]

$$\frac{d^\alpha \theta_1}{dt^\alpha} = -\gamma \left(\frac{1}{p + a_m} \right) u_c e, \tag{11.28}$$

$$\frac{d^\alpha \theta_2}{dt^\alpha} = \gamma \left(\frac{1}{p + a_m} \right) y e, \tag{11.29}$$

where $p = d/dt$, and γ is the adaptation gain, a small positive real number. Equivalently, in frequency domain, (11.28) and (11.29) can be written as

$$\theta_1 = -\frac{\gamma}{s^\alpha} \left(\frac{1}{s + a_m} \right) u_c e, \tag{11.30}$$

$$\theta_2 = \frac{\gamma}{s^\alpha} \left(\frac{1}{s + a_m} \right) y e. \tag{11.31}$$

Clearly, the conventional MRAC [143] is the case when $\alpha = 1$.

A block diagram for the above MRAC scheme for adjusting the unknown parameters θ_1 and θ_2 is shown in Figure 11.7.

In Figure 11.8, simulation results for $a = 1, b = 0.5, a_m = b_m = 2, \gamma = 3$ are presented. Two cases are considered for $\alpha = 1$ and $\alpha = 1.25$. As can be observed from Figure 11.8, under the same conditions, compared to the case when $\alpha = 1$, the updating of the unknown parameters is faster when $\alpha = 1.25$. The benefit due to the use of a slightly higher order of the derivatives is clearly demonstrated in Figure 11.8.

11.2.3.2 Stability Considerations

Equation 11.21, usually known as the MIT rule, performs well if the adaptation gain is small. The allowable value depends on both the magnitude of

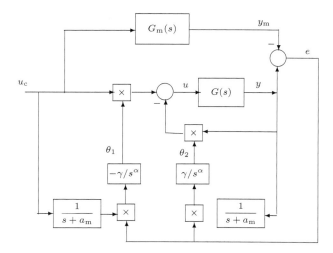

Figure 11.7 Block diagram of a simple MRAC scheme

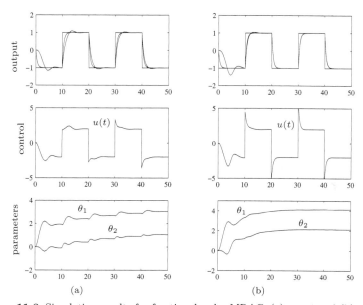

Figure 11.8 Simulation results for fractional-order MRAC: (a) $\alpha = 1$ and (b) $\alpha = 1.25$

the reference signal, u_c, and the process gain. So, if not properly handled, the MIT rule may give an unstable closed-loop system.

As an example, consider the MRAS scheme in Figure 11.9 in which the problem is to adjust a feedforward gain, θ, to the value θ_0 [143]. Consider the transfer function of the system

$$G(s) = \frac{1}{s^2 + a_1 s + a_2}. \tag{11.32}$$

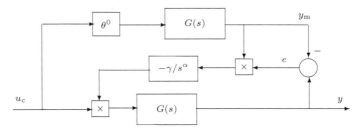

Figure 11.9 A simple MRAC scheme for feedforward gain adjustment

The MIT rule gives

$$\frac{d\theta}{dt} = -\gamma e y_m, \tag{11.33}$$

where

$$e = G(s)(\theta - \theta_0) u_c. \tag{11.34}$$

The governing differential equation for the overall adaptive system is

$$\frac{d^3 y}{dt^3} + a_1 \frac{d^2 y}{dt^2} + a_2 \frac{dy}{dt} + \gamma u_c y_m y = \theta \frac{du_c}{dt} + \gamma u_c y_m^2. \tag{11.35}$$

Some insight into the behavior of the system can be obtained by assuming that the adaptation mechanism is connected when the equilibrium is reached. That is, when $u_c = u_c^0 = y_m = y_m^0$, the time-varying differential equation (11.35) is transformed into a differential equation with constant coefficients that describes an LTI system with its characteristic equation given by

$$s^3 + a_1 s^2 + a_2 s + \gamma u_c^0 y_m^0 = 0. \tag{11.36}$$

It is easy to test the stability condition by using the Routh test which yields

$$a_1 a_2 > \gamma \left(u_c^0\right)^2. \tag{11.37}$$

So, if the adaptation gain γ or the reference signal u_c are sufficiently large, the system may become unstable.

As an illustrative simulation example, let $a_1 = a_2 = \theta^0 = 1$ and $\gamma = 0.1$. For reference signal amplitude $u_c = 0.1, 1$, and 3.5, the results are shown in Figure 11.10. As can be observed from Figure 11.10, when $\left(u_c^0\right)^2 > 0.1$, the system becomes unstable.

Here we adopt an alternative adjustment rule using the FOC as follows:

$$\frac{d^\alpha \theta}{dt^\alpha} = -\gamma e y_m, \quad 0 < \alpha < 1. \tag{11.38}$$

With the flexibility in selecting both the derivative order and the adaptation gain, one can expect an enlarged range of reference signal magnitude with which the system is stable.

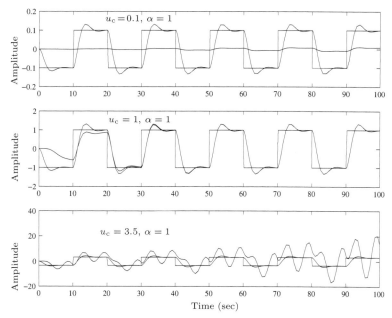

Figure 11.10 Unstable behavior in conventional MRAC with respect to the magnitude of reference input signal

For example, with $\gamma = 0.1$ and $\alpha = 0.75$ the simulation results shown in Figure 11.11 demonstrate clearly that for $u_c = 0.1$ and $u_c = 1$, the behavior of the adaptive system is similar to the conventional case when $\alpha = 1$ shown in Figure 11.10. However, with $\alpha = 0.75$, the overall adaptive system is still stable even when $u_c = 5$.

To obtain some insight into the above beneficial fact, it is noted that for different choices of the design parameter-pair (γ, α) in the operator γ/s^α in order to achieve similar transient performances on the adjustment rate, the induced phase response depends only on α. This could be a desired behavior that the conventional MRAC cannot have.

11.2.3.3 Fractional-order Reference Model

Now we will introduce another modification to MRAC problem by introducing fractional-order system as the reference model. In the simplest MRAC problems, the usual reference models are first-order or the second-order dynamic systems. Clearly, the set of candidates of the reference models can be enlarged by using fractional-order systems. In addition, transient response of MRAC systems can be improved. This is illustrated by a simulation example.

Consider a system described by the transfer function

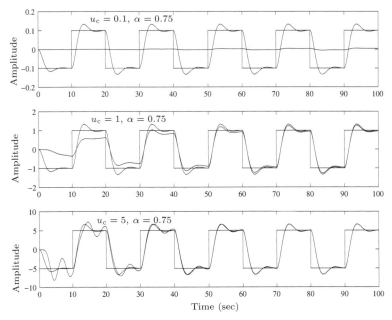

Figure 11.11 Improved stability behavior of MRAC with respect to the magnitude of reference input signal using FOC

$$G(s) = \frac{1}{s+1}. \tag{11.39}$$

The adaptive scheme shown in Figure 11.9 is used to adjust the feedforward gain in order to track the reference model output

$$y_\mathrm{m} = \frac{1}{s^{0.25}+1} u_\mathrm{c}. \tag{11.40}$$

With $\alpha = 1$ in the parameter adjusting rule (11.38), it would be very difficult, if not impossible, to track the reference output even after a significant time interval. However, when a fractional-order reference model is used, it is an easy task if we choose $\alpha \in (0,1)$.

Again, as an illustrative simulation example, in Figure 11.12 the results using the pairs $(\gamma_1, \alpha_1) = (0.2, 1)$ and $(\gamma_2, \alpha_2) = (15, 0.25)$ are shown. We can observe quite a large transient for $(\gamma_1, \alpha_1) = (0.2, 1)$ as shown in the top subplot of Figure 11.12. However, when we choose $(\gamma_2, \alpha_2) = (15, 0.25)$, *i.e.*, a fractional-order reference model is used, the tracking performance is almost perfect as shown in the bottom subplot of Figure 11.12. Note that in the latter case, the adjustment gain γ can be chosen as large as 15.

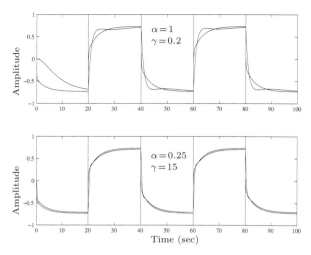

Figure 11.12 The effect of using fractional-order reference model in MRAC

11.2.4 Concluding Remarks

In this section, we have presented two ideas to extend the conventional MRAC by using the fractional-order parameter adjustment rule and the employment of fractional-order reference model. Through examples, benefits from the use of FOC are illustrated.

11.3 Reset Control

11.3.1 Introduction

An integrator is an element found in every control toolbox helping to provide zero steady-state errors to constant exogenous signals. However, this benefit comes at the expense of 90° of phase lag at all frequencies, which means a loss in relative stability (phase margin or overshoot).

The reset or Clegg integrator was introduced in [145] to reduce this phase lag while retaining the integrators desirable magnitude slope frequency response. The potential advantages of using Clegg integrators to meet stringent design specifications, and some stability analysis have been presented in the literature [146–150].

On the other hand, from the very beginnings of the use of fractional calculus in control [15,16], the fractional-order integrator has been considered as an alternative reference system for control purposes in order to obtain closed-loop controlled systems robust to gain changes. From another point of view, the fractional-order integrator can be used in feedback control in order to introduce both a constant phase lag and magnitude slope proportional to the integration order. Thus, fractional-order integrators can be used with the same purposes as the reset integrator.

The only purpose of this section is to present, compare, and combine both integrators in an easy way in order to open new possibilities of use. The rest of the section is organized as follows. Section 11.3.2 presents the fundamentals of the two kinds of integrators. Section 11.3.3 proposes a fractional-order reset integrator. In Section 11.3.4 the analyzed integrators are compared for the control of two first-order plants. Section 11.3.5 concludes the section and gives some insights into future work.

11.3.2 Reset Integrator

The reset or Clegg integrator represents an attempt to synthesize a non-linear circuit possessing the magnitude-frequency characteristic of a linear integrator while avoiding the 90° degrees phase lag associated with the linear transfer function. A functional diagram of the Clegg integrator, which switches on input zero crossings, is illustrated in Figure 11.13 [151, 152]. Basically, operation consists of the input being gated through one of two integrators (the input of the other is simultaneously reset) in accordance with zero-crossing detector (ZCD) commands. An analog implementation of this integrator is given in [149], including four diodes, four R-C networks, and two operational amplifiers.

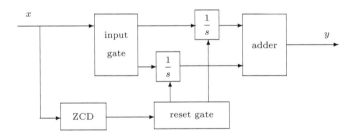

Figure 11.13 Functional diagram of the Clegg integrator

In the interval $0 < \varphi < \pi$, $\varphi = \omega t$, for the input $x(t) = A \sin \varphi$, the output is

$$y(\varphi) = \int_0^{\varphi/\omega} x(\varphi) \mathrm{d} \left(\frac{\varphi}{\omega}\right) = \frac{1}{\omega} \int_0^{\varphi} A \sin \varphi \mathrm{d}\varphi = \frac{A}{\omega} (1 - \cos \varphi.) \qquad (11.41)$$

The input and output waveforms are shown in Figure 11.14.

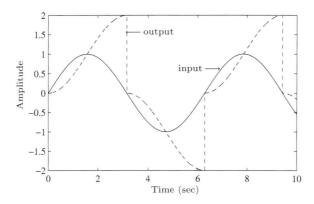

Figure 11.14 Input and output of the Clegg integrator

The describing function (DF) is given by

$$N(A,\omega) = \frac{2j}{\pi A} \int_0^\pi \frac{A}{\omega} (1 - \cos\varphi)\, e^{-j\varphi} d\varphi = \frac{4}{\pi\omega} \left(1 - j\frac{\pi}{4}\right), \tag{11.42}$$

$$|N(A,\omega)| = \frac{1}{\omega}\sqrt{1 + \left(\frac{4}{\pi}\right)^2}, \quad \arg N(A,\omega) = -\arctan\frac{\pi}{4} \approx -38.15°. \tag{11.43}$$

So, the DF is independent of the input amplitude, has a constant phase of $-38.15°$, and a magnitude slope of $-20\,\text{dB/dec}$. A disadvantage of the Clegg integrator is that it may induce oscillations [153].

11.3.3 Generalized Reset Integrator

If, for $0 < \varphi < \pi$, the input and output of a generalized device are

$$x(t) = A\sin\varphi, \quad y(t) = \frac{A}{\omega}(1 - \cos(\varphi + \theta)), \quad \varphi = \omega t, \tag{11.44}$$

θ being a constant, then the DF will be given by

$$N(A,\omega) = \frac{2j}{\pi A} \int_0^\pi \frac{A}{\omega} (1 - \cos(\varphi + \theta))\, e^{-j\varphi} d\varphi = \frac{4}{\pi\omega} \left(1 - j\frac{\pi}{4} e^{j\theta}\right),$$

$$\arg N(A,\omega) = -\arctan\frac{\frac{\pi}{4}\sin\left(\theta + \frac{\pi}{2}\right)}{1 - \frac{\pi}{4}\cos(\theta + \frac{\pi}{2})}. \tag{11.45}$$

The input and output waveforms are represented in Figure 11.15. The argument of the DF is represented in Figure 11.16 as a function of θ. As can

be observed, such a device may introduce a tunable phase-lag ranging from
$0°$ approximately, and could be denominated *generalized reset integrator*. On
the other hand, it could be a candidate for nonlinear implementation of an
approximated fractional-order integrator of order from 0 to 0.5.

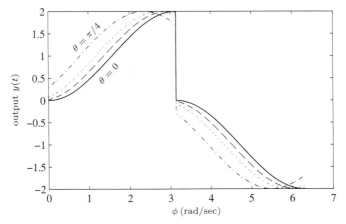

Figure 11.15 Output waveforms for values of $\theta = 0, \pi/16, \pi/8, \pi/4$

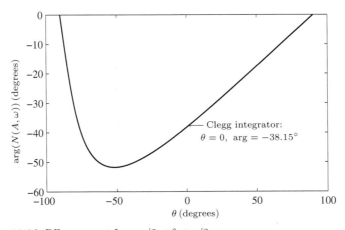

Figure 11.16 DF argument for $-\pi/2 < \theta < \pi/2$

11.3.4 Fractional Reset Integrator

Assuming that all the functions are causal and $\beta > -1$, we have [3]

$$\mathscr{D}^{\beta} \left[\sin (\omega t) \right] = \omega^{\beta} \sin \left(\omega t + \beta \frac{\pi}{2} \right), \tag{11.46}$$

$\mathscr{D}^{\beta}$ being the Riemann–Liouville fractional-order derivative of order β. With $\beta = -\alpha$, we obtain the fractional-order integral of order α as

$$\mathscr{I}^{\alpha}\left[\sin\left(\omega t\right)\right] = \omega^{-\alpha}\sin\left(\omega t - \alpha\frac{\pi}{2}\right) \tag{11.47}$$

So, if the input to a fractional-order reset integrator is $x(t) = A\sin\omega t$, the output can be obtained as

$$y(t) = \mathscr{I}^{\alpha}x(t)\Big|_0^t = \omega^{-\alpha}\sin\left(\omega t - \frac{\alpha\pi}{2}\right)\Big|_0^t = A\omega^{-\alpha}\left[\sin\left(\omega t - \frac{\alpha\pi}{2}\right) + \sin\left(\frac{\alpha\pi}{2}\right)\right]. \tag{11.48}$$

An implementation of the device in Simulink is given in Figure 11.17, where the fractional-order differentiator block is defined in Section 13.3, with the Oustaloup recursive filter.

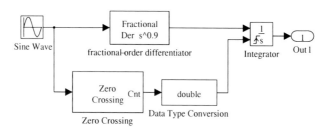

Figure 11.17 Fractional-order reset integrator implementation in Simulink

The output waveforms for different values of α are given in Figures 11.18 and 11.19.

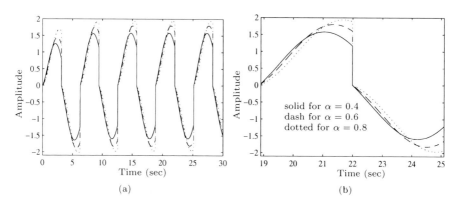

Figure 11.18 Output waveforms of the fractional-order reset integrator for different values of α: (a) when $\alpha < 1$ and (b) zoomed plots

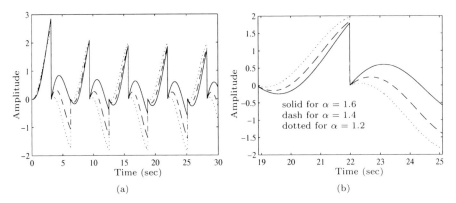

Figure 11.19 Output waveforms of the fractional-order reset integrator for different values of α: Detail after transient: (a) when $\alpha > 1$ and (b) zoomed plots

By following the methodology used in previous sections, the DF of the system can be obtained as

$$
\begin{aligned}
N(A,\omega) &= \frac{2\mathrm{j}}{\pi A}\int_0^\pi \left[A\omega^{-\alpha}\left[\sin\left(\varphi-\alpha\frac{\pi}{2}\right)+\sin\left(\alpha\frac{\pi}{2}\right)\right]\right]\mathrm{e}^{-\mathrm{j}\varphi}\mathrm{d}\varphi \\
&= \frac{4}{\pi\omega^\alpha}\left[\sin\alpha\frac{\pi}{2}+\frac{\pi}{4}\mathrm{e}^{-\mathrm{j}\alpha\frac{\pi}{2}}\right],
\end{aligned}
\tag{11.49}
$$

its argument being given by

$$
\arg N(A,\omega)=\arctan\frac{\dfrac{\pi}{4}\sin\dfrac{\alpha\pi}{2}}{\sin\dfrac{\alpha\pi}{2}+\dfrac{\pi}{4}\cos\dfrac{\alpha\pi}{2}}=\arctan\frac{\dfrac{\pi}{4}}{1+\dfrac{\pi}{4}\cot\dfrac{\alpha\pi}{2}}.
\tag{11.50}
$$

This argument is represented in Figure 11.20 for $\alpha \in [0, 1.5]$. As can be observed, we can obtain a phase lag ranging from $0°$ to $75°$ for $\alpha \in [0, 1.5]$, being the phase lag introduced by the reset integrator for a particular case, $\alpha = 1$. However, in this case the magnitude slope is linked to the phase lag, with a value of -20α dB/dec.

11.3.5 Simulation Results

In this section the integrators are compared in the control of two first-order plants. The first consists of an integrator, and for comparing the reset and fractional-order integrators with the same phase lag, the fractional-order integrator is of order $\alpha = 38.15/90 = 0.424$, and has been approximated by Oustaloup's method. The Bode plots of the approximation is given in Figure 11.21, and the step responses of the controlled plant in Figure 11.22.

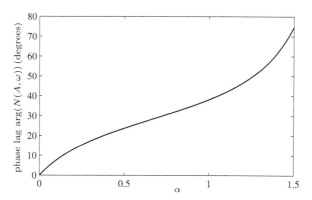

Figure 11.20 Phase lag $(\arg(N(A,\omega))$ introduced by the fractional-order reset integrator

As can be observed, the rise time is the same for both integrators, and the fractional-order integrator has an overshoot in agreement with the integration order [48].

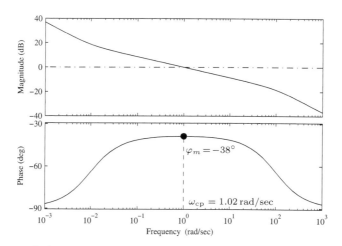

Figure 11.21 Bode plot of the approximated fractional-order integrator

The second plant considered has the transfer function

$$G(s) = \frac{5}{s+1}.$$

The unit step responses are given in Figure 11.23. The fractional-order integrators are of order $\alpha = 0.8$. As can be observed, fractional-order integrators, even the reset one, do not induce permanent oscillations.

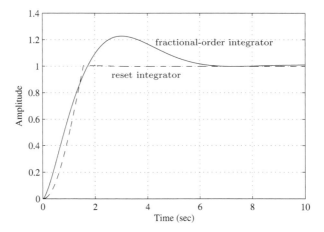

Figure 11.22 Unit step responses of reset and fractional-order integrators

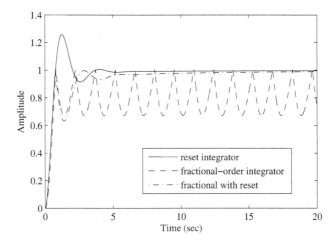

Figure 11.23 Unit step response of the first-order plant

11.4 Summary

The only purpose of this chapter has been to give some insights over the possibilities of applying the fractional-order operators in order to extend not only linear control strategies based on the basic control actions, but also more sophisticated nonlinear strategies such as sliding mode control, adaptive control, and reset control. By means of simple examples easy to understand, we have tried to show how, by using the fractional-order operators, the control designer can obtain more powerful and flexible tools in order to enlarge the dynamical richness of the controlled systems.

Part V
Implementations of Fractional-order Controllers: Methods and Tools

Chapter 12
Continuous-time and Discrete-time Implementations of Fractional-order Controllers

In the previous chapters, different types of fractional-order controllers are addressed. The most difficult problem yet to be solved is how to implement them. Although some work has been performed with hardware devices for fractional-order integrator, such as fractances (*e.g.*, RC transmission line circuit and Domino ladder network) [154] and fractors [155], there are restrictions, since these devices are difficult to tune. An alternative feasible way to implement fractional-order operators and controllers is to use finite-dimensional integer-order transfer functions.

Theoretically speaking, an integer-order transfer function representation to a fractional-order operator s^α is infinite-dimensional. However it should be pointed out that a band-limit implementation of fractional-order controller (FOC) is important in practice, *i.e.*, the finite-dimensional approximation of the FOC should be done in a proper range of frequencies of practical interest [17,51]. Moreover, the fractional-order can be a complex number as discussed in [51]. In this book, we focus on the case where the fractional order is a real number.

For a single term s^α with α a real number, there are many approximation schemes proposed. In general, we have analog realizations [156–159] and digital realizations.

This chapter describes different approximations or implementations of fractional-order operators and systems. When fractional-order controllers have to be implemented or simulations have to be performed, FOTFs are usually replaced by integer-order transfer functions with a behavior close enough to that desired, but much easier to handle.

There are many different ways of finding such approximations but unfortunately it is not possible to say that one of them is the best, because even though some of them are better than others in regard to certain characteristics, the relative merits of each approximation depend on the differentiation order, on whether one is more interested in an accurate frequency behavior or in accurate time responses, on how large admissible transfer functions may be, and other factors like these. A good review of these approximations can be found in [126, 154, 160].

This chapter is organized as follows. In Section 12.1, continuous-time implementation to fractional-order operators are studied. Several continued fraction expansion based approximation methods are presented first. Then the well-established Oustaloup filter is given and generalized. A modified Oustaloup filter is also demonstrated. Generally speaking, Oustaloup's approximation to fractional-order operators are good enough in most cases. In Section 12.2, discrete-time implementation schemes to fractional-order operators are presented. In particular, FIR filter design and Tustin discretization methods are presented, and the step and impulse response invariants retaining methods are demonstrated. Another practical category of implementation methods, the frequency response identification based methods, for fractional-order systems are presented in Section 12.3. With the use of these methods, the linear fractional-order system of any complexity can be easily approximated, when a suitable frequency range of interest is selected. In Section 12.4, a sub-optimal $\mathcal{H}_2$ pseudo-rational approximation algorithm [161] for FOTFs is presented with a demonstration of its applications in PID controller design.

12.1 Continuous-time Implementations of Fractional-order Operators

For the fractional-order operator, its Laplace representation is s^γ, which exhibits straight lines in both magnitude and phase Bode plots. Thus it is not possible to find a finite-order filter to fit the straight lines for all the frequencies. However, it is useful to fit the frequency responses over a frequency range of interest (ω_b, ω_h).

Different continuous-time filters have been studied in [154, 160], and some of the approximations can be constructed by relevant MATLAB functions in N-Integer Toolbox [162].

12.1.1 Continued Fraction Approximations

Continued fraction expansion (CFE) is often regarded as a useful type of rational-function approximation to a given function $f(s)$. It usually has better convergence than the power series functions such as Taylor series expansions. For the fractional-order operator $G(s) = s^\alpha$, the continued fraction expansion can be written as

$$G(s) = \cfrac{b_0(s)}{a_0(s) + \cfrac{b_1(s)}{a_1(s) + \cfrac{b_2(s)}{a_2(s) + \cdots}}}, \tag{12.1}$$

where $a_i(s)$ and $b_i(s)$ can be expressed by rational functions of s. One should first find the continued fraction expansion to the original fractional-order operator, then get the integer-order transfer function, *i.e.*, rational function, representation.

There are several well-established continued fraction expansion based approximation method to the fractional-order operator $G(s) = s^{\alpha}$. The N-integer Toolbox provides a `nid()` function for finding the rational-function approximation.

Example 1 Consider the fractional-order integrator with $\alpha = 0.5$. The rational function approximation using different continued fraction expansion based methods can be found in [160] as

Low-frequency CFE: $H_1(s) = \dfrac{0.351s^4 + 1.405s^3 + 0.843s^2 + 0.157s + 0.009}{s^4 + 1.333s^3 + 0.478s^2 + 0.064s + 0.002844}$,

High-frequency CFE: $H_2(s) = \dfrac{s^4 + 4s^3 + 2.4s^2 + 0.448s + 0.0256}{9s^4 + 12s^3 + 4.32s^2 + 0.576s + 0.0256}$,

Carlson's method: $H_3(s) = \dfrac{s^4 + 36s^3 + 126s^2 + 84s + 9}{9s^4 + 84s^3 + 126s^2 + 36s + 1}$,

Matsuda's method: $H_4(s) = \dfrac{0.08549s^4 + 4.877s^3 + 20.84s^2 + 12.99s + 1}{s^4 + 13s^3 + 20.84s^2 + 4.876s + 0.08551}$.

The Bode plots with different approximations can be obtained as shown in Figure 12.1. It can be seen that the fitting ranges are rather small and the quality of fit is not satisfactory. □

12.1.2 Oustaloup Recursive Approximations

Oustaloup filter approximation to a fractional-order differentiator is a widely used one in fractional calculus [50, 51]. A generalized Oustaloup filter can be designed as

$$G_f(s) = K \prod_{k=1}^{N} \frac{s + \omega'_k}{s + \omega_k}, \tag{12.2}$$

where the poles, zeros, and gain are evaluated from

$$\omega'_k = \omega_b \omega_u^{(2k-1-\gamma)/N}, \quad \omega_k = \omega_b \omega_u^{(2k-1+\gamma)/N}, \quad K = \omega_h^{\gamma}, \tag{12.3}$$

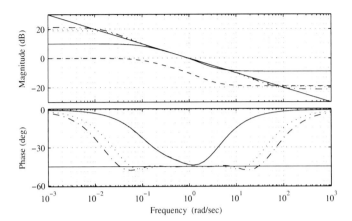

Figure 12.1 Bode plots comparisons with different approximations, with *solid lines* for $H_1(s)$, *dashed lines* for $H_2(s)$, *dotted lines* for $H_3(s)$, and *dashed dotted lines* for $H_4(s)$. Also the *straight lines* are the theoretical results

where $\omega_u = \sqrt{\omega_h/\omega_b}$. We used the term "generalized" because N here can be either odd or even integers.

Based on the above algorithm, the following function can be written:

```
function G=ousta_fod(gam,N,wb,wh)
k=1:N; wu=sqrt(wh/wb);
wkp=wb*wu.^((2*k-1-gam)/N);  wk=wb*wu.^((2*k-1+gam)/N);
G=zpk(-wkp,-wk,wh^gam);  G=tf(G);
```

and the Oustaloup filter can be designed with $G=$ousta_fod$(\gamma,N,\omega_b,\omega_h)$, where γ is the order of derivative, and N is the order of the filter.

Example 2 To illustrate the method, the approximation of the fractional-order integrator of order 0.45 can be obtained. In this particular case, the orders of the approximation are selected as 4 and 5, respectively, with $\omega_h = 1000$ rad/sec and $\omega_b = 0.01$ rad/sec. The filters can be designed with the following MATLAB commands:

```
>> G1=ousta_fod(-0.45,4,1e-2,1e3);
   G2=ousta_fod(-0.45,5,1e-2,1e3);
   bode(G1,'-',G2,'--',{1e-3,1e4})
```

and the two filters respectively obtained are

$$G_1(s) = \frac{0.04467s^4 + 21.45s^3 + 548.2s^2 + 783.2s + 59.57}{s^4 + 131.5s^3 + 920.3s^2 + 360.1s + 7.499},$$

and

$$G_2(s) = \frac{0.04467s^5 + 26.35s^4 + 1413s^3 + 7500s^2 + 3942s + 188.4}{s^5 + 209.3s^4 + 3982s^3 + 7500s^2 + 1399s + 23.71}.$$

The Bode plots are shown in Figure 12.2. It can be seen that the Bode plots of the two filters are relatively close to that of the theoretical one over the frequency range of interest. It can be seen that the fitting quality is much superior to those obtained with continued fraction based approaches.   □

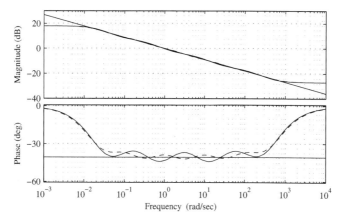

Figure 12.2 Bode plots of $H_{\text{Oust}}(s)$, corresponding to the approximation of a fractional-order integrator of order 0.45 with the Oustaloup method, with *solid lines* for $G_1(s)$, *dashed lines* for $G_2(s)$, and *dotted lines* for the theoretical Bode plot

12.1.3 Modified Oustaloup Filter

In practical applications, it is frequently found that the filter from using the ousta_fod() function cannot exactly fit the whole expected frequency range of interest. A new improved filter for a fractional-order derivative in the frequency range of interest (ω_b, ω_h), which is shown to perform better, is introduced in this section. The modified filter is [163]

$$s^\gamma \approx \left(\frac{d\omega_h}{b}\right)^\gamma \left(\frac{ds^2 + b\omega_h s}{d(1-\gamma)s^2 + b\omega_h s + d\gamma}\right) \prod_{k=-N}^{N} \frac{s + \omega_k'}{s + \omega_k}, \tag{12.4}$$

and the filter is stable for $\gamma \in (0, 1)$, and

$$\omega_k' = \omega_b \omega_u^{(2k-1-\gamma)/N}, \quad \omega_k = \omega_b \omega_u^{(2k-1+\gamma)/N}, \tag{12.5}$$

with $\omega_u = \sqrt{\omega_h/\omega_b}$.

Through a number of experiment confirmations and theoretical analyses, the modified filter achieves good approximation when $b = 10$ and $d = 9$. With the above algorithm, a MATLAB function new_fod() is written as

```
function G=new_fod(r,N,wb,wh,b,d)
if nargin==4, b=10; d=9; end
k=1:N; wu=sqrt(wh/wb); K=(d*wh/b)^r;
wkp=wb*wu.^((2*k-1-r)/N); wk=wb*wu.^((2*k-1+r)/N);
G=zpk(-wkp',-wk',K)*tf([d,b*wh,0],[d*(1-r),b*wh,d*r]);
```

with the syntax G_f=new_fod$(\gamma,N,\omega_b,\omega_h,b,d)$. Again here, the modified Oustaloup filter is also extended which can be used for handling odd and even order.

Example 3 Consider an FOTF model $G(s) = \dfrac{s+1}{10s^{3.2} + 185s^{2.5} + 288s^{0.7} + 1}$.

The approximations to the 0.2th-order derivative using Oustaloup's filter and the modified Oustaloup's filter can be obtained as shown in Figure 12.3 (a). The frequency range of good fitting is larger with the improved filter. The exact Bode plot can be obtained with the bode() function. Also the two approximations to the $G(s)$ model is shown in Figure 12.3 (b). In the following commands, function fotf() is used to define an FOTF object, and will be fully presented in the next chapter:

```
>> b=[1 1]; a=[10,185,288,1]; nb=[1 0]; na=[3.2,2.5,0.7,0];
   w=logspace(-4,4,200); G0=fotf(a,na,b,nb); H=bode(G0,w);
   s=zpk('s'); N=4; w1=1e-3; w2=1e3; b=10; d=9;
   g1=ousta_fod(0.2,N,w1,w2); g2=ousta_fod(0.5,N,w1,w2);
   a1=g1; g3=ousta_fod(0.7,N,w1,w2);
   G1=(s+1)/(10*s^3*g1+185*s^2*g2+288*g3+1);
   g1=new_fod(0.2,N,w1,w2,b,d); g2=new_fod(0.5,N,w1,w2,b,d);
   g3=new_fod(0.7,N,w1,w2,b,d); bode(g1,a1); figure
   G2=(s+1)/(10*s^3*g1+185*s^2*g2+288*g3+1); bode(H,G1,G2)
```

It can be seen that the modified method provided a much better fit. Thus for certain fractional-order differentiators, the modified filter may be more appropriate.   □

12.2 Discrete-time Implementation of Fractional-order Operators

The key step in digital implementation of an FOC is the numerical evaluation or discretization of the fractional-order differentiator s^{α}. In general, there are two classes of discretization methods: *direct discretization* and *indirect discretization*. In *indirect discretization* methods [51], two steps are required, *i.e.*, frequency domain fitting in continuous time domain first and then discretizing

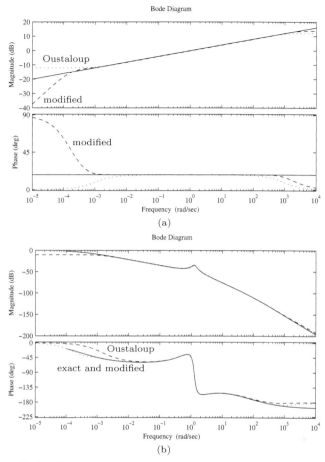

Figure 12.3 Bode plot comparisons, *straight lines* for exact ones, *dashed lines* for Oustaloup filters, and *dotted lines* for modified Oustaloup filters: (a) $s^{0.2}$ fittings and (b) Bode plot comparisons

the fit s-transfer function. Other frequency-domain fitting methods can also be used but without guaranteeing the stable minimum-phase discretization. Existing *direct discretization* methods include the application of the direct power series expansion (PSE) of the Euler operator [78,93,160,164], continued fraction expansion (CFE) of the Tustin operator [78, 148, 160, 164, 165], and numerical integration based method [93, 148, 166]. However, as pointed out in [167–169], the Tustin operator based discretization scheme exhibits large errors in high frequency range. A new mixed scheme of Euler and Tustin operators is proposed in [148] which yields the so-called Al-Alaoui operator [167]. These discretization methods for s^{α} are in infinite impulse response (IIR) form. Recently, there have been some reported methods to obtain directly the digital fractional-order differentiators in FIR form [170,171]. However, using an FIR filter to approximate s^α may be less efficient due to very high order of the FIR filter. So, discretizing fractional-order differentiators in IIR forms is perferred [148,165,166,172].

In this section, FIR filter approximation and Tustin discretization method are, in particular, presented. Then an introduction is made on finite-dimensional integer-order approximations retaining step and impulse response invariants of the fractional-order operators.

12.2.1 FIR Filter Approximation

FIR filter is a class of widely used filters in signal processing [170]. Ivo Petráš proposed a MATLAB function filt(), which can be used in FIR filter approximation of fractional-order differentiators [173]. The kernel part of the function is

```
function H=dfod2(n,T,r)
if r>0
    bc=cumprod([1,1-((r+1)./[1:n])]); H=filt(bc,[T^r],T);
elseif r<0
    bc=cumprod([1,1-((-r+1)./[1:n])]); H=filt([T^(-r)],bc,T);
end
```

where n is the expected order of the filter, T is the sampling period, and r is the expected order of differentiation. With the use of the function, the FIR filter can be designed. Normally to achieve good approximation results, the order n must be assigned to a very high number, $i.e.$, $n = 50$.

12.2.2 Discretization Using The Tustin Method with Prewarping

As is known, the Tustin method [118] relates the s and z domains with the following substitution formula:

$$s = \frac{2}{T}\frac{z-1}{z+1}, \tag{12.6}$$

where T is the sampling period. In signal processing literature the Tustin method is frequently denoted the *bilinear transformation method*. The term bilinear is related to the fact that the imaginary axis in the complex s-plane for continuous-time systems is mapped or transformed onto the unity circle for the corresponding discrete-time system. In addition, the poles are transformed so that the stability property is preserved.

With the substitution formula in (12.6) the discrete version $H_d(z)$ of a continuous transfer function $H_c(s)$ is obtained. In general, the frequency responses of $H_c(s)$ and $H_{disc}(z)$ are not equal at the same frequencies. Tustin method can be modified or enhanced so that a similar frequency response can be obtained for both $H_c(s)$ and $H_d(z)$ at one or more user-defined critical frequencies. This is done by modifying (*prewarping*) the critical frequencies of $H_c(s)$ so that the frequency responses are equal after the discretization [118].

In our case, MATLAB function c2d() [174] is used to obtain the discrete transfer function of a continuous system, whose syntax is H_d=c2d(H_c, T,METHOD), where H_d is the resulting discrete transfer function, H_c the continuous transfer function to discretize, and T the sampling period. The string METHOD selects the discretization method among the following:

- 'zoh': Zero-order hold on the inputs.
- 'foh': Linear interpolation of inputs.
- 'tustin': Bilinear approximation.
- 'prewarp': Tustin approximation with frequency prewarping. The critical frequency ω_c (in rad/sec) is specified as fourth input by H_d=c2d(H_c, T,'prewarp',ω_c). In our case, the critical frequency will be the gain crossover frequency, that is, $\omega_c = \omega_{cg}$.
- 'matched': Matched pole-zero method (for SISO systems only).
- The default option is 'zoh' when METHOD is omitted.

Example 4 To illustrate this method, the discrete-time transfer function $H_{invf}(z)$ corresponding to the continuous approximation $H_1(s)$ from the previous section is obtained with the following statements:

```
>> H1=ousta_fod(-0.5,4,1e-2,1e2);
   H2=c2d(H1,0.1,'prewarp',1), bode(H1,'-',H2,'--')
```

resulting in respectively

$$H_1(s) = \frac{0.1s^4 + 6.248s^3 + 35.45s^2 + 19.76s + 1}{s^4 + 19.76s^3 + 35.45s^2 + 6.248s + 0.1}$$

and

$$H_2(z) = \frac{0.2425z^4 - 0.491z^3 + 0.2033z^2 + 0.106z - 0.06079}{z^4 - 2.875z^3 + 2.802z^2 - 0.974z + 0.0478},$$

with $\omega_c = 1$ rad/sec and $T = 0.001$ sec. From the Bode plots of Figure 12.4, the similarity between the frequency responses of $H_2(z)$ and $H_1(s)$ in Figure 12.6 can be observed. □

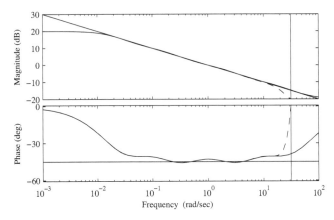

Figure 12.4 Bode plots of the transfer functions $H_1(s)$ and $H_2(z)$, with *solid lines* for $H_1(s)$ and *dashed lines* for $H_2(z)$. The *straight lines* are for the theoretical results

12.2.3 Discrete-time Implementation with Step or Impulse Response Invariants

A set of MATLAB functions on discrete-time implementation of fractional-order differentiator, integrator, as well as complicated transfer functions with non-integer powers has been developed and given in Table 12.1, based on step response invariants and impulse response invariants fitting. These functions can be downloaded for free and used directly in establishing the discrete-time implementations [175]. With the use of the functions, a discrete-time implementation to the fractional-order terms can easily be constructed.

Table 12.1 MATLAB functions for discrete-time implementations

Function	Syntax	Descriptions
irid_fod()	G=irid_fod(α,T,N)	s^α fitting with impulse response invariants
srid_fod()	G=srid_fod(α,T,N)	s^α fitting with step response invariants
irid_folpf()	G=irid_folpf(τ,α,T,N)	$(\tau s + 1)^{-\alpha}$ fitting with impulse response invariants

Example 5 Selecting a sampling period of $T = 0.1\,\text{sec}$, and the order of 5, the 0.5th-order integrator can be implemented with the step response invariants and impulse response invariants with the following statements:

```
>> G1=irid_fod(-0.5,0.1,5); G2=srid_fod(-0.5,0.1,5);
   bode(G1,'--',G2,':')
```

with the results

$$G_1(z) = \frac{0.09354z^5 - 0.2395z^4 + 0.2094z^3 - 0.06764z^2 + 0.003523z + 0.0008224}{z^5 - 3.163z^4 + 3.72z^3 - 1.966z^2 + 0.4369z - 0.02738},$$

$$G_2(z) = \frac{2.377 \times 10^{-6}z^5 + 0.1128z^4 - 0.367z^3 + 0.4387z^2 - 0.2269z + 0.04241}{z^5 - 3.671z^4 + 5.107z^3 - 3.259z^2 + 0.882z - 0.05885},$$

and the Bode plot comparisons are given in Figure 12.5. It can be seen that the fittings are satisfactory.   □

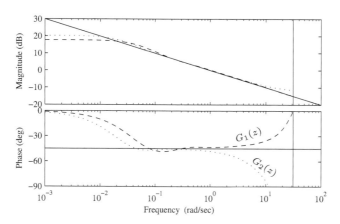

Figure 12.5 Bode plots comparisons with discrete-time implementations

12.3 Frequency Response Fitting of Fractional-order Controllers

12.3.1 Continuous-time Approximation

In general, any available method for frequency domain identification can be applied in order to obtain a rational function, whose frequency response fits that corresponding to the original transfer function. For example, a minimization of the cost function of the ISE form is generally aimed, *i.e.*,

$$J = \int W(\omega) \left| G(\omega) - \hat{G}(\omega) \right|^2 d\omega, \tag{12.7}$$

where $W(\omega)$ is a weighting function, $G(\omega)$ is the original frequency response, and $\hat{G}(\omega)$ is the frequency response of the approximated rational function.

MATLAB function `invfreqs()` [176] follows this criterion, with the next syntax: $[\boldsymbol{B}, \boldsymbol{A}] = \text{invfreqs}(H, \boldsymbol{w}, n_b, n_a)$. This function gives real numerator and denominator coefficients $\boldsymbol{B}$ and $\boldsymbol{A}$ of orders n_b and n_a, respectively. H

is the desired complex frequency response of the system at frequency points w, and w contains the frequency values in rad/sec. Function invfreqs() yields a filter with real coefficients. This means that it is sufficient to specify positive frequencies only.

The approximation of the fractional-order integrator of order 0.5 has been obtained using this method. The order of the approximation is 4, that is $n_b = n_a = 4$, and the frequency range w goes from 0.01 rad/sec to 100 rad/sec. The identified model can be obtained with the following statements:

```
>> w=logspace(-2,2,100); H=1./(sqrt(-1)*w).^0.5;
   [n,d]=invfreqs(H,w,4,4); G=tf(n,d), bode(G).
```

The resulting transfer function is

$$G(s) = \frac{B(s)}{A(s)} = \frac{0.02889s^4 + 17.08s^3 + 1102s^2 + 1.027e004s + 4567}{s^4 + 172.1s^3 + 4378s^2 + 1.148e004s + 459.8},$$

and the Bode plots are shown in Figure 12.6.

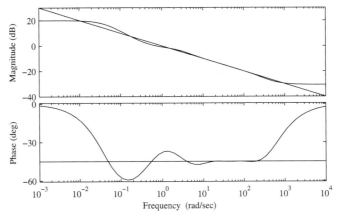

Figure 12.6 Bode plots of $G(s)$, corresponding to the approximation of a fractional-order integrator of order 0.5 with MATLAB function invfreqs()

12.3.2 Discrete-time Approximation

If the frequency response of a fractional-order operator is given, discrete-time implementation can also be obtained. There are several way for finding the discrete-time IO transfer function approximations to the fractional-order controllers. One may use MATLAB function invfreqz() for a direct approximation to the given frequency response data. On the other hand, the continuous-time approximation can be obtained first, then with the use of c2d() function, the discrete-time implementation can be obtained. One may also use special algorithms for specific types of fractional-order controllers. For instance, the impulse response invariant function $G=$irid_folpf(τ,α,T,N) given in Table 12.1 can be used for fitting controllers of $(\tau s + 1)^{-\alpha}$.

Example 6 Consider a fractional-order model given by $G(s) = (3s+2)^{-0.4}$. One may simple rewrite the model by $G(s) = 2^{-0.4}(1.5s + 1)^{-0.4}$. It can be seen that $\tau = 1.5$ and $\alpha = 0.4$. Selecting sampling periods as $T = 0.1\sec$, with order $N = 4$, the discrete-time implementation using impulse response invariants can be obtained as

```
>> tau=1.5; a=0.4; T=0.1; N=4;
   G1=2^(-0.4)*irid_folpf(tau,a,T,N);
```

The approximation model is

$$G_1(z) = \frac{0.2377z^4 - 0.4202z^3 + 0.2216z^2 - 0.02977z - 0.00138}{z^4 - 2.222z^3 + 1.663z^2 - 0.4636z + 0.03388}.$$

The Bode plot comparisons of the fitting model and the original model can be shown in Figure 12.7. It can be seen that the fitting results are good for this example. □

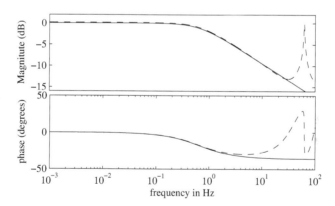

Figure 12.7 Bode plots comparisons, with *solid lines* for exact controller and *dashed lines* for the discrete-time implementation model

12.3.3 Transfer Function Approximations to Complicated Fractional-order Controllers

In control applications, sometimes the fractional-order controller designed may be rather complicated. For instance, the QFT controller designed in Section 10.2 may consist of several series connected terms such as $[(as+b)/(cs+d)]^\alpha$. To implement the controllers in continuous-time form, the following steps should be taken:

1. Get the exact frequency response of the fractional-order controller.
2. Select appropriate orders for the numerator and denominator of the integer-order filters.
3. Identify the continuous-time integer-order controllers with the use of `invfreqs()` function.
4. Verify frequency response fitting. If the fitting is not satisfactory, go back to Step 2 to select another set of orders, or another frequency range of interest, until satisfactory approximations can be obtained.

Example 7 Consider again the QFT controller presented in (10.20). For ease of presentation, the controller is given again here as

$$G_c(s) = 1.8393 \left(\frac{s+0.011}{s} \right)^{0.96} \left(\frac{8.8 \times 10^{-5}s+1}{8.096 \times 10^{-5}s+1} \right)^{1.76} \frac{1}{(1+s/0.29)^2}.$$

It should be noted that the filter is too complicated to implement with impulse response invariant fitting method given earlier. With the use of MATLAB, the function `frd()` can be used to get the frequency response of an integer-order block, and the `RespnseData` membership of the frequency response object can be used to extract the frequency response data. Then dot multiplications and dot powers in MATLAB can be used to evaluate the exact frequency response data. Selecting the orders of numerator and denominator as 4 for continuous-time fitting, and the fitting frequency range of $\omega \in (10^{-4}, 10^0)$ rad/sec, the following commands can be used:

```
>> w=logspace(-4,0); G1=tf([1 0.011],[1 0]); F1=frd(G1,w);
   G2=tf([8.8e-5 1],[8.096e-5 1]); F2=frd(G2,w);
   s=tf('s'); G3=1/(1+s/0.29)^2; F3=frd(G3,w); F=F1;
   h1=F1.ResponseData; h2=F2.ResponseData; h3=F3.ResponseData;
   h=1.8393*h1.^0.96.*h2.^1.76.*h3; F.ResponseData=h; %exact
   [n,d]=invfreqs(h(:),w,4,4); G=tf(n,d);
```

The continuous-time approximate integer-order controller can be obtained as

$$G(s) = \frac{2.213 \times 10^{-7}s^4 + 1.732 \times 10^{-6}s^3 + 0.1547s^2 + 0.001903s + 2.548 \times 10^{-6}}{s^4 + 0.5817s^3 + 0.08511s^2 + 0.000147s + 1.075 \times 10^{-9}}.$$

To verify the controller from the point of view of frequency response fitting, we should compare the original and fitted controller over a larger frequency interval. The following commands can be used to compare the two controller in the frequency range of $(10^{-6}, 10^2)$ rad/sec.

```
>> w=logspace(-6,2,200); F1=frd(G1,w); F2=frd(G2,w); F=F1;
   F3=frd(G3,w); h1=F1.ResponseData; h2=F2.ResponseData;
   h3=F3.ResponseData; h=1.8393*h1.^0.96.*h2.^1.76.*h3;
```

```
F.ResponseData=h; bode(F,'-',G,'--',w)
```

The Bode plots of both controllers over the new frequency range are shown in Figure 12.8. It can be seen that the frequency response of the controller is satisfactory, albeit there is small discrepancy at very-low frequency range. If such an extremely low-frequency range is to be fitted, we should go to Step 2 to generate more frequency response points in the range.   □

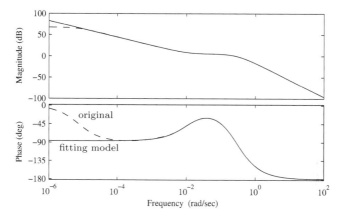

Figure 12.8 Bode plot comparisons for fractional-order QFT controller

12.4 Sub-optimal Approximation of FOTFs

In this section, we consider the general fractional-order FO-LTI systems with non-commensurate fractional orders as follows:

$$G(s) = \frac{b_m s^{\gamma_m} + b_{m-1} s^{\gamma_{m-1}} + \cdots + b_1 s^{\gamma_1} + b_0}{a_n s^{\eta_n} + a_{n-1} s^{\eta_{n-1}} + \cdots + a_1 s^{\eta_1} + a_0}. \qquad (12.8)$$

Using the aforementioned approximation schemes for a single s^r and then for the general FO-LTI system (12.8) could be very tedious, leading to a very high order model. In this section, we propose to use a numerical algorithm to achieve a good approximation of the overall transfer function (12.8) using finite-dimensional integer-order rational transfer function with a possible time delay term and illustrate how to use the approximated integer-order model for integer-order controller design.

Our target now is to find an approximate integer-order model with a relative low order, possibly with a time delay in the following form:

$$G_{r/m,\tau}(s) = \frac{\beta_1 s^r + \cdots + \beta_r s + \beta_{r+1}}{s^m + \alpha_1 s^{m-1} + \cdots + \alpha_{m-1} s + \alpha_m} e^{-\tau s}. \qquad (12.9)$$

An objective function for minimizing the $\mathcal{H}_2$-norm of the reduction error signal $e(t)$ can be defined as

$$J = \min_{\theta} \|\widehat{G}(s) - G_{r/m,\tau}(s)\|_2, \tag{12.10}$$

where θ is the set of parameters to be optimized such that

$$\theta = [\beta_1, \ldots, \beta_r, \alpha_1, \cdots, \alpha_m, \tau]. \tag{12.11}$$

For an easy evaluation of the criterion J, the delayed term in the reduced order model $G_{r/m,\tau}(s)$ can be further approximated by a rational function $\widehat{G}_{r/m}(s)$ using the Padé approximation technique [177]. Thus, the revised criterion can then be defined by

$$J = \min_{\theta} \|\widehat{G}(s) - \widehat{G}_{r/m}(s)\|_2, \tag{12.12}$$

and the $\mathcal{H}_2$ norm computation can be evaluated recursively using the algorithm in [178].

Suppose that for a stable transfer function type $E(s) = \widehat{G}(s) - \widehat{G}_{r/m}(s) = B(s)/A(s)$, the polynomials $A_k(s)$ and $B_k(s)$ can be defined such that

$$A_k(s) = a_0^k + a_1^k s + \cdots + a_k^k s^k, \; B_k(s) = b_0^k + b_1^k s + \cdots + b_{k-1}^k s^{k-1} \tag{12.13}$$

The values of a_i^{k-1} and b_i^{k-1} can be evaluated recursively from

$$a_i^{k-1} = \begin{cases} a_{i+1}^k, & i \text{ even} \\ a_{i+1}^k - \alpha_k a_{i+2}^k, & i \text{ odd} \end{cases} \quad i = 0, \cdots, k-1 \tag{12.14}$$

and

$$b_i^{k-1} = \begin{cases} b_{i+1}^k, & i \text{ even} \\ b_{i+1}^k - \beta_k a_{i+2}^k, & i \text{ odd} \end{cases} \quad i = 1, \cdots, k-1, \tag{12.15}$$

where $\alpha_k = a_0^k/a_1^k$ and $\beta_k = b_1^k/a_1^k$.

The $\mathcal{H}_2$-norm of the approximate reduction error signal $\hat{e}(t)$ can be evaluated from

$$J = \sum_{k=1}^{n} \frac{\beta_k^2}{2\alpha_k} = \sum_{k=1}^{n} \frac{(b_1^k)^2}{2a_0^k a_1^k}. \tag{12.16}$$

The sub-optimal $\mathcal{H}_2$-norm reduced order model for the original high-order fractional-order model can be obtained using the following procedure [177]:

1. Select an initial reduced model $\widehat{G}_{r/m}^0(s)$.
2. Evaluate an error $\|\widehat{G}(s) - \widehat{G}_{r/m}^0(s)\|_2$ from (12.16).
3. Use an optimization algorithm (for instance, Powell's algorithm [179]) to iterate one step for a better estimated model $\widehat{G}_{r/m}^1(s)$.
4. Set $\widehat{G}_{r/m}^0(s) \leftarrow \widehat{G}_{r/m}^1(s)$, go to Step 2 until an optimal reduced model $\widehat{G}_{r/m}^*(s)$ is obtained.
5. Extract the delay from $\widehat{G}_{r/m}^*(s)$, if any.

Based on the above approach, a MATLAB function opt_app() can be designed, with the syntax G_r=opt_app$(G,r,d,$key$,G_0)$, where key is the indicator whether delay is required in the reduced order model. G_0 is the initial reduced order model, optional. The listings of the function is

```
function G_r=opt_app(G_Sys,r,k,key,G0)
GS=tf(G_Sys); num=GS.num{1}; den=GS.den{1};
Td=totaldelay(GS); GS.ioDelay=0;
GS.InputDelay=0;GS.OutputDelay=0;
if nargin<5,
    n0=[1,1]; for i=1:k-2, n0=conv(n0,[1,1]); end
    G0=tf(n0,conv([1,1],n0));
end
beta=G0.num{1}(k+1-r:k+1); alph=G0.den{1}; Tau=1.5*Td;
x=[beta(1:r),alph(2:k+1)]; if abs(Tau)<1e-5, Tau=0.5; end
dc=dcgain(GS); if key==1, x=[x,Tau]; end
y=opt_fun(x,GS,key,r,k,dc);
x=fminsearch('opt_fun',x,[],GS,key,r,k,dc);
alph=[1,x(r+1:r+k)]; beta=x(1:r+1); if key==0, Td=0; end
beta(r+1)=alph(end)*dc;
if key==1, Tau=x(end)+Td; else, Tau=0; end
G_r=tf(beta,alph,'ioDelay',Tau);
```

Two lower-level MATLAB function should also be designed as

```
function y=opt_fun(x,G,key,nn,nd,dc)
ff0=1e10; alph=[1,x(nn+1:nn+nd)];
beta=x(1:nn+1); beta(end)=alph(end)*dc; g=tf(beta,alph);
if key==1,
    tau=x(end); if tau<=0, tau=eps; end
    [nP,dP]=pade(tau,3); gP=tf(nP,dP);
else, gP=1; end
G_e=G-g*gP;
G_e.num{1}=[0,G_e.num{1}(1:end-1)];
[y,ierr]=geth2(G_e);
if ierr==1, y=10*ff0; else, ff0=y; end
%---sub function geth2
function [v,ierr]=geth2(G)
G=tf(G); num=G.num{1}; den=G.den{1}; ierr=0;
n=length(den); v=0;
if abs(num(1))>eps
    disp('System not strictly proper'); ierr=1; return
else, a1=den; b1=num(2:end); end
```

```
for k=1:n-1
  if (a1(k+1)<=eps), ierr=1; v=0; return
  else,
      aa=a1(k)/a1(k+1); bb=b1(k)/a1(k+1);
      v=v+bb*bb/aa; k1=k+2;
      for i=k1:2:n-1
          a1(i)=a1(i)-aa*a1(i+1); b1(i)=b1(i)-bb*a1(i+1);
end, end, end
v=sqrt(0.5*v);
```

We call the above procedure sub-optimal since Oustaloup's method is used for each single term s^γ in (12.8), and the Padé approximation is used for pure delay terms.

Example 8 Consider the following non-commensurate FO-LTI system:

$$G(s) = \frac{5s^{0.6} + 2}{s^{3.3} + 3.1s^{2.6} + 2.89s^{1.9} + 2.5s^{1.4} + 1.2}.$$

Using the following MATLAB scripts:

```
>> N=5; w1=1e-3; w2=1e3;
   g1=ousta_fod(0.3,N,w1,w2); g2=ousta_fod(0.6,N,w1,w2);
   g3=ousta_fod(0.9,N,w1,w2); g4=ousta_fod(0.4,N,w1,w2);
   s=tf('s');
   G=(5*g2+2)/(s^3*g1+3.1*s^2*g2+2.89*s*g3+2.5*s*g4+1.2);
   G=minreal(G)
```

an extremely high-order model can be obtained with Oustaloup's filter, such that

$$G(s) = \frac{\begin{matrix} 39.97s^{20} + 7.68 \times 10^4 s^{19} + 5.16 \times 10^7 s^{18} + 1.53 \times 10^{10} s^{17} + 2.06 \times 10^{12} s^{16} \\ +1.339 \times 10^{14} s^{15} + 4.388 \times 10^{15} s^{14} + 7.32 \times 10^{16} s^{13} + 6.053 \times 10^{17} s^{12} \\ +2.515 \times 10^{18} s^{11} + 5.422 \times 10^{18} s^{10} + 6.149 \times 10^{18} s^9 + 3.597 \times 10^{18} s^8 \\ +1.067 \times 10^{18} s^7 + 1.671 \times 10^{17} s^6 + 1.41 \times 10^{16} s^5 + 6.229 \times 10^{14} s^4 \\ +1.344 \times 10^{13} s^3 + 1.459 \times 10^{11} s^2 + 7.703 \times 10^8 s + 1.577 \times 10^6 \end{matrix}}{\begin{matrix} s^{23} + 2211s^{22} + 1.782 \times 10^6 s^{21} + 6.524 \times 10^8 s^{20} + 1.122 \times 10^{11} s^{19} \\ +9.336 \times 10^{12} s^{18} + 4.013 \times 10^{14} s^{17} + 9.167 \times 10^{15} s^{16} + 1.114 \times 10^{17} s^{15} \\ +7.328 \times 10^{17} s^{14} + 2.739 \times 10^{18} s^{13} + 6.03 \times 10^{18} s^{12} + 8.058 \times 10^{18} s^{11} \\ +6.695 \times 10^{18} s^{10} + 3.651 \times 10^{18} s^9 + 1.462 \times 10^{18} s^8 + 3.952 \times 10^{17} s^7 \\ +6.472 \times 10^{16} s^6 + 6.007 \times 10^{15} s^5 + 2.934 \times 10^{14} s^4 + 6.775 \times 10^{12} s^3 \\ +7.745 \times 10^{10} s^2 + 4.275 \times 10^8 s + 9.103 \times 10^5 \end{matrix}},$$

and the order of rational approximation to the original order model is the 23th for $N = 5$. For larger values of N, the order of rational approximation may be much higher. For instance, the order of the approximation may reach the 30th- and 40th-order respectively for the selections $N = 7$ and $N = 9$, with extremely large coefficients. Thus the model reduction algorithm should be used with the following MATLAB statements:

```
>> G2=opt_app(G,2,3,0); G3=opt_app(G,3,4,0);
   G4=opt_app(G,4,5,0); step(G,G2,G3,G4,60)
```

The step responses can be compared in Figure 12.9 and it can be seen that the seventh-order approximation is satisfactory and the fourth order fitting gives a better approximation. The obtained optimum approximated results are listed in the following:

$$G_2(s) = \frac{0.41056s^2 + 0.75579s + 0.037971}{s^3 + 0.24604s^2 + 0.22176s + 0.021915},$$

$$G_3(s) = \frac{-4.4627s^3 + 5.6139s^2 + 4.3354s + 0.15330}{s^4 + 7.4462s^3 + 1.7171s^2 + 1.5083s + 0.088476},$$

$$G_4(s) = \frac{1.7768s^4 + 2.2291s^3 + 10.911s^2 + 1.2169s + 0.010249}{s^5 + 11.347s^4 + 4.8219s^3 + 2.8448s^2 + 0.59199s + 0.0059152}.$$

It can be seen that with the lower-order models obtained, the system response of the system may not change much. The sub-optimum fitting algorithm presented may be useful in a class of linear fractional-order system approximation.   □

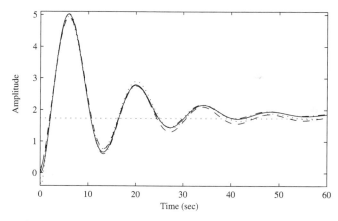

Figure 12.9 Step responses comparisons: *solid lines* for the original system, and the other lines are respectively for $G_2(s)$, $G_3(s)$, and $G_4(s)$

Example 9 Let us consider the following FO-LTI plant model:

$$G(s) = \frac{1}{s^{2.3} + 3.2s^{1.4} + 2.4s^{0.9} + 1}.$$

Let us first approximate it with Oustaloup's method and then fit it with a fixed model structure known as first-order lag plus deadtime (FOLPD) model, where $G_r(s) = \dfrac{K}{Ts+1}\mathrm{e}^{-Ls}$. The following MATLAB scripts:

```
>> N=5; w1=1e-3; w2=1e3; g1=ousta_fod(0.3,N,w1,w2);
   g2=ousta_fod(0.4,N,w1,w2); g3=ousta_fod(0.9,N,w1,w2);
   s=tf('s'); G=1/(s^2*g1+3.2*s*g2+2.4*g3+1);
   G2=opt_app(G,0,1,1); step(G,'-',G2,'--')
```

can perform this task and the optimal FOLPD model obtained is given as follows:

$$G_r(s) = \frac{0.9951}{3.5014s+1}\mathrm{e}^{-1.634s}.$$

The comparison of the open-loop step response is shown in Figure 12.10. It can be observed that the approximation is fairly effective.

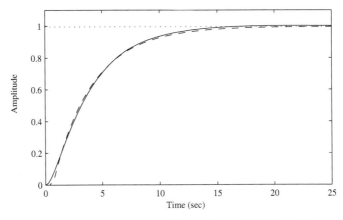

Figure 12.10 Step response comparison of the optimum FOLPD and the original model

Designing a suitable feedback controller for the original FO-LTI system G can be a formidable task. Now let us consider designing an integer-order PID controller for the optimally reduced model $G_r(s)$ and let us see if the designed controller still works for the original system.

The integer-order PID controller to be designed is in the following form:

$$G_c(s) = K_p\left(1 + \frac{1}{T_i s} + \frac{T_d s}{T_d s/N + 1}\right). \qquad (12.17)$$

The optimum ITAE criterion-based PID tuning formula [180] can be used:

$$K_{\mathrm{p}} = \frac{(0.7303 + 0.5307T/L)(T + 0.5L)}{K(T + L)},$$ (12.18)

$$T_{\mathrm{i}} = T + 0.5L, \quad T_{\mathrm{d}} = \frac{0.5LT}{T + 0.5L}.$$ (12.19)

Based on this tuning algorithm, a PID controller can be designed for $G_{\mathrm{r}}(s)$ as follows:

```
>> L=0.63; T=3.5014; K=0.9951; N=10; Ti=T+0.5*L;
   Kp=(0.7303+0.5307*T/L)*Ti/(K*(T+L));
   Td=(0.5*L*T)/(T+0.5*L); [Kp,Ti,Td]
   Gc=Kp*(1+1/Ti/s+Td*s/(Td/N*s+1))
```

The parameters of the PID controller are then $K_{\mathrm{p}} = 3.4160$, $T_{\mathrm{i}} = 3.8164$, $T_{\mathrm{d}} = 0.2890$, and the PID controller can be written as

$$G_{\mathrm{c}}(s) = \frac{1.086s^2 + 3.442s + 0.8951}{0.0289s^2 + s}.$$

Finally, the step response of the original FO-LTI with the above designed PID controller is shown in Figure 12.11. A satisfactory performance can be clearly observed. Therefore, we believe the method presented can be used for integer-order controller design for general FO-LTI systems. □

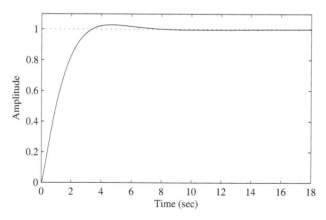

Figure 12.11 Closed-loop step response of the fractional-order plant model under the integer-order PID controller

12.5 Summary

In this chapter, several finite-dimensional integer-order approximations to fractional-order operators as well as systems are explored. For fractional-order differentiators and integrators, the frequency domain fitting using Oustaloup's recursive method is usually good enough over the frequency range of interest. The modified version is an extension to Oustaloup's fitting filter. The approximation method can be used to construct a Simulink block for fractional-order operators, and details will be given in the next chapter. Based on the approximation, a block diagram based simulation approach will be presented in the next chapter to handle fractional-order systems with any complexity. Discrete-time implementation to fractional-order operators are also presented in the chapter. Further, finite-dimensional integer-order identification methods based on given frequency domain response data are also presented and fractional-order controllers with complicated structures are explored. For FOTF models, an $\mathcal{H}_2$ norm minimization approach for model fitting is presented, and an illustrative example is given to demonstrate its use in controller design applications.

Chapter 13
Numerical Issues and MATLAB Implementations for Fractional-order Control Systems

There are several relevant MATLAB toolboxes which can be used to handle fractional-order systems. The N-integer toolbox [162] is the one used widely by the researchers. We also developed useful MATLAB code and an object-oriented toolbox in [181–183] which solves similar problems. This chapter is designed to be self-contained and is presented in the sequence of modeling, analysis, and design of fractional-order systems. Readers can run the examples on their own computer and obtain the same results. Many illustrative examples are given in the chapter to demonstrate the modeling, analysis, and design problems in fractional-order systems. Also the code can easily be applied to tackle other problems in the other chapters.

In Chapter 2, the theoretical aspects of fractional calculus were given. In this chapter, numerical issues of fractional calculus and fractional-order control are discussed. In Section 13.1, the computation aspect of fractional calculus problems are discussed. First, the evaluation of Mittag–Leffler functions is presented with MATLAB implementations. Then the fractional-order differentiations of a given function are given and closed-form solutions to linear fractional-order differential equations can be obtained. Analytical solutions are also explored in some special cases. In Section 13.2, a MATLAB object FOTF is designed and, based on the object, overload functions are developed for model simplifications, time and frequency domain analysis, as well as stability assessment. The concepts and ideas follow the Control System Toolbox of MATLAB, which is very easy for users to start with. In Section 13.3, filter approximations to fractional-order differentiators are explored and a Simulink block is designed to implement it. A block diagram based simulation strategy is illustrated for fractional-order nonlinear systems, including delay systems, through examples. Section 13.4 concentrates on optimum design of controllers for fractional-order systems. A meaningful control criterion, the finite-time ITAE criterion, is introduced and explored.

Then a controller design approach using an optimization problem solver is presented through several examples.

13.1 Computations in Fractional Calculus

In this section, numerical solutions to typical problems in fractional calculus are presented. The evaluation of Mittag–Leffler functions of different kinds is explored first, then computation of fractional-order derivatives using Grünwald–Letnikov's and Caputo's definitions are presented. A closed-form solution algorithm for linear fractional-order differential equations together with MATLAB implementations are given. Then analytical solutions to some special forms of fractional-order differential equations are explored.

13.1.1 Evaluation of Mittag–Leffler functions

In fractional calculus and control, Mittag–Leffler functions play very important roles. These can be considered analogous to the use of exponential functions in integer-order systems.

In Section 2.2, Mittag–Leffler functions in one and two parameters can respectively be expressed as

$$\mathscr{E}_{\alpha}(z) = \sum_{k=0}^{\infty} \frac{z^k}{\Gamma(\alpha k + 1)} \tag{13.1}$$

and

$$\mathscr{E}_{\alpha,\beta}(z) = \sum_{k=0}^{\infty} \frac{z^k}{\Gamma(\alpha k + \beta)}, \tag{13.2}$$

where $\alpha, \beta \in \mathbb{C}$, and $\Re(\alpha) > 0, \Re(\beta) > 0$.

It can be seen that the Mittag–Leffler function in one parameter is a special case of that in two parameters, with $\beta = 1$, *i.e.*,

$$\mathscr{E}_{\alpha,1}(z) = \mathscr{E}_{\alpha}(z). \tag{13.3}$$

Also, the Mittag–Leffler function in one parameter is an extension of the exponential function e^z, where

$$\mathscr{E}_1(z) = \sum_{k=0}^{\infty} \frac{z^k}{k!} = \mathrm{e}^z. \tag{13.4}$$

It can be seen that, in most applications, Mittag–Leffler functions can be evaluated directly by the truncation method, if MATLAB is used. The MATLAB function thus defined can be written. And in some applications, truncation methods may fail due to the nature of poor convergence. The numerical integration based method implemented by Podlubny [184] can be embedded. The listings of the MATLAB function will be given later.

The syntaxes of the function are respectively y=ml_func$(\alpha, z, n, \epsilon_0)$, and y=ml_func$([\alpha, \beta], z, n, \epsilon_0)$ for Mittag–Leffler functions in one and two parameters, where n is the order of derivative, with a default value of 0, while ϵ_0 is the error tolerance. It should be noted that, in the case where the truncation method is not convergent, the Podlubny's code can be called automatically instead. However, unfortunately, the speed of the embedded code is extremely slow.

Example 1 The curves of Mittag–Leffler functions $\mathscr{E}_1(-t)$, $\mathscr{E}_{3/2,3/2}(-t)$, and $\mathscr{E}_{1,2}(-t)$ can be drawn as shown in Figure 13.1, with the following MATLAB statements:

```
>> t=0:0.1:5; y1=ml_func(1,-t); y2=ml_func([1,2],-t);
   y3=ml_func([3/2,3/2],-t); plot(t,y1,t,y2,t,y3)
```

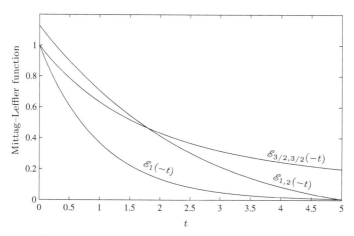

Figure 13.1 Curves of some Mittag–Leffler functions

It should be noted that $\mathscr{E}_1(-t)$ is in fact the exponential function e^{-t}. The decay of the other two curves is slower than the exponential function. □

In some applications, generalized Mittag–Leffler functions, *i.e.*, Mittag–Leffler functions in three or four parameters, have to be used. These functions are defined respectively as [185]

$$\mathscr{E}^{\gamma}_{\alpha,\beta}(z) = \sum_{k=0}^{\infty} \frac{(\gamma)_k}{\Gamma(\alpha k + \beta)} \frac{z^k}{k!} \tag{13.5}$$

and

$$\mathscr{E}_{\alpha,\beta}^{\gamma,q}(z) = \sum_{k=0}^{\infty} \frac{(\gamma)_{kq}}{\Gamma(\alpha k + \beta)} \frac{z^k}{k!}, \qquad (13.6)$$

where $\alpha, \beta, \gamma \in \mathbb{C}$, $\Re(\alpha) > 0$, $\Re(\beta) > 0$, $\Re(\gamma) > 0$, and $q \in \mathbb{N}$;

$$(\gamma)_0 = 1, \text{ and } (\gamma)_k = \gamma(\gamma+1)(\gamma+2)\cdots(\gamma+k-1) = \frac{\Gamma(k+\gamma)}{\Gamma(\gamma)}. \qquad (13.7)$$

It is easily seen, by comparing the definitions, that

$$\mathscr{E}_{\alpha,\beta}(z) = \mathscr{E}_{\alpha,\beta}^1(z), \quad \mathscr{E}_{\alpha,\beta}^{\gamma,1}(z) = \mathscr{E}_{\alpha,\beta}^\gamma(z). \qquad (13.8)$$

The above-mentioned ml_func() can still be used to deal with the generalized Mittag–Leffler function evaluation problems. The syntaxes for the evaluation of generalized Mittag–Leffler functions in three and four parameters are y=ml_func([α,β,γ],z,n,ϵ_0) and y=ml_func([α,β,γ,q],z,n,ϵ_0), respectively. The listing of the function is given below, where the function mlf() is the embedded code to be downloaded separately from [184]:

```
function f=ml_func(aa,z,n,eps0)
aa=[aa,1,1,1]; a=aa(1); b=aa(2); c=aa(3); q=aa(4);
f=0; k=0; fa=1; aa=aa(1:4); if nargin<4, eps0=eps; end
if nargin<3, n=0; end
if n==0
  while norm(fa,1)>=eps0
    fa=gamma(k*q+c)/gamma(c)/gamma(k+1)/gamma(a*k+b) *z.^k;
    f=f+fa; k=k+1;
  end
  if ~isfinite(f(1))
    if c*q==1
      f=mlf(a,b,z,round(-log10(eps0))); f=reshape(f,size(z));
    else, error('Error: truncation method failed'); end, end
  else
    aa(2)=b+n*a; aa(3)=c+q*n;
    f=gamma(q*n+c)/gamma(c)*ml_func(aa,z,0,eps0);
  end, end
```

Example 2 The curves of Mittag–Leffler functions in more parameters, $\mathscr{E}_{1,2}^{1/3}(-t)$, $\mathscr{E}_{2,2}^{1/3,2}(-t)$ are shown in Figure 13.2, together with $\mathscr{E}_{2,2}(-t)$:

```
>> t=0:0.01:5; y1=ml_func([1,2,1/3],-t);
   y2=ml_func([2,2,1/3,2],-t);
   y3=ml_func([2,2],-t); plot(t,y1,t,y2,t,y3)
```

□

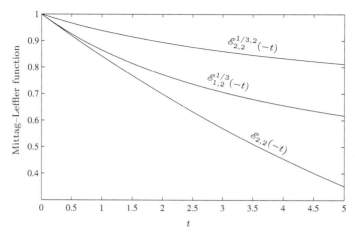

Figure 13.2 Curves of some Mittag–Leffler functions in more parameters

Also the nth-order derivative of the Mittag–Leffler function in four parameters can be evaluated from [185]

$$\frac{\mathrm{d}^n}{\mathrm{d}t^n}\mathscr{E}_{\alpha,\beta}^{\gamma,q}(z) = (\gamma)_{qn}\mathscr{E}_{\alpha,\beta+n\alpha}^{\gamma+qn,q}(z), \tag{13.9}$$

and, in particular, the integer nth-order derivative of the Mittag–Leffler function in two parameters can be evaluated from [3]

$$\frac{\mathrm{d}^n}{\mathrm{d}z^n}\mathscr{E}_{\alpha,\beta}(z) = \sum_{j=0}^{\infty} \frac{(j+n)!}{j!\,\Gamma(\alpha j + \alpha n + \beta)} z^j. \tag{13.10}$$

13.1.2 Evaluations of Fractional-order Derivatives

There are several definitions on fractional-order differentiations, such as Riemann–Liouville's definition, Grünwald–Letnikov's definition, Caputo's definition, and others, as shown in Section 2.1. It has been shown that Riemann–Liouville's definition and Grünwald–Letnikov's definition are equivalent, while there exist discrepancies between the Grünwald–Letnikov's definition and Caputo's definition due to differences in the description at initial time instances. The computation via Grünwald–Letnikov's definition and Caputo's definition is discussed here. Among the definitions, the Grünwald–Letnikov's definition is the most straightforward from the numerical implementation point of view.

13.1.2.1 Grünwald–Letnikov's Definition

Recall the approximate Grünwald–Letnikov's definition given below, where the step size of h is assumed to be very small:

$$_a\mathscr{D}_t^\alpha f(t) \approx \frac{1}{h^\alpha} \sum_{j=0}^{[(t-a)/h]} w_j^{(\alpha)} f(t - jh), \qquad (13.11)$$

where the binomial coefficients can recursively be calculated with the following formula:

$$w_0^{(\alpha)} = 1, \quad w_j^{(\alpha)} = \left(1 - \frac{\alpha + 1}{j}\right) w_{j-1}^{(\alpha)}, \; j = 1, 2, \cdots. \qquad (13.12)$$

Based on the above algorithm, the γth-order derivative of a given function can be evaluated, and the syntax of $\boldsymbol{y_1}$=glfdiff$(\boldsymbol{y},\boldsymbol{t},\gamma)$, where $\boldsymbol{y}$ and $\boldsymbol{t}$ are signal and time vectors, respectively, and $\boldsymbol{y_1}$ is a vector of γth-order derivative of $f(t)$:

```
function dy=glfdiff(y,t,gam)
h=t(2)-t(1); dy(1)=0; y=y(:); t=t(:); w=1;
for j=2:length(t), w(j)=w(j-1)*(1-(gam+1)/(j-1)); end
for i=2:length(t), dy(i)=w(1:i)*[y(i:-1:1)]/h^gam; end
```

Example 3 Consider a sinusoidal function $f(t) = \sin(3t + 1)$. It is known from Cauchy's formula that the kth-order derivative of the function is

$$f^{(k)}(t) = 3^k \sin\left(3t + 1 + k\pi/2\right),$$

and the formula also works for non-integer values of k. It is known from integer-order calculus that the integer-order derivatives can only be sinusoidal functions with phase shift of multiples of $\pi/2$. The fractional-order derivatives may provide more intermediate information, since the phase shifts are no longer integer multiples of $\pi/2$. The 3D plot of the fractional-order integrals and derivatives is shown in Figure 13.3 (a), with the following MATLAB commands:

```
>> t=0:0.1:pi; y=sin(3*t+1); Y=[]; n_vec=[-1:0.2:1];
   for n=n_vec, Y=[Y; glfdiff(y,t,n)]; end
   surf(t,n_vec,Y), shading flat
```

With Grünwald–Letnikov's definition, the 0.75th-order derivative of function $f(t)$ can be obtained as shown in Figure 13.3 (b), while that with the Cauchy formula above can also be shown:

```
>> t=0:0.01:pi; y=sin(3*t+1); y1=3^0.75*sin(3*t+1+0.75*pi/2);
   y2=glfdiff(y,t,0.75); plot(t,y1,t,y2)
```

It can be seen that there exist some differences only initially, since in Grünwald–Letnikov's definition, the initial values of function $f(t)$, for $t \leqslant 0$, are assumed to be zero, while in the Cauchy formula, the initial values of the function $f(t)$ is still assumed to be obtainable from $f(t) = \sin(3t + 1)$. Thus one must be careful with the differences in the definitions.   □

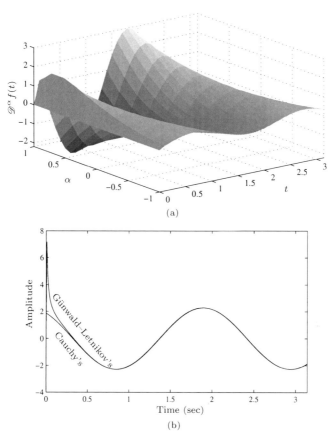

(a)

(b)

Figure 13.3 Fractional-order derivatives and integrals: (a) fractional-order derivatives and integrals for different orders and (b) comparisons under different definitions

Example 4 It is well known in the field of integer-order calculus that the derivative of a step function is a straight line. Now let us investigate the case for fractional-order derivatives and integrals. With the following MATLAB statements, the derivatives and integrals of selected orders can be obtained as shown in Figure 13.4:

```
>> t=0:0.01:1; u=ones(size(t));
   n_vec=[-0.5,0,0.5,1,1.5]; Y=[];
```

```
for n=n_vec, Y=[Y; glfdiff(u,t,n)]; end
plot(t,Y), ylim([-2 2])
```

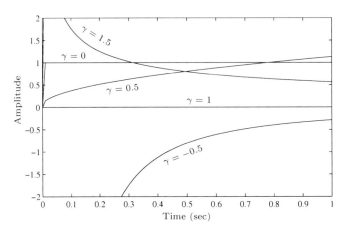

Figure 13.4 Fractional-order differentiations of a unit step function

It can be seen that, when fractional calculus is involved, the fractional-order derivatives and integrals of a step function may not be straight lines. □

13.1.2.2 Evaluating Caputo's Derivatives

Caputo's fractional-order differentiation is defined by

$$_0\mathscr{D}_t^\alpha f(t) = \frac{1}{\Gamma(1-\alpha)} \int_0^t \frac{f^{(m+1)}(\tau)}{(t-\tau)^\alpha} \, d\tau, \tag{13.13}$$

where $\alpha = m + \gamma$, m is an integer, and $0 < \gamma \leqslant 1$. Similarly, by Caputo's definition, the integral is described by

$$_0\mathscr{D}_t^{-\gamma} f(t) = \frac{1}{\Gamma(\gamma)} \int_0^t \frac{f(\tau)}{(t-\tau)^{1-\gamma}} \, d\tau, \quad \gamma > 0. \tag{13.14}$$

With the help of Symbolic Math Toolbox, a MATLAB function `caputo()` can be written as

```
function dy=caputo(t0,f,gam)
m=floor(gam); a=gam-m; dy=0;
if gam>0, syms t; fd=diff(f,t,m+1); else; a=-gam; end
for t1=t0(2:end)
    if gam>0,
        f=@(x)subs(fd,t,x)./(t1-x).^a/gamma(1-a);
    else
```

```
        f=@(x)subs(f,t,x)./(t1-x).^(1-a)/gamma(a);
    end
    dy=[dy; quadl(f,0,t1)];
end
```

and the syntax of the function is y_1=caputo(t_0,f,γ), where f is a symbolic function of variable t, t_0 is a numerical vector of evenly distributed time instances. The function returns a numerical vector y_1 of γth-order derivatives.

Example 5 Again consider the original function in Example 3, where $f(t) = \sin(3t + 1)$. The numerical solutions of 0.3th-order derivatives by Caputo's definition and Grünwald–Letnikov's definition can be obtained with the following statements:

```
>> syms t; f=sin(3*t+1); t0=0:0.01:5; dy=caputo(t0,f,0.3);
   y0=subs(f,t,t0); y1=glfdiff(y0,t0,0.3); plot(t0,dy,t0,y1)
```

and the comparison of the two derivatives is shown in Figure 13.5, and it can be seen that there are differences between the two definitions, due to non-zero initial values.   □

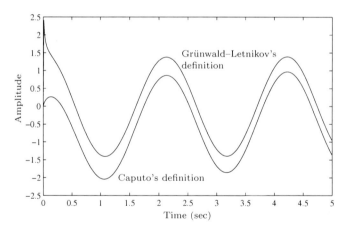

Figure 13.5 Fractional-order differentiations of step functions

13.1.3 Closed-form Solutions to Linear Fractional-order Differential Equations

As in the case for conventional linear systems, for linear fractional-order systems, linear fractional-order differential equations are the fundamental governing equations. The linear fractional-order differential equation is defined as

$$a_1 \mathscr{D}_t^{\beta_1} y(t) + a_2 \mathscr{D}_t^{\beta_2} y(t) + \cdots + a_n \mathscr{D}_t^{\beta_n} y(t)$$
$$= b_1 \mathscr{D}_t^{\gamma_1} v(t) + b_2 \mathscr{D}_t^{\gamma_2} v(t) + \cdots + b_m \mathscr{D}_t^{\gamma_m} v(t). \tag{13.15}$$

Denoting the left hand side of the equation by

$$u(t) = b_1 \mathscr{D}^{\gamma_1} v(t) + \cdots + b_m \mathscr{D}^{\gamma_m} v(t), \tag{13.16}$$

where for a given function $v(t)$, the signal $u(t)$ can easily be evaluated with the above-mentioned algorithms, the original fractional-order differential equation can be rewritten in the form

$$a_1 \mathscr{D}_t^{\beta_1} y(t) + a_2 \mathscr{D}_t^{\beta_2} y(t) + \cdots + a_n \mathscr{D}_t^{\beta_n} y(t) = u(t). \tag{13.17}$$

Substituting (13.11) into the above equation, one may find that

$$\frac{a_1}{h^{\beta_1}} \sum_{j=0}^{[(t-a)/h]} w_j^{(\beta_1)} y(t - jh) + \cdots + \frac{a_n}{h^{\beta_n}} \sum_{j=0}^{[(t-a)/h]} w_j^{(\beta_n)} y(t - jh) = u(t), \tag{13.18}$$

where the binomial coefficients $w_j^{(\beta_i)}$ can still be evaluated recursively with

$$w_0^{(\beta_i)} = 1, \quad w_j^{(\beta_i)} = \left(1 - \frac{\beta_i + 1}{j}\right) w_{j-1}^{(\beta_i)}, \ j = 1, 2, \cdots. \tag{13.19}$$

By slight rearrangement of the terms, the closed-form solution of the fractional-order differential equation can be obtained as

$$y(t) = \frac{1}{\sum_{i=1}^{n} \frac{a_i}{h^{\beta_i}}} \left[u(t) - \sum_{i=1}^{n} \frac{a_i}{h^{\beta_i}} \sum_{j=1}^{[(t-a)/h]} w_j^{(\beta_i)} y(t - jh) \right]. \tag{13.20}$$

Based on the above algorithm, the numerical solutions to the linear fractional-order differential equation can be obtained with a MATLAB function of

```
function y=fode_sol(a,na,b,nb,u,t)
h=t(2)-t(1); D=sum(a./[h.^na]); nT=length(t);
vec=[na nb]; W=[]; D1=b(:)./h.^nb(:); nA=length(a);
y1=zeros(nT,1); W=ones(nT,length(vec));
for j=2:nT, W(j,:)=W(j-1,:).*(1-(vec+1)/(j-1)); end
for i=2:nT,
    A=[y1(i-1:-1:1)]'*W(2:i,1:nA);
    y1(i)=(u(i)-sum(A.*a./[h.^na]))/D;
end
for i=2:nT, y(i)=(W(1:i,nA+1:end)*D1)'*[y1(i:-1:1)]; end
```

whose syntax is y=fode_sol$(a,n_{\mathrm{a}},b,n_{\mathrm{b}},u,t)$, where $a = [a_1, \cdots, a_n]$ and $b = [b_1, \cdots, b_m]$ are respectively the coefficients of y and u of the equation, while $n_{\mathrm{a}} = [\beta_1, \cdots, \beta_n]$ and $n_{\mathrm{b}} = [\gamma_1, \cdots, \gamma_m]$ are, respectively, the orders. The vectors t and u contain the time instances and the input sequence. The function returns the solution vector y of the equation.

Example 6 Consider a linear fractional-order differential equation
$$\mathscr{D}^{1.6}y(t) + 10\mathscr{D}^{1.2}y(t) + 35\mathscr{D}^{0.8}y(t) + 50\mathscr{D}^{0.4}y(t) + 24y(t)$$
$$= \mathscr{D}^{1.2}u(t) + 3\mathscr{D}^{0.4}u(t) + 5u(t),$$
with zero initial conditions. It is also known that the input signal is a step function. To solve the differential equation numerically, one should specify the coefficients and orders in the equation; then fode_sol() can be called. The following MATLAB statements can be issued:

```
>> a=[1,10,35,50,24]; na=[1.6 1.2 0.8 0.4 0];
   b=[1 3 5]; nb=[1.2 0.4 0]; t=0:0.01:10; u=ones(size(t));
   y=fode_sol(a,na,b,nb,u,t); plot(t,y)
```

The step response of the system is obtained as Figure 13.6. The most important step in numerical solutions is the validation process of the solutions. A simple way to validate the solution is to select a smaller step-size, and see whether consistent results can be obtained. If not, one should select smaller step-sizes and try again. For the problem in this example, the step-size of 0.01 is good enough.

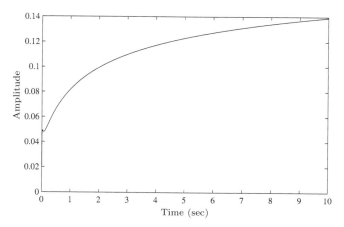

Figure 13.6 Solution of the fractional-order differentiations

As in the case of integer-order systems, only very limited classes of fractional-order systems can be studied analytically. □

13.1.4 Analytical Solutions to Linear Fractional-order Differential Equations

As in the case of integer-order systems, only very limited classes of fractional-order systems can be studied analytically. Laplace transforms of some particular functions are presented first. Based on the properties, step and impulse responses for commensurate-order systems are presented. Then for ordinary fractional-order systems, in particular, three-term fractional-order systems, analytical solutions to step responses are given.

13.1.4.1 Some Useful Laplace Transform Formulas

One of the most important Laplace transform formula for finding the analytical solutions of a class of fractional-order systems is presented first. Based on the formula, some of the other particular properties will be presented.

The general form of the Laplace transform formula is given by [185, 186]

$$\mathscr{L}^{-1}\left[\frac{s^{\alpha\gamma-\beta}}{(s^\alpha+a)^\gamma}\right] = t^{\beta-1}\mathscr{E}_{\alpha,\beta}^\gamma\left(-at^\alpha\right), \tag{13.21}$$

and with the use of the property, many useful formulas can be derived:

- When $\gamma = 1$ and $\alpha\gamma = \beta$, then the inverse Laplace transform can be interpreted as the analytical response of an FOTF $1/(s^\alpha + a)$ driven by an impulse input. In this case, one has $\beta = \alpha$, and the Laplace transform can be expressed as

$$\mathscr{L}^{-1}\left[\frac{1}{s^\alpha+a}\right] = t^{\alpha-1}\mathscr{E}_{\alpha,\alpha}\left(-at^\alpha\right). \tag{13.22}$$

- When $\gamma = 1$, and $\alpha\gamma - \beta = -1$, then the inverse Laplace transform can be interpreted as the analytical solution of a fractional-order transfer function $1/(s^\alpha + a)$ driven by a step input. In this case, one has $\beta = \alpha + 1$, and the Laplace transform can be expressed as

$$\mathscr{L}^{-1}\left[\frac{1}{s(s^\alpha+a)}\right] = t^\alpha\mathscr{E}_{\alpha,\alpha+1}\left(-at^\alpha\right). \tag{13.23}$$

It can also be shown that the inverse Laplace transform of the function can alternatively be written as

$$\mathscr{L}^{-1}\left[\frac{1}{s(s^\alpha+a)}\right] = \frac{1}{a}\left[1 - \mathscr{E}_\alpha\left(-at^\alpha\right)\right]. \tag{13.24}$$

- When $\gamma = k$ is an integer and $\alpha\gamma = \beta$, then the inverse Laplace transform can be interpreted as the analytical solution of an FOTF $1/(s^\alpha + a)^k$ driven by an impulse input. In this case, one has $\beta = \alpha k$, and the Laplace transform can be expressed as

$$\mathscr{L}^{-1}\left[\frac{1}{(s^\alpha + a)^k}\right] = t^{\alpha k - 1}\mathscr{E}_{\alpha,\alpha k}^k\left(-at^\alpha\right).\tag{13.25}$$

- When $\gamma = k$ is an integer and $\alpha\gamma - \beta = -1$, then the inverse Laplace transform can be interpreted as the analytical solution of a fractional-order transfer function $1/(s^\alpha + a)^k$ driven by a step input. In this case, one has $\beta = \alpha k + 1$, and the Laplace transform can be expressed as

$$\mathscr{L}^{-1}\left[\frac{1}{s(s^\alpha + a)^k}\right] = t^{\alpha k}\mathscr{E}_{\alpha,\alpha k+1}^k\left(-at^\alpha\right).\tag{13.26}$$

13.1.4.2 Analysis of Commensurate-order Systems

Consider the orders in (13.15). If one can find a greatest common divisor α to them, then the original equation can be rewritten as

$$
\begin{aligned}
a_1\mathscr{D}_t^{n\alpha}y(t) &+ a_2\mathscr{D}_t^{(n-1)\alpha}y(t) + \cdots + a_n\mathscr{D}_t^\alpha y(t) + a_{n+1}y(t)\\
&= b_1\mathscr{D}_t^{m\alpha}v(t) + b_2\mathscr{D}_t^{(m-1)\alpha}v(t) + \cdots + b_m\mathscr{D}_t^\alpha v(t) + b_{m+1}v(t).
\end{aligned}\tag{13.27}
$$

If zero initial conditions are assumed for the input and output signals, then the Laplace transform can be applied to the fractional-order differential equation, and the FOTF

$$G(s) = \frac{b_1 s^{m\alpha} + b_2 s^{(m-1)\alpha} + \cdots + b_m s^\alpha + b_{m+1}}{a_1 s^{n\alpha} + a_2 s^{(n-1)\alpha} + \cdots a_n s^\alpha + a_{n+1}}\tag{13.28}$$

can be obtained. This kind of system is usually referred to as commensurate-order systems. For commensurate-order systems, denoting $\lambda = s^\alpha$, the transfer function can be re-expressed as an integer-order rational function of variable of λ, such that

$$G(\lambda) = \frac{b_1\lambda^m + b_2\lambda^{m-1} + \cdots + b_m\lambda + b_{m+1}}{a_1\lambda^n + a_2\lambda^{n-1} + \cdots a_n\lambda + a_{n+1}}.\tag{13.29}$$

Suppose there are no repeated poles in $G(\lambda)$, then partial fraction expansion can be made to the original system such that

$$G(\lambda) = \sum_{i=1}^n \frac{r_i}{\lambda + p_i} = \sum_{i=1}^n \frac{r_i}{s^\alpha + p_i}.\tag{13.30}$$

With the use of the Laplace transforms defined in (13.22) and (13.23), the analytical solutions to impulse and step input can be obtained directly such that

$$\mathscr{L}^{-1}\left[\sum_{i=1}^n \frac{r_i}{s^\alpha + p_i}\right] = \sum_{i=1}^n r_i t^{\alpha-1}\mathscr{E}_{\alpha,\alpha}\left(-p_i t^\alpha\right),\tag{13.31}$$

$$\mathscr{L}^{-1}\left[\sum_{i=1}^{n}\frac{r_i}{s\left(s^\alpha+p_i\right)}\right]=\sum_{i=1}^{n}r_it^\alpha\mathscr{E}_{\alpha,\alpha+1}\left(-p_it^\alpha\right), \qquad (13.32)$$

or, alternatively,

$$\mathscr{L}^{-1}\left[\sum_{i=1}^{n}\frac{r_i}{s\left(s^\alpha+p_i\right)}\right]=\sum_{i=1}^{n}\frac{r_i}{p_i}\left[1-\mathscr{E}_\alpha\left(-p_it^\alpha\right)\right]. \qquad (13.33)$$

Example 7 The fractional-order differential equation is described as

$$G(s)=\frac{s^{1.2}+3s^{0.4}+5}{s^{1.6}+10s^{1.2}+35s^{0.8}+50s^{0.4}+24}.$$

Denoting $\lambda=s^{0.4}$, a commensurate-order system can be obtained:

$$G(\lambda)=\frac{\lambda^3+3\lambda+5}{\lambda^4+10\lambda^3+35\lambda^2+50\lambda+24}.$$

With MATLAB commands

```
>> n=[1,0,3,5]; d=[1,10,35,50,24];
   [r,p,k]=residue(n,d)
```

it is easily found that the partial fraction expansion of the system can be written as

$$G(\lambda)=\frac{71}{6}\times\frac{1}{\lambda+4}-\frac{31}{2}\times\frac{1}{\lambda+3}+\frac{9}{2}\times\frac{1}{\lambda+2}+\frac{1}{6}\times\frac{1}{\lambda+1},$$

from which the impulse response of the system can be written as

$$y_1(t)=\frac{71}{6}t^{-0.6}\mathscr{E}_{0.4,0.4}\left(-4t^{0.4}\right)-\frac{31}{2}t^{-0.6}\mathscr{E}_{0.4,0.4}\left(-3t^{0.4}\right)$$
$$+\frac{9}{2}t^{-0.6}\mathscr{E}_{0.4,0.4}\left(-2t^{0.4}\right)+\frac{1}{6}t^{-0.6}\mathscr{E}_{0.4,0.4}\left(-t^{0.4}\right).$$

Also, with (13.23) and (13.24), the step responses of the system can be written as

$$y_2(t)=\frac{71}{6}t^{0.4}\mathscr{E}_{0.4,1.4}\left(-4t^{0.4}\right)-\frac{31}{2}t^{0.4}\mathscr{E}_{0.4,1.4}\left(-3t^{0.4}\right)$$
$$+\frac{9}{2}t^{0.4}\mathscr{E}_{0.4,1.4}\left(-2t^{0.4}\right)+\frac{1}{6}t^{0.4}\mathscr{E}_{0.4,1.4}\left(-t^{0.4}\right),$$

or alternatively

$$y_3(t)=\frac{5}{24}-\frac{71}{24}\mathscr{E}_{0.4}\left(-4t^{0.4}\right)+\frac{31}{6}\mathscr{E}_{0.4}\left(-3t^{0.4}\right)-\frac{9}{4}\mathscr{E}_{0.4}\left(-2t^{0.4}\right)-\frac{1}{6}\mathscr{E}_{0.4}\left(-t^{0.4}\right).$$

The impulse response of the system can be obtained as shown in Figure 13.7 (a), with the following MATLAB statements:

```
>> t=0:0.02:2;
   y1=t.^(-0.6).*(71/6*ml_func([0.4,0.4],-4*t.^(0.4))...
                 -31/2*ml_func([0.4,0.4],-3*t.^(0.4))...
                 +9/2*ml_func([0.4,0.4],-2*t.^(0.4))...
```

```
                    +1/6*ml_func([0.4,0.4],-t.^(0.4)));
     plot(t,y1)
```

The numerical solutions of step responses can alternatively be evaluated using one of the above two formulas, with the following MATLAB statements and the step response is shown in Figure 13.7 (b). It should be noted that the truncation method does not converge for this example, and Podlubny's code is called automatically, which makes the computation extremely slow:

```
>> y3=5/24-71/24*ml_func(0.4,-4*t.^(0.4))...
            +31/6*ml_func(0.4,-3*t.^(0.4))...
             -9/4*ml_func(0.4,-2*t.^(0.4))...
             -1/6*ml_func(0.4,-t.^(0.4));
   y2=(71/6*ml_func([0.4,1.4],-4*t.^(0.4))-...
       31/2*ml_func([0.4,1.4],-3*t.^(0.4))+...
       9/2*ml_func([0.4,1.4],-2*t.^(0.4))+...
       1/6*ml_func([0.4,1.4],-t.^(0.4))).*t.^(0.4);
   plot(t,y2,t,y3)
```

By comparing the results obtained numerically in the previous example, it can be seen that the step responses with the two analytical formulas are almost the same as that using numerical methods, which validates the results.

If there are repeated poles of s^α with multiplicity of k, for instance, the related partial fraction expansion can be written as

$$\frac{r_j}{s^\alpha + p_i} + \frac{r_{j+1}}{(s^\alpha + p_i)^2} + \cdots + \frac{r_{j+m-1}}{(s^\alpha + p_i)^m} = \sum_{k=1}^{m} \frac{r_{j+k-1}}{(s^\alpha + p_i)^k}, \qquad (13.34)$$

then using the inverse Laplace transform properties, the step response and impulse response can be obtained as

$$\mathscr{L}^{-1}\left[\sum_{k=1}^{m} \frac{r_{j+k-1}}{(s^\alpha + p_i)^k}\right] = \sum_{k=1}^{m} r_{j+k-1} t^{\alpha k-1} \mathscr{E}_{\alpha,\alpha k}^k(-p_i t^\alpha), \qquad (13.35)$$

$$\mathscr{L}^{-1}\left[\sum_{k=1}^{m} \frac{r_{j+k-1}}{s(s^\alpha + p_i)^k}\right] = \sum_{k=1}^{m} r_{j+k-1} t^{\alpha k} \mathscr{E}_{\alpha,\alpha k+1}^k(-p_i t^\alpha). \qquad (13.36)$$

Example 8 Consider the fractional-order differential equation given by

$$\mathscr{D}^{1.2}y(t) + 5\mathscr{D}^{0.9}y(t) + 9\mathscr{D}^{0.6}y(t) + 7\mathscr{D}^{0.3}y(t) + 2y(t) = u(t),$$

where $u(t)$ is a step input. Assume that the initial conditions of $y(t)$ and its derivatives are zero. Again, by selecting $\lambda = s^{0.3}$, the original transfer function can be rewritten as

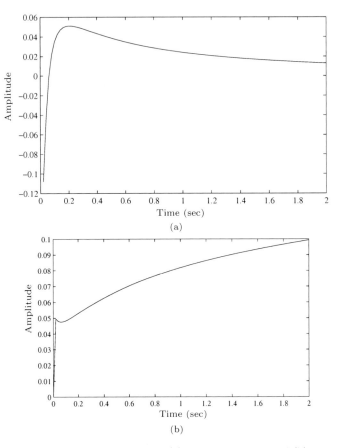

Figure 13.7 Impulse and step responses: (a) impulse response and (b) step response

$$G(\lambda) = \frac{1}{\lambda^4 + 5\lambda^3 + 9\lambda^2 + 7\lambda + 2}.$$

The following MATLAB statements can be used to find the partial fraction expansion of the related transfer function:

```
>> num=1; den=[1 5 9 7 2]; [r,p]=residue(num,den)
```

from which it is immediately seen that the transfer function of variable λ can be rewritten as

$$G(\lambda) = \frac{1}{\lambda + 2} + \frac{1}{\lambda + 1} - \frac{1}{(\lambda + 1)^2} + \frac{1}{(\lambda + 1)^3}.$$

With the use of the above formula, the analytical solutions to the impulse response of the system can be written as

$$y_1(t) = -t^{-0.7}\mathscr{E}_{0.3,0.3}\left(-2t^{0.3}\right) + t^{-0.7}\mathscr{E}_{0.3,0.3}\left(-t^{0.3}\right)$$
$$- t^{0.4}\mathscr{E}_{0.3,0.6}^2\left(-t^{0.3}\right) + t^{0.1}\mathscr{E}_{0.3,0.9}^3\left(-t^{0.3}\right),$$

and the analytical solution to the step response can be written as

$$y_2(t) = -t^{0.3}\mathscr{E}_{0.3,1.3}\left(-2t^{0.3}\right) + t^{0.3}\mathscr{E}_{0.3,1.3}\left(-t^{0.3}\right)$$
$$- t^{0.6}\mathscr{E}_{0.3,1.6}^2\left(-t^{0.3}\right) + t^{0.9}\mathscr{E}_{0.3,1.9}^3\left(-t^{0.3}\right).$$

It can be seen that, with the use of partial fraction expansion technique, the analytical solutions can be found:

```
>> t=0:0.05:2; t1=0:0.001:2;
   y2=-t.^(0.3).*ml_func([0.3,1.3],-2*t.^(0.3))+...
      t.^(0.3).*ml_func([0.3,1.3,1],-t.^(0.3))-...
      t.^(0.6).*ml_func([0.3,1.6,2],-t.^(0.3))+...
      t.^(0.9).*ml_func([0.3,1.9,3],-t.^(0.3));
   y0=fode_sol(den,[1.2:-0.3:0],1,0,ones(size(t1)),t1);
   plot(t,y2,t1,y0)
```

Since the numerical solution to step responses of the system can also be obtained, as shown in Figure 13.8, the responses can be compared with the analytical solutions and it can be seen that both results agree well. □

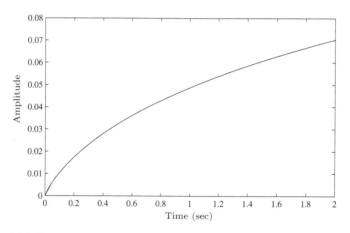

Figure 13.8 Step responses comparison

13.1.4.3 Analytical Solutions with Mittag–Leffler Functions for a Class of Linear Fractional-order Differential Equations

The Mittag–Leffler function plays an important part in the analytic solutions of linear fractional-order systems.

Consider an $(n+1)$-term fractional-order differential equation given by

$$a_n \mathscr{D}_t^{\beta_n} y(t) + a_{n-1} \mathscr{D}_t^{\beta_{n-1}} y(t) + \cdots + a_0 \mathscr{D}_t^{\beta_0} y(t) = u(t), \qquad (13.37)$$

whose analytical solution to step input function can be obtained as [3]

$$
y(t) = \frac{1}{a_n} \sum_{m=0}^{\infty} \frac{(-1)^m}{m!} \sum_{\substack{k_0+k_1+\cdots+k_{n-2}=m \\ k_0 \geqslant 0, \ \cdots, \ k_{n-2} \geqslant 0}} (m; k_0, k_1, \cdots, k_{n-2})
$$

$$
\prod_{i=0}^{n-2} \left(\frac{a_i}{a_n}\right)^{k_i} t^{(\beta_n-\beta_{n-1})m+\beta_n+\sum_{j=0}^{n-2}(\beta_{n-1}-\beta_j)k_j-1} \tag{13.38}
$$

$$
\mathscr{E}^{(m)}_{\beta_n-\beta_{n-1}, \ \beta_n+\sum_{j=0}^{n-2}(\beta_{n-1}-\beta_j)k_j} \left(-\frac{a_{n-1}}{a_n}t^{\beta_n-\beta_{n-1}}\right),
$$

where $(m; k_0, k_1, \cdots, k_{n-2})$ is a multinomial coefficient defined as

$$
(m; k_0, k_1, \cdots, k_{n-2}) = \frac{m!}{k_0!k_1!\cdots k_{n-2}!}. \tag{13.39}
$$

Unfortunately, for general fractional-order differential equations, the above formulas are very difficult, if not impossible, to solve. One usually considers only special cases, *i.e.*, systems with three terms only in the denominator. The transfer function thus studied is given below:

$$
G(s) = \frac{1}{a_2 s^{\beta_2} + a_1 s^{\beta_1} + a_0}, \tag{13.40}
$$

where $\hat{a}_0 = a_0/a_2, \hat{a}_1 = a_1/a_2$. The step response to such a system can be written as

$$
y(t) = \frac{1}{a_2} \sum_{k=0}^{\infty} \frac{(-1)^k \hat{a}_0^k t^{-\hat{a}_1+(k+1)\beta_2}}{k!} \mathscr{E}^{(k)}_{\beta_2-\beta_1, \beta_2+\beta_1 k+1} \left(-\hat{a}_1 t^{\beta_2-\beta_1}\right), \tag{13.41}
$$

where $\hat{a}_0 = a_0/a_2, \hat{a}_1 = a_1/a_2$. A MATLAB function ml_step is written to implement the step response of the system. The syntax y=ml_step(a,b,t,ε) can be used to find the numerical solution of the three-term system, where $a = [a_0, a_1, a_2]$, and $b = [\beta_1, \beta_2]$. The argument ε is the error tolerance:

```
function y=ml_step(a,b,t,eps0)
a0=a(1); a1=a(2); a2=a(3); b1=b(1); b2=b(2);
y=0; k=0; ya=1; a0=a0/a2; a1=a1/a2;
if nargin==3, eps0=eps; end
while max(abs(ya))>=eps0
    ya=(-1)^k/gamma(k+1)*a0^k*t.^((k+1)*b2).*...
        ml_fun(b2-b1,b2+b1*k+1,-a1*t.^(b2-b1),k,eps0);
    y=y+ya; k=k+1;
end
y=y/a2;
```

Example 9 For the FOTF

$$G(s) = \frac{1}{s^{0.8} + 0.75s^{0.4} + 0.9}$$

it can be seen that the parameters can be expressed as $a_0 = 0.9, a_1 = 0.75, a_2 = 1, \beta_1 = 0.4, \beta_2 = 0.8$. The step response can be obtained as shown in Figure 13.9, and the curve agrees well with that obtained by closed-form solution algorithm:

```
>> t=0:0.001:5; u=ones(size(t));
   y=ml_step([0.9,0.75,1],[0.4,0.8],t);
   y1=fode_sol([1,0.75,0.9],[0.8,0.4,0],1,0,u,t);
   plot(t,y,t,y1)
```

whose analytical solution of step response is

$$y(t) = \sum_{k=0}^{\infty} \frac{(-0.9)^k}{k!} t^{-0.75+0.8(k+1)} \mathscr{E}_{0.4, 1.8+0.4k}^{(k)} \left(-0.75t^{0.4}\right).$$ □

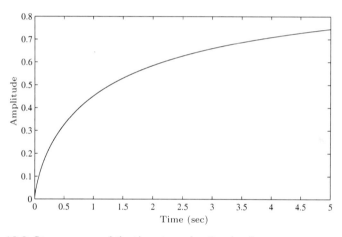

Figure 13.9 Step response of the three-term fractional-order system

13.2 Fractional-order Transfer Functions

It is well known that the Control System Toolbox of MATLAB provides a very simple and straightforward way for the modeling, analysis, and design of linear transfer function models.

With the use of the property of Laplace transform for signal $y(t)$ at rest

$$\mathscr{L}[\mathrm{d}^\alpha y(t)/\mathrm{d}t^\alpha] = s^\alpha \mathscr{L}[y(t)], \tag{13.42}$$

the FOTF can be defined as

$$G(s) = \frac{b_1 s^{\gamma_1} + b_2 s^{\gamma_2} + \cdots + b_m s^{\gamma_m}}{a_1 s^{\beta_1} + a_2 s^{\beta_2} + \cdots + a_{n-1} s^{\beta_{n-1}} + a_n s^{\beta_n}}. \tag{13.43}$$

In order to deal better with control systems, the above FOTF is extended to include a pure time delay as

$$G(s) = \frac{b_1 s^{\gamma_1} + b_2 s^{\gamma_2} + \cdots + b_m s^{\gamma_m}}{a_1 s^{\beta_1} + a_2 s^{\beta_2} + \cdots + a_{n-1} s^{\beta_{n-1}} + a_n s^{\beta_n}} e^{-\tau s}. \tag{13.44}$$

It can be seen that, like integer-order transfer functions, one can uniquely describe the FOTF, with the four vectors, *i.e.*, coefficient vectors of numerator and denominator, and the vectors of orders, and a delay constant τ. In this section, a MATLAB object for FOTFs is designed and, based on the object, different modeling and analysis routines are defined.

13.2.1 Design of an FOTF Object

To implement an object in MATLAB, for instance, the FOTF object, `fotf`, one should create a folder `@fotf`, and in the folder, at least two functions should be written. One is `fotf.m` file which defines the object, and the other, `display.m`, which displays the object.

- **FOTF object creation file**: The file `fotf.m` listed below can be used to define FOTFs:

```
function G=fotf(a,na,b,nb,T)
if nargin==0,
   G.a=[]; G.na=[]; G.b=[]; G.nb=[]; G.ioDelay=0;
   G=class(G,'fotf');
elseif isa(a,'fotf'), G=a;
elseif nargin==1 & isa(a,'double'), G=fotf(1,0,a,0,0);
elseif nargin==1 & a=='s', G=fotf(1,0,1,1,0);
else, ii=find(abs(a)<eps); a(ii)=[]; na(ii)=[];
   ii=find(abs(b)<eps); b(ii)=[]; nb(ii)=[];
   if nargin==5, G.ioDelay=T; else, G.ioDelay=0; end
   G.a=a; G.na=na; G.b=b; G.nb=nb; G=class(G,'fotf');
end
```

One may define an FOTF object with a command in the syntax $G=$`fotf`$(\boldsymbol{a}, \boldsymbol{n_a}, \boldsymbol{b}, \boldsymbol{n_b}, \tau)$, where $\boldsymbol{a} = [a_1, \cdots, a_n]$, $\boldsymbol{b} = [b_1, \cdots, b_m]$, $\boldsymbol{n_a} = [\beta_1, \cdots, \beta_n]$, and $\boldsymbol{n_b} = [\gamma_1, \cdots, \gamma_m]$ are respectively the coefficients and orders of the numerator and denominator, and τ is the delay constant. If there is no delay in the system, the argument can be omitted. Also a fractional-order operator s can alternatively be defined as $s=$`fotf`$('s')$.

- **Object display file**: Another file which is essential in the folder is the display file, `display.m`, listed below. The function can be called automatically once an FOTF object is defined:

```
function display(G)
strN=polydisp(G.b,G.nb); strD=polydisp(G.a,G.na);
nn=length(strN); nd=length(strD); nm=max([nn,nd]);
disp([char(' '*ones(1,floor((nm-nn)/2))) strN]), ss=[];
T=G.ioDelay; if T>0, ss=[' exp(-' num2str(T) 's)']; end
disp([char('-'*ones(1,nm)), ss]);
disp([char(' '*ones(1,floor((nm-nd)/2))) strD])
function strP=polydisp(p,np)
P=''; [np,ii]=sort(np,'descend'); p=p(ii);
for i=1:length(p),
    P=[P,'+',num2str(p(i)),'s^{',num2str(np(i)),'}'];
end
P=P(2:end); P=strrep(P,'s^{0}','');
P=strrep(P,'+-','-'); P=strrep(P,'^{1}','');
P=strrep(P,'+1s','+s');
strP=strrep(P,'-1s','-s'); nP=length(strP);
if nP>=2 & strP(1:2)=='1s', strP=strP(2:end); end
```

Example 10 An FOTF

$$G(s) = \frac{0.8s^{1.2} + 2}{1.1s^{1.8} + 1.9s^{0.5} + 0.4} e^{-0.5s}$$

can be entered into MATLAB environment with the following command:

```
>> G=fotf([1.1,1.9,0.4],[1.8,0.5,0],[0.8,2],[1.2,0],0.5);
```

and the object can be displayed as follows:

```
    0.8s^{1.2}+2
------------------------ exp(-0.5s)
1.1s^{1.8}+1.9s^{0.5}+0.4
```

In later examples, mathematical ways of display rather than direct MATLAB display will be given in the chapter. □

13.2.2 Modeling Using FOTFs

Based on the newly defined FOTF class, the functions `plus()`, `mtimes()` and `feedback()` can be written to implement interconnections of FOTF objects.

In this way, complicated systems under different connections can be obtained easily with simple commands:

- **Multiplication function** `mtimes()` for blocks in series connections. If two FOTF object G_1 and G_2 are multiplied together, *i.e.*, performing series connections to the two blocks, one can simply multiply the numerators of the blocks together to form the numerator of overall system, and multiply the denominator together, to form the denominator of the overall system. In mathematical terms,

$$G(s) = G_1(s)G_2(s) = \frac{N_1(s)N_2(s)}{D_1(s)D_2(s)}, \qquad (13.45)$$

where $N_i(s)$ and $D_i(s)$ are the numerator and denominator polynomials respectively. The MATLAB implementation of the above manipulation is given below:

```
function G=mtimes(G1,G2)
G1=fotf(G1); G2=fotf(G2); na=[]; nb=[];
a=kron(G1.a,G2.a); b=kron(G1.b,G2.b);
for i=1:length(G1.na), na=[na,G1.na(i)+G2.na]; end
for i=1:length(G1.nb), nb=[nb,G1.nb(i)+G2.nb]; end
G=simple(fotf(a,na,b,nb,G1.ioDelay+G2.ioDelay));
```

If the above-listed file is included in the @fotf folder, one can then use $*$ operator in MATLAB for FOTF objects. This kind of programming technique is the so-called overload function design in object-oriented programming. A lower level function `simple()` is also programmed and will be explained later.

- **Plus function** `plus()` for blocks in parallel connections. Similarly, the $+$ operator can be redefined for FOTF objects. The overall system can be expressed as

$$G(s) = G_1(s) + G_2(s) = \frac{N_1(s)D_2(s) + N_2(s)D_1(s)}{D_1(s)D_2(s)}, \qquad (13.46)$$

where the listing of the function is

```
function G=plus(G1,G2)
G1=fotf(G1); G2=fotf(G2); na=[]; nb=[];
if G1.ioDelay==G2.ioDelay
    a=kron(G1.a,G2.a); b=[kron(G1.a,G2.b),kron(G1.b,G2.a)];
    for i=1:length(G1.a),
        na=[na G1.na(i)+G2.na]; nb=[nb, G1.na(i)+G2.nb];
    end
    for i=1:length(G1.b), nb=[nb G1.nb(i)+G2.na]; end
```

```
        G=simple(fotf(a,na,b,nb,G1.ioDelay));
    else, error('cannot handle different delays'); end
```

It should be noted that, if the two blocks have different time delay constants, the overall model cannot be expressed with FOTF block. In this case, an error message is given by the function.

- **Feedback function** feedback() for blocks in negative feedback connections. The overall system is then described as

$$G(s) = \frac{G_1(s)}{1 \pm G_1(s)G_2(s)} = \frac{N_1(s)D_2(s)}{D_1(s)D_2(s) \pm N_1(s)N_2(s)}, \qquad (13.47)$$

and the system can be implemented with

```
function G=feedback(F,H)
F=fotf(F); H=fotf(H); na=[]; nb=[];
if F.ioDelay==H.ioDelay
    b=kron(F.b,H.a); a=[kron(F.b,H.b), kron(F.a,H.a)];
    for i=1:length(F.b),
        nb=[nb F.nb(i)+H.nb]; na=[na,F.nb(i)+H.nb];
    end
    for i=1:length(F.a), na=[na F.na(i)+H.na]; end
    G=simple(fotf(a,na,b,nb,F.ioDelay));
else, error('cannot handle different delays'); end
```

For positive feedback systems, the feedback $-H(s)$ should be used instead. Again, if the two blocks have different delay constants, the feedback system cannot be expressed by FOTF object, and an error message will be given.

- **Division of FOTFs.** The division of FOTFs, $G(s) = G_1(s)/G_2(s)$, can be evaluated by taking multiplications as $G(s) = G_1(s)G_2^{-1}(s)$. Thus an overload function can be written as

```
function G=mrdivide(G1,G2)
G1=fotf(G1); G2=fotf(G2); G=G1*inv(G2);
G.ioDelay=G1.ioDelay-G2.ioDelay;
if G.ioDelay<0, warning('block with positive delay'); end
```

- **Simplification function** simple(). Polynomial collection simplifications as well as other simplification method, including removing the minus terms of s, can be obtained. The function can be written as

```
function G=simple(G1)
[a,n]=polyuniq(G1.a,G1.na); G1.a=a; G1.na=n; na=G1.na;
[a,n]=polyuniq(G1.b,G1.nb); G1.b=a; G1.nb=n; nb=G1.nb;
nn=min(na(end),nb(end)); nb=nb-nn; na=na-nn;
```

```
G=fotf(G1.a,na,G1.b,nb,G1.ioDelay);
% local function polyuniq for collecting polynomial terms
function [a,an]=polyuniq(a,an)
[an,ii]=sort(an,'descend'); a=a(ii); ax=diff(an); key=1;
for i=1:length(ax)
    if ax(i)==0,
        a(key)=a(key)+a(key+1); a(key+1)=[]; an(key+1)=[];
    else, key=key+1; end
end
```

- **Power of FOTF**. This file only works for integer power of FOTF or taking the power of the Laplace operator. This facility is also useful in specifying fractional-order PID controllers:

```
function G1=mpower(G,n)
if n==fix(n),
    if n>=0, G1=1; for i=1:n, G1=G1*G; end
    else, G1=inv(G^(-n)); end
    G.ioDelay=n*G.ioDelay;
elseif length(G.a)*length(G.b)==1 & G.na==0 & G.nb==1,
    G1=fotf(1,0,1,n);
else, error('mpower: power must be an integer.'); end
```

- **Other supporting functions**. Other functions should also be designed, such as the minus(), uminus(), inv(), and the files should be placed in the @fotf directory to overload the existing ones. The listings of the three functions are given below:

```
function G=uminus(G1), G=G1; G.b=-G.b;
function G=minus(G1,G2), G=G1+(-G2);
function G=inv(G1)
G=fotf(G1.b,G1.nb,G1.a,G1.na,-G.ioDelay);
```

Example 11 Fractional-order PID controller $G_c(s) = 5 + 2s^{-0.2} + 3s^{0.6}$ can easily be entered into MATLAB with the following statements:

```
>> s=fotf('s'); Gc=5+2*s^(-0.2)+3*s^0.6
```

which returns the $G_c(s) = \dfrac{3s^{0.8} + 5s^{0.2} + 2}{s^{0.2}}$.   □

Example 12 The transfer function $G(s) = \dfrac{(s^{0.3} + 3)^2}{(s^{0.2} + 2)(s^{0.4} + 4)(s^{0.4} + 3)}$

can easily be entered into the MATLAB environment with the following statements:

```
>> s=tf('s'); G=(s^0.3+3)^2/(s^0.2+2)/(s^0.4+4)/(s^0.4+3)
```

which gives $G(s) = \dfrac{s^{0.6} + 6s^{0.3} + 9}{s + 2s^{0.8} + 7s^{0.6} + 14s^{0.4} + 12s^{0.2} + 24}$.   □

Example 13 Suppose in the unity negative feedback system, the models are given by

$$G(s) = \frac{0.8s^{1.2} + 2}{1.1s^{1.8} + 0.8s^{1.3} + 1.9s^{0.5} + 0.4}, \quad G_c(s) = \frac{1.2s^{0.72} + 1.5s^{0.33}}{3s^{0.8}}.$$

The plant and controller can easily be entered and the closed-loop system can be directly obtained with the following commands:

```
>> G=fotf([1.1,0.8 1.9 0.4],[1.8 1.3 0.5 0],[0.8 2],[1.2 0]);
   Gc=fotf([3],[0.8],[1.2 1.5],[0.72 0.33]);
   GG=feedback(G*Gc,1)
```

and the result is obtained as

$$G(s) = \frac{0.96s^{1.59} + 1.2s^{1.2} + 2.4s^{0.39} + 3}{3.3s^{2.27} + 2.4s^{1.77} + 0.96s^{1.59} + 1.2s^{1.2} + 5.7s^{0.97} + 1.2s^{0.47} + 2.4s^{0.39} + 3}.$$

It can be seen from the above illustrations that, although both the plant and controller models are relatively simple, extremely complicated closed-loop models may be obtained. This makes the analysis and design of a fractional-order system a difficult task.   □

13.2.3 Stability Assessment of FOTFs

The stability assessment of fractional-order systems is different from those in integer-order systems. The only known stability assessment method is for commensurate-order systems. For commensurate-order systems with no time delay, where $\lambda = s^{\alpha}$, if the absolute values of the angles of all the poles of λ are larger than $\alpha\pi/2$, the system is stable [14]. The stable region of commensurate-order systems is shown in Figure 13.10.

A MATLAB function isstable() is written for assessing the stability of FOTFs. For a given FOTF, the greatest common divisor for the orders are found and denoted by α. To prevent extremely high-order system being obtained, the minimum value of α is assigned to 0.01, and approximate commensurate-order model can be obtained automatically. Then for $\lambda = s^{\alpha}$, the roots of polynomial of λ can be evaluated and validated. The stability assessments based on the positions of roots of λ are made. The syntax of the

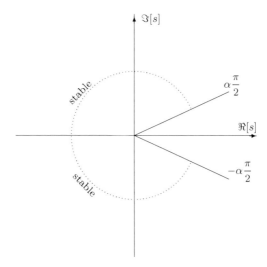

Figure 13.10 Stable region illustration for commensurate-order systems

function is $[K,\alpha,\varepsilon]$=isstable(G,a_0), where G is an FOTF object, a_0 is the minimum allowed greatest common divisor, with a default of 0.01. If the return variable K is one, the system G is stable, otherwise it is not. ε returns the error norm when the roots are substituted back to the polynomials:

```
function [K,alpha,err,apol]=isstable(G,a0)
if G.ioDelay~=0, error('delay system cannot be assessed'); end
if nargin==1, a0=0.01; end
a=G.na; a1=fix(a/a0); n=gcd(a1(1),a1(2));
for i=3:length(a1), n=gcd(n,a1(i)); end
alpha=n*a0; a=fix(a1/n); b=G.a; c(a+1)=b; c=c(end:-1:1);
p=roots(c); p=p(abs(p)>eps); err=norm(polyval(c,p));
plot(real(p),imag(p),'x',0,0,'o')
apol=min(abs(angle(p))); K=apol>alpha*pi/2;
xm=xlim; xm(1)=0; line(xm,alpha*pi/2*xm)
```

Example 14 For the FOTF model
$$G(s) = \frac{-2s^{0.63} - 4}{2s^{3.501} + 3.8s^{2.42} + 2.6s^{1.798} + 2.5s^{1.31} + 1.5},$$
the system model can be entered into the MATLAB environment, and the stability of the system can be assessed with the following statements:

```
>> b=[-2,-4]; nb=[0.63,0];
   a=[2,3.8,2.6,2.5,1.5]; na=[3.501,2.42,1.798,1.31,0];
   G=fotf(a,na,b,nb); [key,alpha,err,apol]=isstable(G)
```

With the function call, the commensurate-order description, after assuming automatically by the program for $\alpha = 0.001$, can be rewritten as

$$G(\lambda) = \frac{-2\lambda^{630} - 4}{2\lambda^{3501} + 3.8\lambda^{2420} + 2.6\lambda^{1798} + 2.5\lambda^{1310} + 1.5}.$$

The poles of λ equation can be obtained as shown in Figure 13.11 (a), and the zoomed plot in the interested area is given in Figure 13.11 (b). It can be seen that all the poles of the system are located in the stable area, and the fractional-order system is stable.   □

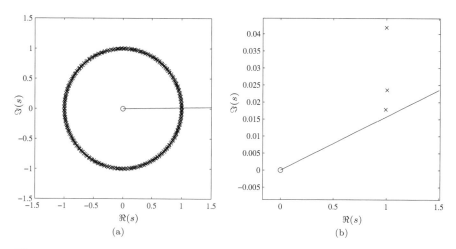

Figure 13.11 Pole locations and stability assessment: (a) pole locations and (b) zoomed plot

13.2.4 Numerical Time Domain Analysis

The solutions of fractional-order differential equations have been presented in the previous section. In particular, a closed-form algorithm has been given for linear fractional-order differential equations, with a MATLAB function fode_sol(). Based on the function, step responses of an FOTF object, and also time response of FOTF subject to arbitrary input signal, can be obtained with the following overload functions:

```
function y=step(G,t)
y=fode_sol(G.a,G.na,G.b,G.nb,ones(size(t)),t);
ii=find(t>G.ioDelay); lz=zeros(1,ii(1)-1);
y=[lz, y(1:end-length(lz))];
```

and

```
function y=lsim(G,u,t)
y=fode_sol(G.a,G.na,G.b,G.nb,u,t);
```

```
ii=find(t>G.ioDelay); lz=zeros(1,ii(1)-1);
y=[lz, y(1:end-length(lz))];
```

The syntaxes of the functions are y=step(G,t) and y=lsim(G,u,t) respectively, where G is an FOTF object, t is an evenly distributed time vector, and u is a vector of input samples. The syntaxes are quite similar those in the Control System Toolbox.

Example 15 Consider the numerical solution problem of the fractional-order differential equation

$$\mathscr{D}_t^{3.5}y(t) + 8\mathscr{D}_t^{3.1}y(t) + 26\mathscr{D}_t^{2.3}y(t) + 73\mathscr{D}_t^{1.2}y(t) + 90\mathscr{D}_t^{0.5}y(t) = 90\sin t^2.$$

From the given fractional-order differential equation, one can easily extract the FOTF object

$$G(s) = \frac{1}{s^{3.5} + 8s^{3.1} + 26s^{2.3} + 73s^{1.2} + 90s^{0.5}},$$

with the input signal $u(t) = 90\sin t^2$. The following MATLAB commands can then be used to find the time response of the system, and the solution and input signals are shown in Figure 13.12:

```
>> a=[1,8,26,73,90]; n=[3.5,3.1,2.3,1.2,0.5];
   G=fotf(a,n,1,0); t=0:0.002:10; u=90*sin(t.^2);
   y=lsim(G,u,t);
   subplot(211), plot(t,y); subplot(212), plot(t,u)
```

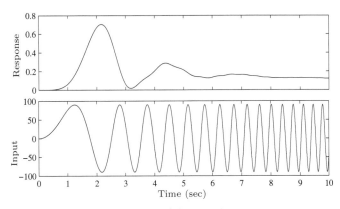

Figure 13.12 Input signal and solution of the equation

One can change the step-size of the solution process to different values and check whether consistent results can be obtained. If so, one may say that the solution is accurate. However, if they are not consistent, one has to reduce the step-sizes and perform the validation process again. □

13.2.5 Frequency Domain Analysis

For the FOTF $G(s)$ defined in (13.44), if one substitutes operator s by $j\omega$, complex gain $G(j\omega)$ can be found. To compare Bode plots, an overload function bode() for FOTF objects is written and placed in the @fotf directory such that

```
function H=bode(G,w)
if nargin==1, w=logspace(-4,4); end
j=sqrt(-1); H1=freqresp(j*w,G); H1=frd(H1,w);
if nargout==0, bode(H1); else, H=H1; end
```

with a low-level function given by

```
function H1=freqresp(w,G)
a=G.a; na=G.na; b=G.b; nb=G.nb; j=sqrt(-1);
for i=1:length(w)
    P=b*(w(i).^nb.); Q=a*(w(i).^na.); H1(i)=P/Q;
end
if G.ioDelay>0,
    A=abs(H1); B=angle(H1)-w1*G.ioDelay; H1=A.*exp(j*B);
end
```

Similarly, the Nyquist plot and Nichols chart of FOTFs can also be obtained easily with the following overload functions:

```
function nyquist(G,w)
if nargin==1, w=logspace(-4,4); end
H=bode(G,w); nyquist(H);
```

and

```
function nichols(G,w)
if nargin==1, w=logspace(-4,4); end
H=bode(G,w); nichols(H);
```

The facilities of the Control System Toolbox, such as the M and N circles, can be drawn immediately with the grid command.

Example 16 Consider again the FOTF model given in Example 14. The Bode plot of the system can be obtained as shown in Figure 13.13, with the following statements:

```
>> b=[-2,-4]; nb=[0.63,0]; w=logspace(-2,2);
   a=[2,3.8,2.6,2.5,1.5]; na=[3.501,2.42,1.798,1.31,0];
   G=fotf(a,na,b,nb); bode(G,w)
```

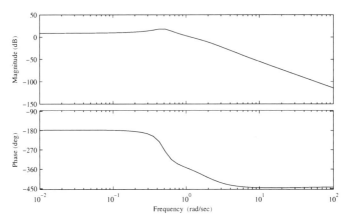

Figure 13.13 Bode plot of FOTF

It can be seen that the exact Bode plot of a fractional-order system can easily be drawn. One can also draw other frequency domain responses and easily use mouse facilities as in the Control System Toolbox functions. □

13.2.5.1 Norm Evaluations of FOTFs

Norm measures of systems are very important in robust controller design problems. Like in the cases of integer-order systems, the $\mathcal{H}_2$ and $\mathcal{H}_\infty$ norms of an FOTF $G(s)$ are defined, respectively, as

$$||G(s)||_2 = \frac{1}{2\pi j} \int_{-j\infty}^{j\infty} G(s)G(-s)\mathrm{d}s \tag{13.48}$$

and

$$||G(s)||_\infty = \sup_\omega |G(j\omega)|. \tag{13.49}$$

It can be seen that the evaluation of $||G(s)||_2$ involves a numerical integration problem, while the evaluation of $||G(s)||_\infty$ involves an optimization problem. An overload function `norm()` can be written, with the syntaxes `norm(G)` and `norm(G,inf)`, for the calculation of the $\mathcal{H}_2$ and $\mathcal{H}_\infty$ norms of the FOTF $G(s)$. Also for $\mathcal{H}_2$ norm evaluation, an error tolerance ε can be introduced, with `norm(G,ε)`:

```
function n=norm(G,eps0)
j=sqrt(-1); dx=1; f0=0;
if nargin==1, eps0=eps; end
if nargin==2 & ~isfinite(eps0) % H∞ norm
    f=@(x)[-abs(freqresp(j*x,G))];
    x=-fminsearch(f,0); n=abs(freqresp(j*x,G));
```

```
    else % ℋ₂ norm
       f=@(x)freqresp(x,G).*freqresp(-x,G);
       while (1)
          n=quadgk(f,-dx*j,dx*j)/(2*pi*j);
          if abs(n-f0)<eps0, break; else, f0=n; dx=dx*1.2;
    end, end, end
```

Example 17 Consider again the FOTF object defined in Example 14. The $\mathcal{H}_2$ and $\mathcal{H}_\infty$ norms of the system can be obtained with the following MATLAB statements:

```
>> b=[-2,-4]; nb=[0.63,0];
   a=[2,3.8,2.6,2.5,1.5]; na=[3.501,2.42,1.798,1.31,0];
   G=fotf(a,na,b,nb); n1=norm(G), n2=norm(G,inf)
```

from which follows that $n_1 = 7.3808$, and $n_2 = 8.6115$, respectively. ☐

13.3 Simulation Studies of Fractional-order Nonlinear Systems with Block Diagrams

A closed-form solution has been presented previously and the method can only be applicable to linear fractional-order systems. Many numerical methods for solving fractional-order nonlinear systems, however, require that the input signal to the system should be given first, before solution can be made. This requirement is problematic since, for a component in a closed-loop control system, the input signal injected to the component is not known. Thus the methods cannot be used in simulation of the closed-loop systems.

In integer-order nonlinear system simulation study, it is known that an integrator is very essential in defining relevant signals in constructing block diagrams. Similarly, for fractional-order systems, fractional-order integrator or differentiator may also be very useful in describing block diagrams. The Oustaloup's recursive filter presented in Section 12.1.2 can be used to define essential signals in the block diagrams. In this section, a Simulink block for fractional-order integrator and differentiator is constructed first, then, based on the block, a block diagram modeling strategy. Hints on validation processes for simulation results are also suggested.

Fractional order differentiators and integrators are essential components in block diagram based simulation. In this section, a fractional-order differentiator/integrator block is created.

13.3.1 Design of a Fractional-order Operator Block in Simulink

It can be seen that the Oustaloup recursive filter and the modified version presented in Section 12.1.2 are effective ways for evaluating fractional-order differentiations. Because of the orders of the numerator and denominator in the ordinary Oustaloup filter, it is likely to cause algebraic loops in simulation. Thus a low-pass filter should be appended to the filter. The block in Figure 13.14 (a) can be used for modeling fractional-order differentiators.

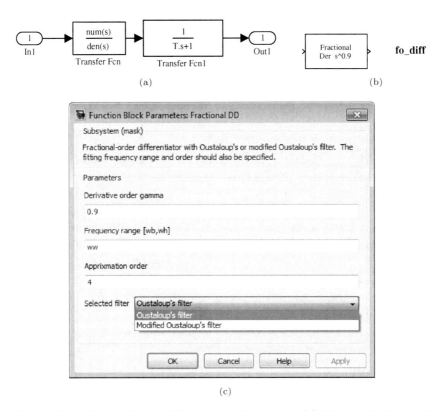

Figure 13.14 Fractional-order differentiator block design: (a) filter for fractional-order differentiator, (b) masked block, and (c) parameter dialog box

With the use of masking technique in Simulink [187], the designed block can be masked as shown in Figure 13.14 (b). The block thus designed is saved in fo_diff.mdl. Double click such a block and a dialog box appears as in Figure 13.14 (c). The corresponding parameters can be filled into the dialog box to complete the fractional-order differentiator block. The following code can be attached to the masked block:

```
wb=ww(1); wh=ww(2);
if key==1, G=ousta_fod(gam,n,wb,wh);
else, G=new_fod(gam,n,wb,wh); end
num=G.num{1}; den=G.den{1}; T=1/wh; str='Fractional\n';
if isnumeric(gam)
    if gam>0, str=[str, 'Der  s^' num2str(gam) ];
    else, str=[str, 'Int  s^{' num2str(gam) '}']; end
else, str=[str, 'Der  s^gam']; end
```

Currently, only two filters, Oustaloup filter and modified Oustaloup filter, can be used in the block. However, the users can add their own filter easily by slightly modifying the masking parameters in the block.

It is strongly recommended to use stiff equation solvers, such as ode15, ode23t, in the Simulink model to get fast simulation results.

13.3.2 Simulation Studies by Examples

In this part, three examples are given. In the first example, a linear fractional-order differential equation is studied again using block diagram method and consistent results can be obtained. In the second example, a nonlinear fractional-order system is studied, and in the third example, a nonlinear fractional-order delay system is simulated. It might be quite difficult, if not impossible, to study the last two examples with other methods.

Example 18 Consider the linear fractional-order differential equations in Example 15, which is again expressed as

$$\mathscr{D}_t^{3.5}y(t) + 8\mathscr{D}_t^{3.1}y(t) + 26\mathscr{D}_t^{2.3}y(t) + 73\mathscr{D}_t^{1.2}y(t) + 90\mathscr{D}_t^{0.5}y(t) = 90\sin t^2.$$

For linear fractional-order differential equations, the block diagram-based method is not as straightforward as the method used in Example 15. An auxiliary variable $z(t) = \mathscr{D}_t^{0.5}y(t)$ can be introduced, and the original differential equation can be rewritten as

$$z(t) = \sin t^2 - \frac{1}{90}\left[\mathscr{D}_t^3 z(t) + 8\mathscr{D}_t^{2.6}z(t) + 26\mathscr{D}_t^{1.8}z(t) + 73\mathscr{D}_t^{0.7}z(t)\right].$$

The Simulink block diagram shown in Figure 13.15 can be established based on the new equation. With stiff ODE solvers, the numerical solution to the problem can be found and the results are exactly the same as those in Figure 13.12.   □

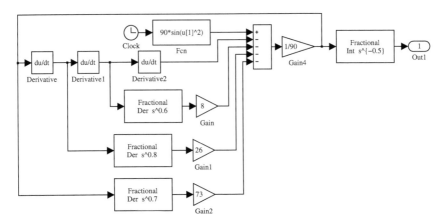

Figure 13.15 Simulink block diagram for the linear system

Example 19 Consider the nonlinear fractional-order differential equation

$$\frac{3\mathscr{D}^{0.9}y(t)}{3 + 0.2\mathscr{D}^{0.8}y(t) + 0.9\mathscr{D}^{0.2}y(t)} + \left|2\mathscr{D}^{0.7}y(t)\right|^{1.5} + \frac{4}{3}y(t) = 5\sin 10t.$$

From the given equation, the explicit form of $y(t)$ can be written as

$$y(t) = \frac{3}{4}\left[5\sin 10t - \frac{3\mathscr{D}^{0.9}y(t)}{3 + 0.2\mathscr{D}^{0.8}y(t) + 0.9\mathscr{D}^{0.2}y(t)} - \left|2\mathscr{D}^{0.7}y(t)\right|^{1.5}\right],$$

and from the explicit expression of $y(t)$, the block diagram in Simulink can be established as shown in Figure 13.16 (a). From the simulation model, the simulation results can be obtained as shown in Figure 13.16 (b). The results are verified by different control parameters in the filter, such as the orders of filters, and they give consistent results. ☐

Example 20 Consider a nonlinear fractional-order delay differential equation described as

$$\mathscr{D}^{1.5}y(t) + 3\mathscr{D}^{0.8}y(t - 0.1)y(t - 0.2) + 2y(t - 0.5) + y(t) = u(t),$$

where $u(t)$ is a step function.

To construct a block diagram model, one should first rewrite the plant model as an explicit expression of $y(t)$ such that

$$y(t) = u(t) - \mathscr{D}^{1.5}y(t) - 3\mathscr{D}^{0.8}y(t - 0.1)y(t - 0.2) - 2y(t - 0.5),$$

then the Simulink model for the closed-loop system can be constructed as shown in Figure 13.17 (a), and the solution to the delay equation is shown in Figure 13.17 (b). ☐

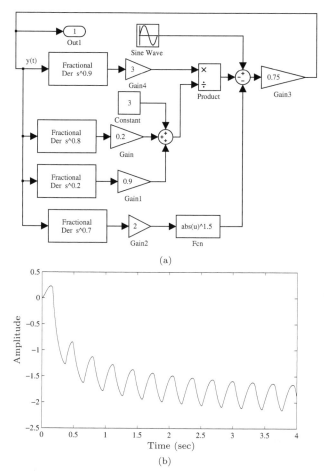

Figure 13.16 Simulink description and simulation results: (a) Simulink model and (b) simulation results

13.3.3 Validations of Simulation Results

It is known that if the parameters in simulation processes are not properly chosen, then the results obtained with Simulink may not be accurate. Sometimes the numerical results may be miseading [182]. Thus the simulation results obtained with the Simulink model must be validated. There are many ways to validate simulation results obtained directly with the block diagram based method. Like the validation process for other Simulink models, the simulation control parameters such as ODE solvers and relative error tolerance can be modified to see whether consistent results can be obtained. If not, one should specify a smaller error tolerance and validate again. Also,

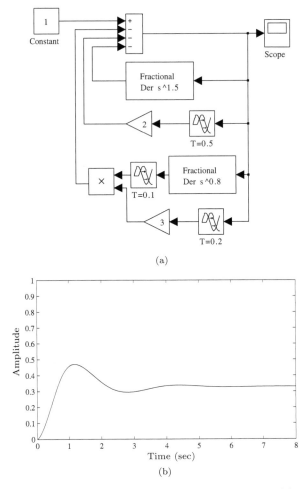

(a)

(b)

Figure 13.17 Simulation analysis of fractional-order delay system: (a) Simulink description of a nonlinear delay differential equation and (b) output signal

for the fractional-order differentiator blocks, the interesting frequency range $(\omega_{\mathrm{b}}, \omega_{\mathrm{h}}])$ and the order N should also be validated.

13.4 Optimum Controller Design for Fractional-order Systems

Many controller design methods have been discussed earlier in Chapters 4–10. In this section, an optimal controller design method and tool are presented.

First we shall pose the key problem of optimality: what kind of controller can be considered "optimal." Then MATLAB based solution methods on optimal controller design are given.

13.4.1 Optimum Criterion Selection

Varieties of optimal criteria for optimal control exist both in control education and real-world controller design. For instance, for linear-quadratical (LQ) optimal control systems, the most widely used criterion is the so-called LQ-criterion, defined as follows:

$$J = \frac{1}{2}\boldsymbol{x}^{\mathrm{T}}(t_{\mathrm{f}})\boldsymbol{S}\boldsymbol{x}(t_{\mathrm{f}}) + \frac{1}{2}\int_{t_0}^{t_{\mathrm{f}}} \left[\boldsymbol{x}^{\mathrm{T}}(t)\boldsymbol{Q}(t)\boldsymbol{x}(t) + \boldsymbol{u}^{\mathrm{T}}(t)\boldsymbol{R}(t)\boldsymbol{u}(t)\right]\mathrm{d}t, \quad (13.50)$$

where the matrices $\boldsymbol{S}$, $\boldsymbol{Q}$, and $\boldsymbol{R}$ are weighting matrices. Since the criterion leads to a Riccati differential equation, which is usually difficult to solve, the infinite-time period optimal control, i.e., setting $t_{\mathrm{f}} \to \infty$, referred to as the regulator problem (LQR), is used. However by introducing regulator problems, the physical meanings of tracking properties are lost.

There are advantages and disadvantages in using LQ criterion. The LQR problem has a good mathematical formula, and usually has closed-form solutions to the problem, which makes LQR a widely used approach. However, by introducing the LQR problem, there is no direct relationship with responses and tracking errors. Also there is no widely acceptable ways in selecting the $\boldsymbol{Q}, \boldsymbol{R}, \boldsymbol{S}$ weighting matrices. Thus the optimality of the problem is superficial and depends heavily upon the selection of the weighting matrices. When the matrices are poorly chosen, the designed controller is useless, and sometimes very bad system behaviors may be observed. Thus LQR-type "optimal" controllers are not suitable for servo control problems.

For servo control problems, the tracking error $e(t)$ is a meaningful specification, and tracking error based criteria, for instance, the ISE criterion, the ITAE criterion, IAE criterion, and finite-time ITAE criterion, should be used to assess the quality of tracking. These criteria are defined as follows:

$$J_{\mathrm{ISE}} = \int_0^\infty e^2(t)\mathrm{d}t, \quad J_{\mathrm{ITAE}} = \int_0^\infty t|e(t)|\mathrm{d}t,$$

$$J_{\mathrm{IAE}} = \int_0^\infty |e(t)|\mathrm{d}t, \quad J_{\mathrm{FT-ITAE}} = \int_0^{t_{\mathrm{f}}} t|e(t)|\mathrm{d}t. \quad (13.51)$$

It should be noted that there is no closed-form solutions to the criteria such as IAE, ITAE, and finite-time ITAE: calculations of these criteria should be made upon simulation results. Although the closed-form solution to the ISE problems exists, there are certain limitations. For instance, if there exists

nonlinear elements, such as actuator saturation, the closed-form solutions
to the ISE criterion do not exist. Again, numerical computations should be
performed instead to evaluate the ISE criterion.

Example 21 Suppose that the plant model is given by $G(s) = \dfrac{1}{(s+1)^5}$,
the best PID controller for ISE criterion can be found as $G_c(s) = 1.5644 + 0.6751s^{-1} + 4.6666s$, while the best for finite-time ITAE criterion is $G_{c2}(s) = 1.1900 + 0.3292s^{-1} + 1.4879s$, with $t_f = 30$. The step responses under the two controllers are obtained as shown in Figure 13.18.

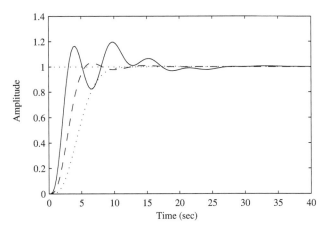

Figure 13.18 Comparisons on closed-loop step responses under the two controllers, *solid line* for optimal ISE controller, *dashed line* for FT-ITAE controller, and *dotted line* for FT-ITAE controller with saturation

It should also be noted that the above controllers may yield an extremely large input signal to the plant at the initial time, which may cause damage to the hardware. In practical control systems, actuator saturation should be considered. A PID controller for finite-time ITAE criterion under $|u(t)| \leqslant 3$ can be redesigned as $G_{c3}(s) = 1.3718 + 0.2863s^{-1} + 1.7850s$, and the closed-loop step response is also shown in Figure 13.18. The redesigned system is a compromise in control quality; however, it is much better than the above designed ISE optimal controller.

It is noted that the ISE criterion has certain disadvantages. Since the errors at any time are treated equally, oscillations in the responses are allowed. However with the use of ITAE criterion, weighting function of t is imposed on the error signal which penalizes the error for large values of time and forces the output signal approaches to the steady-state value as fast as possible. The illustration of the ITAE curve for the example is shown in Figure 13.19.

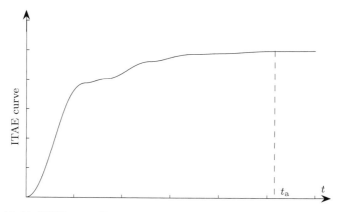

Figure 13.19 ITAE curve illustration

The controller parameters and the ITAE values for different selections of finite-time t_f are given in Table 13.1. Since the integrand is non-negative, the ITAE curve is a monotonically increasing function. Assume that when the ITAE curve goes flat, the time is denoted as t_a. For instance, in this example, $t_a = 25$. Then for the t_f selection of $t_f \in (t_a, 2t_a)$, the optimal controller parameters and ITAE values are quite close. One may say that the selected value of t_f is proper. □

Table 13.1 Controller comparisons for different selections of t_f

t_f	K_p	K_i	K_d	ITAE	t_f	K_p	K_i	K_d	ITAE
15	1.3602	0.27626	1.9773	9.3207	20	1.4235	0.27228	2.0731	9.7011
25	1.3524	0.26555	1.8846	9.945	30	1.3506	0.26454	1.8669	9.9814
35	1.3488	0.26395	1.859	10.008	40	1.3457	0.26363	1.8495	10.016
45	1.346	0.26357	1.8493	10.019	50	1.3452	0.26351	1.8472	10.021
55	1.345	0.26349	1.8467	10.021	60	1.345	0.26349	1.8467	10.021

It is concluded that the finite-time ITAE criterion, with proper choice of finite-time t_f, is the most suitable criterion for servo control problems.

13.4.2 Optimal Controller Design via Optimizations

The mathematical formulation of the unconstrained optimization problem is

$$\min_{x} f(x), \tag{13.52}$$

where $x = [x_1, x_2, \cdots, x_n]^\mathrm{T}$ is referred to as the decision variable and the scalar function $f(x)$ is referred to as the objective function. The interpretation to the formula is finding the vector x such that the objective function $f(x)$ is minimized. If a maximization problem is expected, the objective function can be changed to $-f(x)$, such that it can be converted to a minimization problem. A MATLAB function `fminsearch()` is provided using the well established simplex algorithm [183]. The syntax of the function is `[x,fopt,key,c]=fminsearch(fun,x0, OPT)` where `fun` is a MATLAB function or an anonymous function to describe the objective function. The variable x_0 is the staring point of searching method. The argument `OPT` contains the further control options for the optimization process.

Example 22 For an objective function $z = f(x, y) = (x^2 - 2x)\mathrm{e}^{-x^2-y^2-xy}$, where the minimum point is required, one should first introduce a vector x for the unknown variables x and y. One may select $x_1 = x$ and $x_2 = y$. The objective function can be rewritten as $f(x) = (x_1^2 - 2x_1)\mathrm{e}^{-x_1^2-x_2^2-x_1 x_2}$. The objective function can be expressed as an anonymous function such that

```
>> f=@(x)[(x(1)^2-2*x(1))*exp(-x(1)^2-x(2)^2-x(1)*x(2))];
```

If one selects an initial search point at $[0,0]$, the minimum point can be found with the following statements:

```
>> x0=[0; 0]; x=fminsearch(f,x0)
```

then the solution obtained is $x = [0.6110, -0.3055]^\mathrm{T}$.   □

With the use of the MATLAB Optimization Toolbox, other types of optimization problems, such as constrained optimization problems, can easily be solved with existing functions. Global optimal controllers can also be found with genetic algorithm toolboxes [188, 189].

Example 23 Consider a time varying plant model given by
$$\ddot{y}(t) + \mathrm{e}^{-0.2t}\dot{y}(t) + \mathrm{e}^{-5t}\sin(2t + 6)y(t) = u(t),$$
where $u(t)$ is the signal generated by the preceding controller. One wishes to design an optimal PID controller, which is followed by an actuator saturation with bounds of ± 2. This kind of controller may not be easy to design using traditional methods.

The Simulink model is constructed as shown in Figure 13.20, where the control structure is represented. Also the ITAE integral is expressed as an output port. A MATLAB function is written to describe the objective function as follows, if one selects $t_f = 10$:

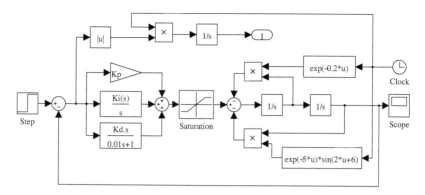

Figure 13.20 PID control for a time varying plant model

```
function y=my_obj(x)
assignin('base','Kp',x(1)); assignin('base','Ki',x(2));
assignin('base','Kd',x(3));
[t,x,y1]=sim('my_sys',[0,5]); y=y1(end);
```

The optimal PID controller can be designed automatically with the following statements:

```
>> x=fminsearch(@my_obj,rand(3,1));
```

The best PID obtained is $G_{\text{PID}}(s) = 121.1324 + 29.4108s$, which is in fact a PD controller. The step response of the closed-loop systems under PD controller is shown in Figure 13.21. It can be seen that the controller can easily be designed with optimization methods.   □

13.4.3 Optimum $PI^\lambda D^\mu$ Controller Design

According to [52], the $PI^\lambda D^\mu$ controller is described as

$$G_c(s) = K_p + K_i s^{-\lambda} + K_d s^\mu, \qquad (13.53)$$

where λ, μ are not necessarily integers. Since there are two more tuning knobs, λ and μ, than the integer-order PID controller, the behavior under $PI^\lambda D^\mu$ controller may have superior properties compared to ordinary PID controllers.

From the previous examples, it can be seen that to design an optimal controller, the following steps should be taken:

- Design a Simulink model, where the feedback control structure should be described. Also the criterion, such as finite-time ITAE criterion, should be included and given in the block diagram, as an output port.

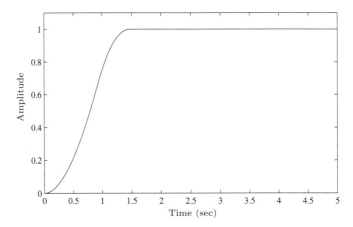

Figure 13.21 Step response with PD controller

- Write an objective function in MATLAB, in the format shown in the previous example.
- Start the optimization process, by calling **fminsearch()**, or others, to find the optimal controllers numerically.

Example 24 Consider again the plant model $G(s) = 1/(s + 1)^5$, with actuator saturation of $|u(t)| \leqslant 2$, studied in Example 21. The Simulink model with a $\text{PI}^\lambda \text{D}^\mu$ controller can be established, as shown in Figure 13.22.

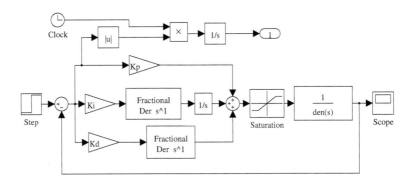

Figure 13.22 Simulink model with fractional-order PID controller

An M-function for the objective function can be written as

```
function y=my_obj1(x)
assignin('base','Kp',x(1)); assignin('base','Ki',x(2));
```

```
assignin('base','Kd',x(3)); assignin('base','L',x(4));
assignin('base','M',x(5));
[t,x,y1]=sim('cxfpid',[0,15]); y=y1(end);
```

In the model, ITAE integration is also implemented and connected to output port 1. In the simulation model, it is suggested to append an integer-order integrator to reduce steady-state error. The following command to find optimal parameters

```
>> x=fminsearch(@my_obj1,rand(5,1))
```

returns a controller $G_c(s) = 0.1138 + 0.5557s^{-0.7807} + 2.1543s^{0.6057}$. Also the obtained x can be used as an initial vector and function `fminsearch()` can be run several times to find better parameters. The step response under such a controller is obtained as shown in Figure 13.23. It can be seen that with fractional-order PID controller, the output of the system is satisfactory. □

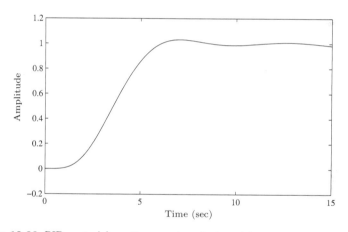

Figure 13.23 PID control for a time varying plant model

It is concluded from this example that one can easily set up a Simulink model for optimal controller design, using the procedure shown in the example, then issue MATLAB commands to tune optimally the parameters of fractional-order PID controllers or other types of fractional-order controllers. A MATLAB graphical user interface OCD [183] can be used directly for controller design problems.

13.5 Summary

In this chapter, the complete process of modeling, analysis, simulation, and design of fractional-order control systems is presented, with reproducible MATLAB code and illustrative examples. This chapter is a self-contained one, with easy-to-follow presentations, which makes it easier for readers to begin research on fractional-order systems.

Computational algorithms and MATLAB implementations of Mittag–Leffler functions, fractional-order differentiations, and linear fractional-order differential equation solutions are presented first, then analytical solutions to certain types of equations are explored.

For linear systems, an FOTF class is designed in MATLAB and a series of overload functions are prepared, which makes the analysis of FOTF systems as easy as the integer-order transfer functions using Control System Toolbox of MATLAB.

For nonlinear fractional-order systems, a block diagram based simulation strategy is presented, with the use of Oustaloup and modified Oustaloup filters. Illustrative examples of linear systems, nonlinear systems, and delay systems with fractional-order terms are presented, again with easy-to-follow procedures. With the use of the strategy, the reader can simulate systems with more complicated structures.

In the final section, a controller parameter tuning methodology based on numerical optimization is adopted. The objective function selection problem is also addressed and it is concluded that finite-time ITAE criterion is suitable for servo control problems. A fractional-order PID controller optimization procedure is demonstrated and again the method can be used for the design of other systems.

Part VI
Real Applications

Chapter 14
Systems Identification

14.1 Introduction

It has been experimentally observed or analytically found that both the time domain and frequency domain behaviors of some linear systems and processes do not fit the standard laws, *i.e.*, exponential evolution in time domain or integer-order slopes in their frequency responses. In the time domain, it has been shown that these complicated dynamics can be described by, (*i.e.*, the solutions of the constitutive equations are) generalized hyperbolic functions, $\mathscr{F}_{\alpha,\beta}^k(z)$, defined as

$$\mathscr{F}_{\alpha,\beta}^k(z) = C \sum_{n=0}^{\infty} \frac{k^n z^{\alpha n + \beta}}{(\alpha n + \beta)!}. \tag{14.1}$$

In particular, the Mittag–Leffler function in two parameters is defined as

$$\mathscr{E}_{\alpha,\beta}(z) = \sum_{n=0}^{\infty} \frac{z^n}{\Gamma(\alpha n + \beta)}, \tag{14.2}$$

from which we can obtain the standard exponential, hyperbolic, or time-scaling functions as particular cases.

On the other hand, and correspondingly, the non-integer-order slopes in the frequency responses can be fitted by irrational-order transfer functions, transfer functions constructed as products of zeros and poles of fractional power, or ratios of polynomials in s^α, $0 < \alpha < 1$, in agreement with the expressions of the Laplace transforms of the mentioned functions.

All these systems and processes have in common the presence of phenomena (anomalous relaxation and diffusion, mainly) that can be seen as incorporating memory into the systems, and, in a formal way, memory can be incorporated into the constitutive equations through a causal convolution. All these ways lead us to consider fractional calculus, as an appropriate tool to describe phenomenologically the richness of dynamic features exhibited by the system, since we know that:

- The solutions of integro-differential equations of fractional-order (FDEs) are the generalized hyperbolic or Mittag–Leffler functions.
- To apply the Laplace transform to these FDEs yields transfer functions that are ratios of polynomials in s^α, $0 < \alpha < 1$, where α is the order of the fractional-order derivative or integral.

Considering these points, we can reformulate the constitutive equations of the systems using fractional calculus in order to agree with the time domain or frequency domain behaviors.

The aim of this chapter is to show how we can generalize some methods and techniques of the control theory in order to manage this class of systems. For this purpose, in Section 14.2, we describe an algorithm for frequency domain identification of commensurate-order systems; in Section 14.3 we apply the identification algorithm to model an electrochemical process, and in Section 14.4 to a flexible structure.

14.2 Frequency Domain Identification of Commensurate-order Systems

Consider a commensurate-order system described by the transfer function

$$H(s, \boldsymbol{\theta}) = \frac{\displaystyle\sum_{k=0}^{n} a_k s^{k\alpha}}{\displaystyle\sum_{k=0}^{m} b_k s^{k\alpha}} , \qquad (14.3)$$

with $m < n$ (the usual assumption for physical causal systems), and $\boldsymbol{\theta}$ the set of parameters to be identified, where

$$\boldsymbol{\theta} = [a_0, a_1, \cdots, a_n, b_0, b_1, \cdots, b_m]. \qquad (14.4)$$

As in [57, 190], we will use the cost function,

$$J = \frac{\displaystyle\int_0^\infty [y(t) - \tilde{y}(t)]^2 \, \mathrm{d}t}{\displaystyle\int_0^\infty [h(t)]^2 \, \mathrm{d}t} = \frac{\displaystyle\int_{-\infty}^\infty \left[H(\omega) - \tilde{H}(\omega)\right]\overline{\left[H(\omega) - \tilde{H}(\omega)\right]}\frac{\mathrm{d}\omega}{\omega^2}}{\displaystyle\int_{-\infty}^\infty H(\omega)\overline{H(\omega)}\,\mathrm{d}\omega} , \qquad (14.5)$$

where $\overline{[\cdot]}$ means the complex conjugate, $h(t)$ is the impulse response of the system with transfer function $H(s)$, $y(t)$ is the step response, $H(\omega)$ the measured frequency response, $\tilde{H}(\omega)$ the frequency response of the fitted model, and the last equality results from the application of Parseval's Theorem. Also, the iterative algorithm proposed by Shanathanan and Koerner [191, 192] will be used for minimizing (14.5). The motivations for the choice of the cost function and the detailed description of the iterative algorithm are in [57, 190], and here we only try to note that these transfer function models have important advantages if compared with the case of $\alpha = 1$. These are:

- We can achieve better fittings of the frequency response data with more simple models, (*i.e.*, with lower n and m) for systems whose frequency responses exhibit non-integer-order slopes.
- We can achieve the same degree of fitting when the frequency response exhibit integer-order slopes.
- The algorithm is numerically more stable than in the case of $\alpha = 1$.

The reasons for the above are:

- In the absence of noise, because these transfer function models can be exact models for that kind of systems, we can find the value of α that fits the data exactly.
- We can choose $\alpha = 1$ in the identification algorithm or, on the other hand, choose $\alpha = 1/q$, q being an integer, only the powers multiples of αq having significant coefficients after identification.
- Because $\alpha < 1$, the condition number of the correlation matrix will be, for the same range of frequency, always smaller than in the usual case of $\alpha = 1$, diminishing the risk of singularity of the matrix.

14.3 Electrochemical Process

One of the most useful methods to study electrochemical processes is the impedance measurement method [193]. This method, shown in Figure 14.1, consists of applying small sinusoidal perturbations (current or voltage) to excite the electrochemical cell system, which brings advantages in terms of the solution of the relevant mathematical equations, since it is possible to use limiting forms of this equations which are normally linear. Measuring the impedance or the admittance (magnitude and phase) allows the analysis of the electrode process in relation to contributions from diffusion, kinetics, double layer, *etc.* Usually, for comparison with the electrochemical cell equivalent electrical circuitsare used whose elements represent the relevant phenomena in the process. Among these equivalent circuits, the most frequently used is the so-called *Randle's equivalent circuit*, shown in Figure 14.2, where C_d is a pure capacitor representing the double layer, R is the uncompensated resistance which is, usually, the solution resistance between the working and reference electrodes, and Z_f is the impedance of the Faradic process.

This last one element, the Faradic impedance, can be subdivided into a purely resistive element representing the resistance to charge transfer and an element representing the difficulty of mass transport of the electroactive species, called the Warburg impedance. We can extend the validity of the Randle's equivalent circuit if we take into account the influence in the double layer of the rugosity and porosity of the electrodes, representing it by the so-called *constant phase element* (CPE) [194–200], thus obtaining the equivalent circuit shown in Figure 14.3.

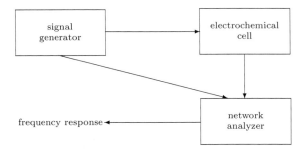

Figure 14.1 Impedance measurement

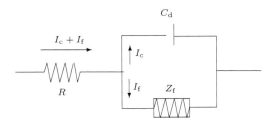

Figure 14.2 Randle's equivalent circuit

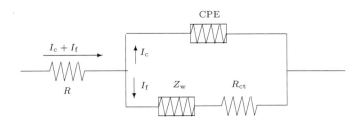

Figure 14.3 Modified Randle's equivalent circuit

In the admittance form, these elements can be described by

$$G_\Omega = \frac{1}{R_\Omega}, \quad G_{\mathrm{ct}} = \frac{1}{R_{\mathrm{ct}}}, \quad Y_{\mathrm{w}} = a\,(\mathrm{j}\omega)^{1/2}, \quad Y_{\mathrm{CPE}} = b\,(\mathrm{j}\omega C)^{\alpha}, \quad (14.6)$$

with $\alpha = 0.5$ for porous electrodes and $\alpha = 1$ for smooth electrodes. As in [201] for viscoelastic models and using the same symbology, we will use the generalized Randle's equivalent circuit shown in Figure 14.4, where FE_i denotes a fractional-order element defined by its admittance as

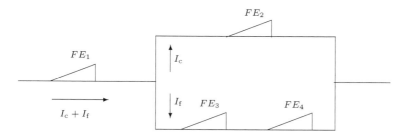

Figure 14.4 Generalized Randle's equivalent circuit

$$FE_i = FE\,(b_i, \tau_i, \alpha_i) = b_i\,(\mathrm{j}\omega\tau_i)^{\alpha_i}, \quad (14.7)$$

from which all the elements of the traditional Randle's equivalent circuit can be obtained as particular cases. For this circuit, and using the given notation for fractional-order operators, the current/voltage relations can be described by the following equations:

$$i_c(t) = b_2 \tau_2^{\alpha_2} \mathscr{D}^{\alpha_2} v_2(t), \quad (14.8)$$

$$b_3 \tau_3^{\alpha_3} \mathscr{D}^{\alpha_3} i_f(t) + b_4 \tau_4^{\alpha_4} \mathscr{D}^{\alpha_4} i_f(t) = b_3 \tau_3^{\alpha_3} b_4 \tau_4^{\alpha_4} \mathscr{D}^{\alpha_3 + \alpha_4} v_2(t), \quad (14.9)$$

$$i(t) = i_c(t) + i_f(t) = b_1 \tau_1^{\alpha_1} \mathscr{D}^{\alpha_1} v_1(t), \quad (14.10)$$

which adequately combined and using the law of exponents for the fractional-order differential operators, gives us the equation that can describe the general relation between the applied voltage and current. That is:

$$i(t) + k_1 \mathscr{D}^{\alpha_4 - \alpha_3} i(t) + k_2 \mathscr{D}^{\alpha_2 - \alpha_1} i(t) + k_3 \mathscr{D}^{\alpha_4 - \alpha_1} i(t) + k_4 \mathscr{D}^{\alpha_4 + \alpha_2 - \alpha_3 - \alpha_1} i(t)$$
$$= k_5 \mathscr{D}^{\alpha_2} v(t) + k_6 \mathscr{D}^{\alpha_4} v(t) + k_7 \mathscr{D}^{\alpha_4 + \alpha_2 - \alpha_3} v(t). \quad (14.11)$$

Now, we will apply the described identification algorithm to show how, by using a transfer function model in $s^\alpha, 0 < \alpha < 1$, we can obtain better fittings to impedance data than by using the traditional s-transfer function model. For this purpose, we simulate the experimental data giving values to the elements of the general equivalent circuit. The values taken from [190] are:

$$FE_1 \rightarrow b_1\tau_1^{\alpha_1} = 100; \alpha_1 = 0 \Longrightarrow \text{resistor,}$$
$$FE_2 \rightarrow b_2\tau_2^{\alpha_2} = 10^4; \alpha_2 = 0.75 \Longrightarrow \text{CPE,}$$
$$FE_3 \rightarrow b_3\tau_3^{\alpha_3} = 10^5; \alpha_3 = 0 \Longrightarrow \text{resistor,}$$
$$FE_4 \rightarrow b_4\tau_4^{\alpha_4} = \sqrt{2} \times 10^3; \alpha_4 = 0.5 \Longrightarrow \text{Warburg impedance.}$$

This corresponds to the following fractional-order differential equation:

$$100\mathscr{D}^{1.25}i(t) + 1.414\mathscr{D}^{0.75}i(t) + 1.001 \times 10^4\mathscr{D}^{0.5}i(t) + 141.4i(t)$$
$$= \mathscr{D}^{1.25}v(t) + 0.0141\mathscr{D}^{0.75}v(t) + 0.1\mathscr{D}^{0.5}v(t),$$

and the following impedance function:

$$Z(s) = \frac{V(s)}{I(s)} = \frac{100s^{1.25} + 1.414s^{0.75} + 1.001 \times 10^4 s^{0.5} + 141.4}{s^{0.5}\left(s^{0.75} + 0.0141s^{0.25} + 0.1\right)}.$$

With the use of MATLAB `invfreq()` function discussed in Chapter 12, a fourth-order model can be obtained, with a warning of a huge condition number of 10^{39}. The identified integer-order model is

$$G_1(s) = \frac{100.2s^4 + 1.218 \times 10^7 s^3 + 1.975 \times 10^{11}s^2 + 2.457 \times 10^{14}s + 1.106 \times 10^{16}}{s^4 + 1.195 \times 10^5 s^3 + 1.812 \times 10^9 s^2 + 1.06 \times 10^{12}s + 3.677 \times 10^{10}}.$$

The Bode plots of the identified model as well as the original model are shown in Figure 14.5. It is immediately seen that the fitting is not good.

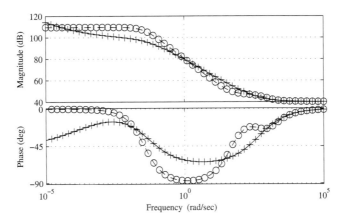

Figure 14.5 Identification for $\alpha = 1$ with `invfreqs()` function, with *circles* for original model, and *plus signs* for identified model

The identification results in the frequency range of 10^{-5} Hz to 10^5 Hz for $\alpha = 1$ shown in Figure 14.6. For $\alpha = 0.25$ and 0.5, the fitting results are virtually the same, as shown in Figure 14.7.

From these results we can observe that:

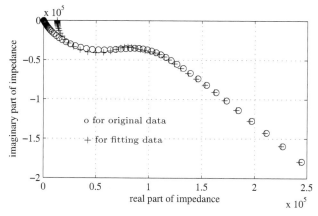

Figure 14.6 Identification for $\alpha = 1$

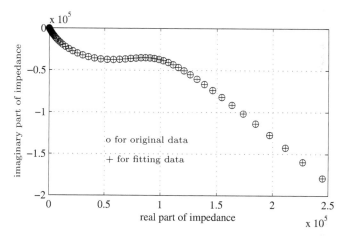

Figure 14.7 Identification for $\alpha = 0.5$ and $\alpha = 0.25$

- The impedance plots agree with the typical shape for electrochemical processes [193] where we can find two different regions:
  - The mass transfer region, characterized by a unit slope curve. For $\omega \to 0$ the reaction is controlled solely by diffusion which predominates the effects of the Warburg impedance.
  - The kinetic control region, characterized by a semicircular shape. For $\omega \to \infty$ the control is purely kinetic and the electrical analogy is a parallel R-C combination.
- In both regions the deviations from the typical shapes are due to the presence of the CPE element instead of a pure or ideal capacitor.

- The fitting is better for $\alpha < 1$. In this case, selecting $\alpha = 0.25$, an exact divisor of 0.5 and 0.75, and the order of the fractional elements (the order of the fractional-order derivatives in the constitutive equation), we can obtain the real transfer function of the system, in the form

$$\widehat{Z}(s) = \frac{\begin{array}{c} 100s^{1.25} - 1.108\times10^{-7}s + 1.414s^{0.75} + 1.001\times10^{4}s^{0.5} \\ +1.282\times10^{-9}s^{0.25} + 141.4 \end{array}}{\begin{array}{c} s^{1.25} - 7.45\times10^{-12}s + 0.01414s^{0.75} + 0.1s^{0.5} \\ +1.179\times10^{-14}s^{0.25} - 2.3\times10^{-16} \end{array}} \approx Z(s).$$

On the other hand, we can verify that the algorithm is more stable numerically for $\alpha < 1$. In our example we verified that the condition numbers of the matrix are 5.39×10^{21} for $\alpha = 1$, 3.6483×10^{13} for $\alpha = 0.5$, and 1.1347×10^{12} for $\alpha = 0.25$.

14.4 Flexible Structure

The system under test is a typical experimental prototype of flexible robotic arms: a beam with transversal distributed flexibility constrained to move on a horizontal plane by the action of piezoelectric crystals, *i.e.*, a clamped-free Euler–Bernouilli beam with piezoelectric actuators. It is schematically illustrated in Figure 14.8, and the relevant parameters and characteristics are given in Table 14.1 for the beam, Table 14.2 for the actuators, and Table 14.3 for the accelerometer.

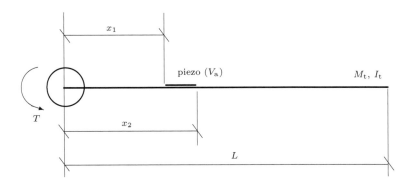

Figure 14.8 Device under test

In the range $0.1 \sim 100\,\mathrm{Hz}$, the frequency response of the system has been measured considering the voltage exciting the crystals as the input and the

Table 14.1 Beam parameters

Tip inertia, I_t	$\rightarrow 0$	Linear mass density, ρ	0.7245 kg/m
Hub inertia, I_h	$\rightarrow \infty$	Stiffness modulus, EI	22.2589 N-m^2
Length, L	1.3 m	Thickness, t_b	4×10^{-3} m
Widht, h	0.07 m		

Table 14.2 Piezoelectric actuator parameters

Type	PZT4D Vernitron
Thickness, t_a	5×10^{-4} m
Width, w	0.07 m
Charge constant, d_{31}	135×10^{-12} m/V
Young modulus, E_a	7.5×10^{10} m
Position in the beam	$x_1 = 8 \times 10^{-3}$ m; $x_2 = 30 \times 10^{-3}$ m

Table 14.3 Accelerometer characteristics

Type	3148 DYTRAN	Frequency range	$1 - 5000$ Hz ($\pm 10\%$)
Weight, M_t	48 g	Sensitivity	100 mV/G ($\pm 5\%$)

acceleration in the free end as the output. The experimental system used for the measurement is shown in Figure 14.9.

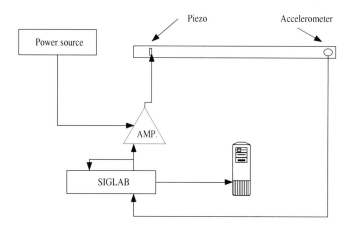

Figure 14.9 Measurement test bench for experiments

After measurement, we have applied the identification algorithm for $\alpha = 1$ and $\alpha = 0.5$ for several values of n and m. The results are shown in Figure 14.10 (for $\alpha = 0.5$ and $n = m = 10$) and Table 14.4.

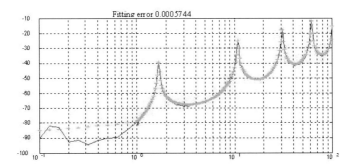

Figure 14.10 Identified response for $\alpha = 0.5$ and measured response

Table 14.4 Identification errors over measurements

n,n	$\alpha = 1$	$\alpha = 0.5$	n,n	$\alpha = 1$	$\alpha = 0.5$
10, 10	3.1×10^{-3}	5.7×10^{-4}	12, 12	1.8×10^{-3}	6.0×10^{-4}
18, 18	1.7×10^{-3}	3.2×10^{-4}	20, 20	2.0×10^{-3}	9.6×10^{-5}

We can verify that the fittings with $\alpha = 0.5$ are better. We will try to show why by analyzing the physical modeling the system.

The physical model of the system can be described by [202, 203]

$$\frac{\partial^2}{\partial x^2}\left[EI\frac{\partial^2 y(x,t)}{\partial x^2} - C_a V_a(x,t)\right] + \rho A\frac{\partial^2 y(x,t)}{\partial t^2} = 0 \qquad (14.12)$$

and the boundary conditions

$$y(0,t) = 0, \qquad\qquad\qquad\qquad\qquad\qquad\qquad (14.13)$$

$$EI\frac{\partial^2 y(0,t)}{\partial x^2} - I_h\frac{\partial^3 y(0,t)}{\partial t^2 \partial x} + T(t) = 0, \qquad (14.14)$$

$$EI\frac{\partial^2 y(L,t)}{\partial x^2} + I_t\frac{\partial^3 y(L,t)}{\partial t^2 \partial x} = 0, \qquad\qquad (14.15)$$

$$EI\frac{\partial^3 y(L,t)}{\partial x^3} - M_t\frac{\partial^2 y(L,t)}{\partial t^2} = 0, \qquad\qquad (14.16)$$

where $C_a V_a(x,t)$ is the moment due to actuators, $C_a = E_a d_{31} w(t_a + t_b)/2$ is the actuator voltage constant, and $V_a(\cdot)$ is the actuator voltage.

Proceeding as in [203] we can obtain the transfer function between the tip acceleration and the actuator voltage (see [57] for more details),

$$G_{\mathrm{a},V_{\mathrm{a}}}(L,s) = \frac{Y(L,s)}{V_{\mathrm{a}}(s)} = \frac{s^2 N(s)}{D(s)}, \tag{14.17}$$

with

$$
\begin{aligned}
N(s) = C_{\mathrm{a}}\beta^2 \Big[&-\cos(\beta L)\cos(\beta x_2) - \sin(\beta L)\sin(\beta x_2) + \cosh(\beta L)\cos(\beta x_1) \\
&+ \sin(\beta L)\sin(\beta x_1) + \sin(\beta L)\sin(\beta x_1) + \cosh(\beta L)\cosh(\beta x_2) \\
&+ \cos(\beta L)\cos(\beta x_2) - \sin(\beta L)\sinh(\beta x_2) - \cosh(\beta L)\cosh(\beta x_1) \\
&- \cosh(\beta L)\cos(\beta x_2) + \sin(\beta L)\sinh(\beta x_1) - \cos(\beta L)\cosh(\beta x_1) \\
&- \sin(\beta L)\sin(\beta x_2) + \sin(\beta L)\sinh(\beta x_1) + \cos(\beta L)\cos(\beta x_1) \\
&- \sin(\beta L)\sin(\beta x_2) \Big],
\end{aligned}
$$

$$
\begin{aligned}
D(s) = 2\rho A\Big[&1 + \cosh(\beta L)\cos(\beta L) \Big] \\
&- 2\beta M_{\mathrm{t}} \Big[\cosh(\beta L)\sin(\beta L) - \cos(\beta L)\sinh(\beta L) \Big],
\end{aligned}
$$

where $\beta^4 = -As^2/(EI)$.

As we can see, this transfer function is a relation between transcendental functions of the variable β, representing an infinite-dimensional system. In Figure 14.11 it is shown that the measurements fit the model.

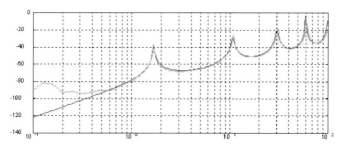

Figure 14.11 Analytic and measured responses, *solid line* for analytic, + for measured data

In order to obtain a transfer function in terms of ratios of polynomials, in [203] the numerator and denominator transcendental functions are rationalized using the Maclaurin series expansion in β^4, *i.e.*, s^2. From the point of view of our identification algorithm this means using $\alpha = 2$. In the same way, we could expand the transcendental functions in the variable β^2 ($s = 1$) or β ($s = 0.5$). Since β is the variable of the transcendental functions, it seems that the expansion in β is a more natural way, which leads us to transfer functions in terms of ratios of polynomials in $s^{1/2}$. This means that we have to model the system using fractional-order derivatives. By fitting the theoretical curve, we obtain the results shown in Table 14.5.

Table 14.5 Identification results over (14.17)

n,n	$\alpha = 2$	$\alpha = 1$	$\alpha = 0.5$	n,n	$\alpha = 2$	$\alpha = 1$	$\alpha = 0.5$
5, 5	1.6×10^{-5}	–	–	10, 10	–	7.2×10^{-5}	3.9×10^{-5}
12, 12	–	2.4×10^{-5}	1.8×10^{-6}	16, 16	–	6.4×10^{-5}	2.6×10^{-7}
18, 18	–	6.6×10^{-5}	8.8×10^{-9}	20, 20	–	–	2.4×10^{-9}

Though by choosing n and m adequately for the different basis functions ($n = m = 5$ for s^2, $n = m = 10$ for s and $s^{1/2}$) we can obtain good fittings in all the cases for the frequency range studied, we can observe that:

- For $n = m = 10$ and $\alpha = 0.5$, the larger power of s in the transfer function is 5 instead of 10. This means that we will need lower-order derivatives to model and control the system.
- Due to both, the nature of the system and the better numerical stability for $\alpha = 0.5$, we can obtain, using larger n and m, better fittings without increasing the order of the derivatives with respect to the cases of $\alpha = 1$ and $\alpha = 2$. Also, we could obtain good fittings for larger frequency ranges. In fact, if we enlarge the frequency range to 200 Hz, only with $\alpha = 0.5$ can we obtain a good fitting, as shown in Figure 14.12.

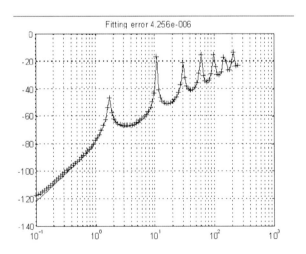

Figure 14.12 Identification for $\alpha = 0.5$

14.5 Summary

In this chapter we have briefly shown in two different ways that by applying the concepts of fractional calculus it is possible to obtain models that fit better the frequency behaviors of some type of distributed-parameter systems. In the first case, the electrochemical process, the convenience of applying fractional calculus has been derived from the presence of phenomena that can be better modeled only using fractional-order derivatives or integrals. In the second, the flexible structure is of infinite-dimensional nature leading us to build an approximate model that explicitly prompts the use of fractional-order derivatives. But, really, both ways are closely related for several reasons:

- It is well known that for infinite-dimensional or distributed-parameter systems a usual and useful electrical equivalent model is the transmission line model, in which the characteristic impedance is a function (algebraic or transcendental) of $s^{1/2}$. In the case of R-C ladder configuration the impedance has the following form:

$$Z_c(s) = \frac{R}{C}\frac{1}{\sqrt{s}}. \tag{14.18}$$

- A fractional-order relation, *i.e.*, a fractional-order element in the equivalent circuits, can be physically realized through transmission line, ladder, trees, or fractal arrangements of traditional elements (resistors and capacitors, springs and dashpots or dampers, *etc.*).
- In certain conditions the Warburg impedance, representing the diffusion phenomena, has an expression corresponding to a distributed-parameter system, that is

$$Z_w(\omega) = R_w \frac{\tanh \sqrt{j\tau_w\omega}}{\sqrt{j\tau_w\omega}}. \tag{14.19}$$

A further justification can be made by noting the fact that the constitutive equation of some materials employed in "smart material systems" (piezoelectrics, electrorheological fluids and solids, fiber optics, *etc.*) could be formulated using fractional-order derivatives.

We can conclude that transfer functions that are ratios of polynomials of s^α, $0 < \alpha < 1$, can be exact "lumped-parameter" models for systems in which diffusion and/or relaxation phenomena are present and, meanwhile, good approximate models for distributed-parameter systems.

Chapter 15
Position Control of a Single-link Flexible Robot

In this chapter fractional-order control is applied to accurate positioning of the tip of a single-link lightweight flexible manipulator. This kind of robot exhibits the advantage of being very lightweight. But they present a drawback in that vibrations appear in the structure when they move that prevent precise positioning of the end effector. Moreover, these vibrations may substantially change their amplitudes and frequencies when the tip payload changes, which is quite usual in robotics. The control of this kind of mechanical structure is nowadays a very challenging and attractive research area. These robots have found application in the aerospace and building construction industries, among others. This chapter develops a fractional-order controller that removes the structural vibrations and is robust to payload changes. The proposed control system is based on Bode's ideal transfer function described in Section 2.3.4. Properties of this transfer function are used to design a controller with the interesting feature that the overshoot of the controlled robot is independent of the tip mass. This allows a constant safety zone to be delimited for any given placement task of the arm, independently of the load carried, thereby making it easier to plan collision avoidance. Other considerations about noise and motor saturation issues are also presented throughout the chapter. To achieve the performance, the overall control scheme proposed consists of three nested control loops. Once the friction and other nonlinear effects have been compensated, the inner-loop is designed to give a fast motor response. The middle-loop simplifies the dynamics of the system, and reduces its transfer function to a double integrator, that allows for Bode's ideal loop transfer function design. Then a fractional-order derivative controller is used to shape the outer-loop into the form of a fractional-order integrator. The result is a constant phase system with, in the time domain, step responses exhibiting constant overshoot, independent of variations in the load. Experimental results are shown when controlling the flexible manipulator with this fractional-order differential operator, that prove the good performance of the system.

15.1 Introduction

Novel robotic applications have demanded lighter robots that can be driven using small amounts of energy, for example, robotic booms in the aerospace industry, where lightweight manipulators with high performance require-ments (high-speed operation, better accuracy, high payload/weight ratio) are required [204]. Another example is the need for lightweight manipulators to be mounted on mobile robots, where power limitations imposed by battery autonomy have to be taken into account.

Unfortunately, the flexibility that appears in these robots as a consequence of minimizing the cross section of their links leads to an undesired oscillatory behavior at the tip of the link, making precise pointing or tip positioning a daunting task that requires complex closed-loop control. In order to address control objectives, such as tip position accuracy and suppression of residual vibrations, many control techniques have been applied to flexible robots (see, for instance, the survey [205]).

In addition, some new robotic designs are being implemented that exploit mechanical flexibility in order to achieve new robotic behaviors. For example, collisions of flexible robots present remarkably less destructive effects than those caused by traditional robots, since the kinetic energy of the movement is transformed into potential energy of deformation at the moment of impact. This fact allows us to perform some control strategy over the actuator before any damage takes place or, at least, to minimize it, which may lead, in the not very distant future, to a robot-human cooperation without the actual dangers in case of malfunction, which is an emergent topic of research interest [206]. Another application is the robotic impedance control or force control, used to carry out dexterous tasks like assembly. This kind of task can be more efficiently carried out if the end effector, the wrist, or even the whole arm, have some degree of structural flexibility.

As a consequence of the preliminary considerations, considerable interest has been attracted to the control of lightweight flexible manipulators during the last 2 decades, becoming a most challenging research area of robotic control. Recently, some reviews in flexible robotics have been published. They divide the previous work into some sort of classification: control schemes [205], modeling [207], overview of main researches [208], *etc.* They are usually comprehensive enumerations of the different approaches and/or techniques used in the diverse fields involving flexible manipulators.

In 1974, W.J. Book provided the first known work dealing with this topic explicitly in his Ph. D. Thesis [209] entitled "Modeling, design and control of flexible manipulators arms" and supervised by D.E. Whitney, who was a professor at MIT Mechanical Engineering Department. In the same department in the very same year Dr. Maizza-Neto also studied the control of flexible manipulator arms but from a modal analysis approach [210]. Fruits of their joint labor, the first work published in a journal in the field of flexible robotics, appeared in 1975, dealing with the feedback control of a two-link-two-joints flexible robot [211]. A recursive, Lagrangian, assumed modes formulation for modeling a flexible arm that incorporates the approach taken by Denavit and Hartenberg [212], to describe in a efficient, complete and straightforward way the kinematics and dynamics of elastic manipulators was proposed by Book in 1984 [213].

This structural flexibility was also intensively studied in satellites and other large spacecraft structures (again spatial purposes and NASA behind the scenes) which generally exhibit low structural damping in the materials used and lack of other forms of damping. The generic studies by M.J. Balas on the control of flexible structures, carried out mainly between 1978 and 1982, deserve a special mention, e.g., [214] and [215]. They established some key concepts such as the influence of highly unmodeled dynamics in system controllability and performance, which is known as "spillover." In addition, the numerical/analytical examples included in his work dealt with controlling and modeling the elasticity of a pinned or cantilevered Euler-Bernoulli beam with a single actuator and a sensor, which is the typical configuration for a one degree of freedom flexible robot as we will discuss in later sections. After these promising origins, the theoretical challenge of controlling a flexible arm (while still very open) turned into the technological challenge of building a real platform testing those control techniques. And there it was, the first known robot exhibiting notorious flexibility to be controlled was built by Schmitz and Cannon [216]. A single-link flexible manipulator was precisely positioned by sensing its tip position while it was actuated on the other end of the link. In this work another essential concept appeared in flexible robots: a flexible robot is a non-collocated system and thus of non-minimum phase nature. Point-to-point motion of elastic manipulators had been studied with remarkable success taking a number of different approaches, but it was not until 1989 that the tracking control problem of the end-point of a flexible robot was properly addressed. This problem was tackled from a mixed open-closed-loop control approach by De Luca and Siciliano in 1989 [217] in the line proposed 2 years before by Bayo [218]. Also in 1989, the passivity concept was used for the first time in this field: D. Wang and M. Vidyasagar studied appropriate outputs of flexible arms in order to attain this property [219].

Book, in his review of 1993 [220], remarks on the exponential growth in the number of publications in this field and also the possibility of corroborating simulation results with experiments, which turns a flexible arm into "one test case for the evaluation of control and dynamics algorithms." And so it was. In the aforementioned [205], a summary of the main control theory contributions to flexible manipulators is shown, such as PD, PID, feedforward, adaptive, intelligent, robust, strain feedback, energy-based, wave-based, etc.

15.2 Problem Statement

One important problem in robotics is the variation of the dynamics as a consequence of changes in the payload to be carried by the manipulator, which causes the untuning of the controllers and degradation on the closed-loop positioning performance. This problem cannot be avoided as most robots are made for this purpose: to pick and place different loads. However this problem is often not relevant in standard robots (rigid industrial robots) because they have reduction gears connected to the actuators that reduce the torques, generated by the payload and robot structure inertias, seen by the actuators, by dividing them by n, this value being the gear reduction ratio. As n is usually high, then changes in the tip payload have relatively little effect on the actuators, and therefore in the actuators controllers performance.

Industrial robots are designed to be heavy and bulky in order to achieve rigidity in the robot mechanical structure. This guarantees that no more dynamics are involved in the robot than the associated rigid body motions and, very important, that controlling only the actuators involves the precise positioning of the tip, as a consequence of applying some trigonometric calculations.

Instead, in a flexible robot, more dynamic rather than the rigid ones appear as a consequence of the flexibility, and tip position cannot be guaranteed by simply controlling the robot actuators dynamics. In this case, payload changes again have little effect on the actuators (because the reduction gears are also present) but flexible link dynamics strongly vary as a consequence of these changes, as the frequencies of the arm vibrations depend (approximately) on the inverse of the square root of the payload mass.

Then if the payload changes from one movement to another, controllers designed for a nominal tip mass become untuned, which produces a very noticeable impairment in the tip trajectory or even instability. Thus the main objective of this chapter is to design a control system for single-link very lightweight flexible arms such that: 1. it achieves a good accuracy in tip trajectory tracking of rapid maneuvers, 2. it is extremely robust to large changes of the tip payload, in the sense of not only keeping the closed-loop system stability but also maintaining the tip tracking performance nearly unaltered, 3. it is very robust to motor friction changes and removes the steady-state error caused by Coulomb friction, 4. it is also robust to high-frequency dynamics not regarded in the controller design (spillover effects), and 5. it is a very simple linear controller that can be tuned in a very straightforward manner. In this chapter a fractional-order controller is developed that removes the structural vibrations and is robust to payload changes. The proposed control system is based on Bode's ideal loop transfer function described in Section 2.3.4. Properties of this transfer function are used to design a controller with the interesting feature that the overshoot of the controlled robot is independent of the tip mass. This allows a constant safety zone to be delimited for any given placement task of the arm, independently of the load being carried, thereby making it easier to plan collision avoidance.

The problem of controlling single-link flexible arms with changing tip payload has already been studied from different points of view. The most efficient strategies to achieve good tip trajectory tracking are introduced next. Adaptive control needs an online parametric estimation to modify, in real time, the controller parameters; [221] is an example of a nonlinear adaptation law applied to a two-link flexible arm in the case of uncertainties in several robot parameters, and [222, 223] deal specifically with the problem of large payload changes. Adaptive control has the drawbacks of 1. the difficulty to guarantee the closed-loop stability in all the circumstances which usually requires some complicated Lyapunov stability analysis, 2. the condition of a persistent excitation condition in order to achieve an accurate estimate of the robot parameters, 3. the deterioration of the transient response of the tip during the time interval from the start of the movement until the manipulator parameters are estimated and the controller retuned, leading to tracking errors that sometimes may be unacceptable. Neural networks have also been used to design controllers robust to load changes for flexible arms [224, 225]. They often implement adaptation laws that change the controller behavior in real time depending on some measured signals. In this technique it is also difficult to guarantee the stability of the control system, and it requires a previous, often laborious, network training process. Another approach is to use sliding control: [226] proposes an adaptive scheme for joint flexibility and [227] a method that adapts the sliding surface depending on initial conditions for a single-link flexible arm. These methods need the fulfilment of the so-called matching condition in order to be applicable, which requires that the uncertainties remain in the space range of the control input to ensure an invariance property of the system behavior during the sliding mode. Moreover, arm dynamic performance is not well controlled during the initial phase to reach the sliding surface. Another approach is to use robust control techniques. In [228], the authors presented a control robust to payload changes composed of feedforward and feedback terms. An energy-based robust controller was designed for multi-link flexible arms in [229], which does not require any information of the system dynamics, and only makes use of the very basic energy relationship of the system. In [230] a very efficient robust to payload changes controller was proposed for the very special case of a single-link flexible arm with only one vibration mode. A robust control combining pole placement and sensitivity function shaping method was developed in [231], where experiments were reported of changes in the tip payload that implied changes of 50% in the moment of inertia of the arm. Another robust controller for the same robot was designed using the quadratic d-stability approach in [232], but changes allowed in the tip payload were smaller. A recent example of design of robust controllers for single-link flexible arms based on the $\mathcal{H}_\infty$ methodology is [233], which splits the arm dynamics into two time scale systems, and imposes some uncertainty bounds mostly to the fast dynamics subsystem. This methodology was mainly designed to control spillover effects but can also be adapted to consider other parameter uncertainties. Finally, robust time-optimal control strategies for the motion of flexible structures in an open-loop mode are studied in [234]. In all these robust control approaches the stability of the controlled system is guaranteed under payload variations or non-negligible spillover effects, but the quality of the reference tracking is not as good as that achieved by a controller designed only for the nominal plant (assuming this controller of the same complexity than the robust one), $i.e.$, the stability margin is increased at the expense of losing system response performance (usually the system becomes slower). Moreover, these controllers exhibit another two drawbacks: 1. they are designed to cope with limited variations of the tip load and 2. the quality of the reference tracking deteriorates considerably as the payload differs more from the nominal value.

In the above context, it will be shown that the fractional-order controller to be developed in this chapter provides a constant stability margin, phase margin, independently of the value of the payload while the gain crossover frequency (speed of response) varies as the payload changes (the speed of response diminishes as the payload increases, keeping approximately constant the maximum of the control signal for the different payloads). These robustness features cannot be achieved by any of the aforementioned techniques. The remainder of this chapter is organized as follows. Section 15.3 outlines the dynamics of the considered flexible arm. Section 15.4 briefly describes the general control scheme composed of three nested loops. Section 15.5 develops the outer control loop which is the one that includes the fractional-order control law and provides the robot with the robustness to payload changes. Section 15.6 studies the effects of the spillover on this control law, and some controller conditions are analytically derived to guarantee stability under these circumstances. Section 15.7 presents the results obtained from the test of the control strategy in an experimental platform. Finally, some relevant concluding remarks are drawn in Section 15.8.

15.3 Dynamic Model of the Single-link Flexible Manipulator

A dynamic model of the single-link flexible arm will be used for the control system design. A lumped mass model, which assumes a massless link and a single mass located at the arm tip (which represents the payload), is used in this work. This is the simplest dynamic model of a flexible arm, and it is well known [230]. Particularly, a single tip mass that can rotate freely (no torque is produced at the tip) will be adopted for the description of the link dynamics [230]. The effect of gravity is assumed negligible since the arm moves in a horizontal plane. The motor has a reduction gear with a reduction relation n. The magnitudes seen from the motor side of the gear will be written with an upper hat, while the magnitudes seen from the link side will be denoted by standard letters.

The dynamics of the link is described by

$$c(\theta_m - \theta_t) = ml^2\ddot{\theta}_t, \tag{15.1}$$

where m is the mass at the end, l and c are the length and the stiffness of the bar, respectively, θ_m is the angle of the motor, and θ_t is the angle of the tip. A scheme of this arm is shown in Figure 15.1.

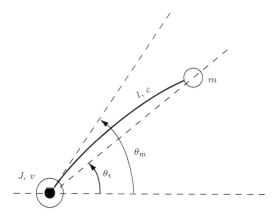

Figure 15.1 Diagram of the single-link flexible arm

The electromechanical actuator is constituted by a DC motor. This DC motor is supplied by a servo-amplifier with a current inner-loop control. The dynamic equation of the system can be written by using Newton's second law:

$$KV = J\ddot{\theta}_m + \nu\dot{\theta}_m + \hat{\Gamma}_{coup} + \hat{\Gamma}_{Coul}, \tag{15.2}$$

where K is the electromechanical constant of the motor servo-amplifier system, J is the motor inertia, ν is the viscous friction coefficient, $\hat{\Gamma}_{coup}$ is the coupling torque between the motor and the link, $\hat{\Gamma}_{Coul}$ is the Coulomb friction, and V is the motor input voltage. This last variable is the control variable of the system, which is the input to a servo-amplifier that controls the input current to the motor by means of an internally PI current controller (see Figure 15.2). This electrical dynamics can be rejected because this is faster than the mechanical dynamics of the motor. Thus, the servo-amplifier can be considered as a constant relation, k_e, between the voltage and the current to the motor: $i_m = Vk_e$ (see Figure 15.3), where i_m is the armature circuit current and k_e includes the gain of the amplifier, $\tilde{k}$, and R the input resistance of the amplifier circuit. The total torque given to the motor, $\hat{\Gamma}$, is directly proportional to the armature circuit in the form $\hat{\Gamma} = k_m i_m$ where k_m is the electromechanical constant of the motor. Thus, the electromechanical constant of the motor servo-amplifier system is $K = k_e k_m$.

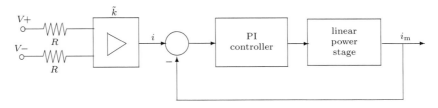

Figure 15.2 Servo-amplifier scheme

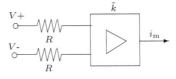

Figure 15.3 Servo-amplifier scheme

From now on it is supposed that the Coulomb friction is negligible or is compensated by a term [235] of the form

$$V_{Coul} = \frac{\bar{\Gamma}_{Coul}}{K}\text{sgn}\left(\ddot{\theta}_m\right), \tag{15.3}$$

as shown in Figure 15.4, where $\bar{\Gamma}_{\text{Coul}}$ is an estimation of the Coulomb friction value.

On the other hand, the coupling torque equation between the motor and the link is

$$\Gamma_{\text{coup}} = c(\theta_{\text{m}} - \theta_{\text{t}}), \tag{15.4}$$

and, finally, the conversion equations $\hat{\theta} = n\theta$ and $\hat{\Gamma}_{\text{coup}} = \Gamma_{\text{coup}}/n$ complete the dynamic model.

Laplace transform is applied to (15.1), leading to the following transfer function for the link:

$$G_{\text{b}}(s) = \frac{\theta_{\text{t}}(s)}{\theta_{\text{m}}(s)} = \frac{\omega_0^2}{s^2 + \omega_0^2}, \tag{15.5}$$

where ω_0 is the natural frequency of the link

$$\omega_0^2 = \frac{c}{ml^2}, \tag{15.6}$$

which is mass dependant.

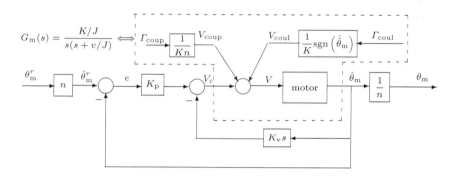

Figure 15.4 Block diagram for the inner-loop with its compensating terms

With all the previous equations, the transfer functions of the robot are

$$\frac{\theta_{\text{m}}(s)}{V(s)} = \frac{Kn\left(s^2 + \omega_0^2\right)}{s\left[Jn^2s^3 + vn^2s^2 + (Jn^2\omega_0^2 + c)s + vn^2\omega_0^2\right]}, \tag{15.7}$$

$$\frac{\theta_{\text{t}}(s)}{V(s)} = \frac{Kn\omega_0^2}{s\left[Jn^2s^3 + vn^2s^2 + (Jn^2\omega_0^2 + c)s + vn^2\omega_0^2\right]}. \tag{15.8}$$

It is evident that these robot equations are mass dependant (so is ω_0) and, therefore, changes in the tip mass (payload) will affect the system behavior.

Table 15.1 shows the parameters of the motor-gear set and the flexible link used for the experimental tests, respectively.

Table 15.1 Data of the motor-gear set and flexible link

Motor-gear set							Flexible link		
γ	K_p	K_v	$J(\text{kg}\cdot\text{m}^2)$	ν (N·m·s)	K (N·m/V)	n	c (N·m)	m (kg)	l (m)
0.022	1	0.025	24.24×10^{-4}	51.66×10^{-4}	3.399	50	443.597	1.9	0.866

15.4 General Control Scheme

This chapter develops a simple control scheme for positioning the tip of single-link flexible arms whose payload changes, based on the use of a fractional-order derivative controller. This scheme uses measurement of the link deflection provided by a strain gauge placed at the base of the link. This sensorial system lets us construct control schemes that are more robust than those based on accelerometer measurements (as was demonstrated in [230], too). This is because strain gauges are placed at the base of the link while accelerometers are located at the tip, where collisions are more likely to happen, so the sensor is more likely to be damaged. Moreover, strain gauges are much simpler to instrument than accelerometers.

The general control scheme proposed here consists of three nested loops (see Figure 15.5). The features of the inner-loops and outer-loops have been previously detailed in [230], while the middle-loop has simplifying purposes, and is included to cope with the fractional-order control strategy. Basically, this scheme allows us to design the loops separately, making the control problem simpler and minimizing the effects of the inaccuracies in the estimation of Coulomb and viscous frictions in control performance (as shown in [235]). The purposes of these three loops are:

- An inner-loop that controls the position of the motor. This loop uses a classical PD controller of high gains in order to give a closed-loop transfer function close to unity (it is detailed in Figure 15.4).
- A simplifying-loop using positive unity-gain feedback. The purpose of this loop is to reduce the dynamics of the system to that of a double integrator.
- An outer-loop that uses a fractional-order derivative controller to shape the loop and to give an overshoot independent of payload changes.

In Figure 15.5, θ_m is the motor angle, θ_t the tip-position angle, $G_b(s)$ the transfer function of the beam, and $R_i(s)$, $R_e(s)$ the inner- and outer-loop controllers, respectively. The design of the first two loops follows [223]. The fractional-order control strategy of the outer-loop utilizes a fractional-order controller and is based on the concept of Bode's ideal loop transfer function design. More details on this control scheme and the fractional-order controller design can be found in [75, 236].

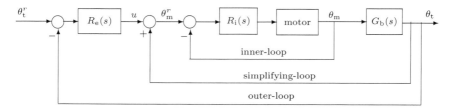

Figure 15.5 Proposed general control scheme

15.4.1 Inner-loop

The inner control loop, shown in Figure 15.4, fixes the dynamic behavior of the motor. This purpose is achieved by means of a standard PD controller with proportional constant K_p and derivative constant K_v, which is tuned to make the motor dynamics critically damped. Moreover two compensating terms are added to this inner-loop: 1. one that cancels the nonlinearity associated with the Coulomb friction by adding a term V_{Coul} calculated from (15.3), and 2. another that compensates the coupling torque existing between the motor and the link, which is expressed as V_{coup}.

This last term is calculated by

$$V_{coup} = \frac{1}{Kn}\Gamma_{coup},\qquad(15.9)$$

that can be implemented because the strain gauge placed at the base of the link provides Γ_{coup}.

The control law uses feedback of the measured motor position $\hat{\theta}_m$, and may be expressed as

$$V(t) = K_p(n\theta_m^r(t) - \hat{\theta}_m(t)) - K_v\dot{\hat{\theta}}_m(t) + V_{Coul} + V_{coup},\qquad(15.10)$$

where $\theta_m^r(t)$ is the reference angle for the motor.

The second-order critically damped equation obtained for the closed-loop motor system, as a consequence of tuning control law (15.10), is

$$M(s) = \frac{\theta_m(s)}{\theta_m^r(s)} = \frac{1}{(1+\gamma s)^2},\qquad(15.11)$$

where γ is the closed-loop motor dynamics constant.

Next we define the PD parameters tuning laws. As a consequence of the two compensating terms (15.3) and (15.9), the motor equation (15.2) reduces to

$$KV_c = J\ddot{\hat{\theta}}_m + \nu\dot{\hat{\theta}}_m,\qquad(15.12)$$

where V_c is a fictitious input to the reduced motor dynamics (see Figure 15.4). The PD parameters easily yield from some algebraic manipulations:

$$K_{\mathrm{p}} = \frac{J}{K\gamma^2}, \tag{15.13}$$

$$K_{\mathrm{v}} = \frac{1}{K}\left(\frac{2J}{\gamma} - \nu\right). \tag{15.14}$$

Theoretically, as $\gamma = \sqrt{J/(K_{\mathrm{p}}K)}$, it is possible to make the motor dynamics as fast as desired ($\gamma \to 0$) by simply making $K_{\mathrm{p}} \to \infty$. But a very demanding speed would saturate the motor, with the subsequent malfunction of the controlled system. This fact implies that, although the motor dynamics can be made quite fast, it cannot be considered negligible in general.

It is important to note that the assumption that the coupling torque may be compensated within the inner-loop by a term of the form (15.9) is well supported. Previous experimental works have proven the correctness of this assumption in direct driven motors, and motors with reduction gears as well [223]. It has also been demonstrated that, in the case of motors with gears, the effect of the coupling torque is very small compared to the motor inertia and friction, as its value is divided by n [230].

15.4.2 Simplifying-loop

As stated previously, the response of the inner-loop (position control of the motor) is significantly faster than the response of the outer-loop (position control of the tip). The motor position is first supposed to track the reference motor position with negligible error and the motor dynamics will be considered later. That is, the dynamics of the inner-loop can be approximated by $M(s) \approx 1$ when designing the outer-loop controller. Taking this into account, a strategy for simplifying the dynamics of the arm, shown in Figure 15.6, is proposed.

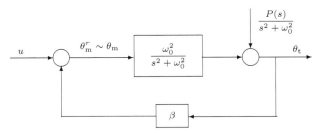

Figure 15.6 Block diagram of the simplifying middle-loop

For the case of a beam with only one vibrational mode, a simplifying-loop can be implemented that reduces the dynamics of the system to a double integrator by simply closing a positive unity-gain feedback loop around the tip position ($\beta = 1$). Then, the equation relating the output and input of the loop is

$$\theta_t(s) = \frac{\omega_0^2}{s^2} u(s) + \frac{1}{s^2} P(s), \tag{15.15}$$

where $P(s)$ represents disturbances with the form of a first-order polynomial in s, which models initial deviations in tip position and tip velocity [75]. In (15.15), the dynamics of the arm has been reduced to a double-integrator dynamics, simplifying the control strategy to be designed for the tip position, as will be seen later.

The stability study by using Nyquist plots shows that the condition $\beta = 1$ is not critical to get stable control systems, being sufficient to implement a feedback gain close to 1.

15.4.3 Outer-loop

The block diagram for the outer-loop used in this work is shown in Figure 15.7.

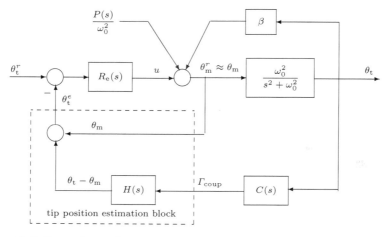

Figure 15.7 Basic scheme of the outer control loop

As observed in the scheme, an estimation of the tip position, θ_t^e, is used to close the loop. We actually feed back the measurements of a strain gauge placed at the base of the link to control the arm. These measurements provide the value of the coupling torque Γ_{coup} between the arm and the motor, which is used to decouple motor and link dynamics and estimate tip position [230]. Combining (15.1), (15.4), and the fundamental frequency definition, we can obtain the relation between Γ_{coup} and the output θ_t, which yields

$$C(s) = \frac{\Gamma_{\text{coup}}(s)}{\theta_t(s)} = c \frac{s^2}{\omega_0^2}. \tag{15.16}$$

Given the values of θ_m and Γ_{coup}, experimentally measured, the value of θ_t can be calculated from (15.4) by

$$\theta_t^e = \theta_m - \Gamma_{coup}/c. \tag{15.17}$$

This estimated tip angle is used to close the control loop, and can be expressed by a block diagram as shown in Figure 15.7, where $H(s) = -1/c$.

The main purpose of this chapter is to design the outer controller $R_e(s)$ of Figure 15.7 so that the time response of the controlled system has an overshoot independent of the tip mass and the effects of disturbances are removed. These specifications will lead to the use of a fractional-order derivative controller, as will be detailed in the next section.

Besides, in this particular case the outer controller will be designed in the frequency domain for the specifications of phase margin (damping of the response), and gain crossover frequency (speed of the response). In order to guarantee a critically damped response (overshoot $M_p = 0$), a phase margin $\varphi_m = 76.5°$ is selected. Besides, the response is desired to have a rise time around 0.3 sec, so the gain crossover frequency is fixed to $\omega_{cg} = 6\,\mathrm{rad/sec}$. The crossover frequency defines the speed of response of the closed-loop system. The practical constraint that limits the speed of response of the arm, and hence the value of ω_{cg}, is the maximum torque provided by the motor. This maximum torque limits the speed of response of the inner motor loop, and hence its bandwidth. The control scheme proposed here works ideally if the dynamics of the inner-loop is negligible. Consequently, its bandwidth must be much larger than the desired gain crossover frequency ω_{cg}. However, it must be taken into account that the experimental results to be presented in this chapter show the behavior of the arm assuming non-negligible inner-loop dynamics, since the value of the torque provided by the motor limits the speed of response of this loop. This fact may change slightly the final frequency specifications found for the system, as will be shown later.

15.5 Design of the Outer-loop Controller $R_e(s)$

With the inner-loops and simplifying-loops closed, the reduced diagram of Figure 15.8 is obtained, which is based on (15.15). From this diagram, the equation for the tip position is

$$\theta_t(s) = \frac{1}{1 + \dfrac{s^2}{R_e(s)\omega_0^2}}\theta_t^r(s) + \frac{1}{1 + \dfrac{s^2}{R_e(s)\omega_0^2}}\frac{P(s)}{R_e(s)\omega_0^2}. \tag{15.18}$$

The controller $R_e(s)$ has a twofold purpose. One objective is to obtain a constant phase margin in the frequency response, in other words, a constant

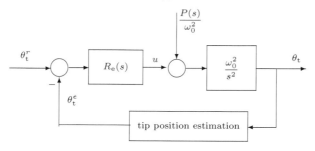

Figure 15.8 Reduced diagram for the outer-loop

overshoot in time response to a step reference for varying payloads. The other is to remove the effects of the disturbance, represented by the initial conditions polynomial, on the steady state. To attain these objectives, most authors propose the use of some kind of adaptive control scheme [223]. We propose here a fractional-order derivative controller with enhanced robustness properties to achieve the above two objectives, without needing any kind of adaptive algorithm. Some methods have been developed in the last few years to tune fractional-order PID controllers with robustness properties to changes in one parameter which is typically the plant gain [237, 238] (note that in our case the process gain is ω_0^2, which changes with the payload according to (15.6)). Some of them have been studied in previous chapters, and were based on optimization procedures. The approach proposed here is based on Bode's ideal loop transfer function studied in Section 2.3.4. It is much simpler than other methods as it is specifically tailored to our particular arm dynamics, leading to very straightforward tuning rules.

15.5.1 Condition for Constant Phase Margin

The condition for a constant phase margin can be expressed as

$$\arg\left[R_e(j\omega)\frac{\omega_0^2}{(j\omega)^2}\right] = \text{constant}, \forall\omega, \tag{15.19}$$

and the resulting phase margin φ_m is

$$\varphi_m = \arg\left[R_e(j\omega)\right]. \tag{15.20}$$

For a constant phase margin $0 < \varphi_m < \pi/2$ the controller that achieves this must be of the form

$$R_e(s) = K_c s^\alpha, \quad \alpha - \frac{2}{\pi}\varphi_m, \tag{15.21}$$

so that $0 < \alpha < 1$. This $R_e(s)$ is a fractional-order derivative controller of order α.

15.5.2 Condition for Removing the Effects of Disturbance

From the Final-Value Theorem, the condition to remove the effects of the disturbance is

$$\lim_{s \to 0} \frac{1}{1 + \dfrac{s^2}{R_e(s)\omega_0^2}} \frac{P(s)}{R_e(s)\omega_0^2} = 0. \tag{15.22}$$

Substituting $R_e(s) = K_c s^\alpha$ and $P(s) = as + b$ (initial tip position and velocity errors different from zero), this condition becomes

$$\lim_{s \to 0} \frac{1}{1 + \dfrac{s^{2-\alpha}}{K_c \omega_0^2}} \frac{b}{K_c \omega_0^2} s^{1-\alpha} = 0, \tag{15.23}$$

which implies that $\alpha < 1$.

15.5.3 Ideal Response to a Step Command

Assuming that the dynamics of the inner-loop can be approximated by unity and that disturbances are absent, the closed-loop transfer function with controller (15.21) is

$$F_{cl}(s) = \frac{\theta_t}{\theta_t^r} = \frac{1}{1 + \dfrac{s^2}{R_e(s)\omega_0^2}} = \frac{K_c \omega_0^2}{s^{2-\alpha} + K_c \omega_0^2}, \tag{15.24}$$

which exhibits the form of Bode's ideal loop transfer function. The corresponding step response is

$$\theta_t(t) = \mathcal{L}^{-1} \left\{ \frac{K_c \omega_0^2}{s(s^{2-\alpha} + K_c \omega_0^2)} \right\} = K_c \omega_0^2 t^{2-\alpha} \mathscr{E}_{2-\alpha,3-\alpha}(-K_c \omega_0^2 t^{2-\alpha}), \tag{15.25}$$

where $\mathscr{E}_{\delta,\delta+1}(-At^\delta)$ is the Mittag–Leffler function in two parameters. The overshoot is fixed by $2 - \alpha$, which is independent of the payload, and the speed by $K_c \omega_0^2$, that is, by the payload and the controller gain. In fact, notice that this equation can be normalized with respect to time by

$$\theta_t(t_n) = t_n^{2-\alpha} \mathscr{E}_{2-\alpha,3-\alpha}(-t_n^{2-\alpha}), \tag{15.26}$$

where $t_n = t(K_c \omega_0^2)^{1/(2-\alpha)}$. This equation shows that the effect of a change in the payload implies a change in ω_0 that only means a time scaling of the response $\theta_t(t)$.

To obtain a required step response, it is necessary to select the values of two parameters. The first is the order α to adjust the overshoot between 0 ($\alpha = 1$) and 1 ($\alpha = 0$), or, equivalently, a phase margin between 90° and 0°. The second is the gain K_c to adjust the gain crossover frequency, or, equivalently, the speed of the response for a nominal payload. Note that increasing α decreases the overshoot but increases the time required to correct the disturbance effects [50].

15.5.4 Controller Design

As stated above, the design of the controller thus involves the selection of two parameters:

- α, the order of the derivative, which determines 1. the overshoot of the step response, 2. the phase margin, or 3. the damping.
- K_c, the controller gain, which determines for a given α 1. the speed of the step response or 2. the gain crossover frequency.

These parameters can be selected by working in the complex plane, the frequency domain, or the time domain. In the frequency domain, the selection of the parameters of the fractional-order derivative controller can be regarded as choosing a fixed phase margin by selecting α, and choosing a gain crossover frequency ω_{cg}, by selecting K_c for a given α. That is,

$$\alpha = \frac{2}{\pi}\varphi_m, \quad K_c = \frac{\omega_{cg}}{\omega_0^2}. \tag{15.27}$$

According to Table 15.1, where the parameters of the flexible manipulator are presented, the fundamental frequency of the system is $\omega_0 = 17.7\,\mathrm{rad/sec}$. The frequency specifications required for the controlled system, stated previously, are phase margin $\varphi_m = 76.5°$ and gain crossover frequency around $\omega_{cg} = 6\,\mathrm{rad/sec}$. Therefore, the parameters of the fractional-order derivative controller are $\alpha = 0.85$ and $K_c = 0.02$. With this controller, and under the assumption of negligible inner-loop dynamics, the Bode plots obtained for the open-loop system are shown in Figure 15.9, where it can be observed that at the gain crossover frequency $\omega_{cg} = 6\,\mathrm{rad/sec}$ the phase margin is $\varphi_m = 76.5°$, fulfilling the design specifications.

The simulated step responses of the controlled system for $m = 0.6\,\mathrm{kg}$, $m = 1.9\,\mathrm{kg}$, $m = 3.2\,\mathrm{kg}$, and $m = 6\,\mathrm{kg}$ are shown in Figure 15.10. It is observed that the overshoot of the response remains constant to payload changes, being $M_p = 0$, fulfilling the robustness purpose. For the nominal mass ($m = 1.9\,\mathrm{kg}$), a rise time $t_r = 0.3\,\mathrm{sec}$ is obtained.

Finally, and for comparison purposes, a standard PD controller, designed for the same frequency specifications as above, has been simulated. In order

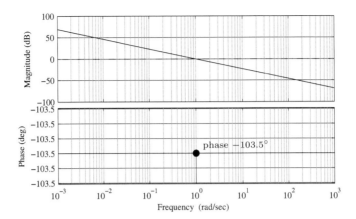

Figure 15.9 Bode plots of the open-loop system with the fractional-order derivative controller, considering negligible inner-loop dynamics

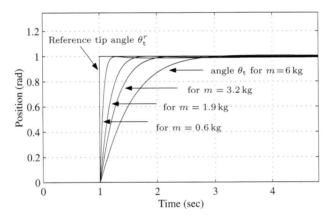

Figure 15.10 Simulated time responses of the system with the fractional-order derivator to a step input for different payloads, considering negligible inner-loop dynamics

to remove an undesirable overshoot that may appear associated with the zero that the controller introduces in the closed-loop system, the PD controller has been modified as shown in Figure 15.11 (velocity is fed back to the controller instead of velocity error). The tuned controller is

$$u(t) = 0.305(\theta_t^r(t) - \theta_t(t)) - 0.013\dot{\theta}_t(t), \tag{15.28}$$

where control signal u was defined in Figure 15.5. Note that the simplifying-loop is also implemented in this case to keep complete equivalence between this controller and the fractional-order one.

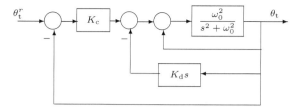

Figure 15.11 Modified PD control system for the outer-loop

Figure 15.12 shows simulated responses of the closed-loop system with this PD controller (15.28) to a step command for different payload values. These plots show that performance deteriorates very noticeably as payload differs from the nominal value: overshoot changes significantly. These results highlight the interest of using a fractional-order controller if a nearly constant overshoot were required.

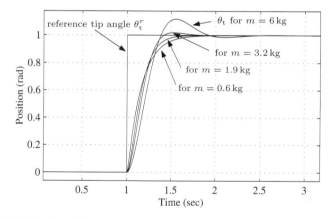

Figure 15.12 Simulated time responses of the system with PD controller (15.28) to a step input for different payloads, considering negligible inner-loop dynamics

15.6 Some Practical Issues

In this section some practical issues related to the effect on the closed-loop performance of the robot dynamics neglected in the controller design procedure are analyzed. In particular two sources of errors and instabilities are studied: spillover effects associated with higher vibration modes of the flexible link that have not been taken into account, and effects of non-negligible motor inner closed-loop dynamics. Moreover, the discrete implementation of the controller is studied.

15.6.1 Robustness to Higher Vibration Modes (Spillover)

The robustness of the developed fractional-order controller to unmodeled higher vibration modes is analyzed here. These modes can influence the closed-loop system in two ways: 1. they are fed back to the controller and, if they were not taken into account in the controller design, the global system can become unstable, 2. the estimator of the tip position based on (15.17) no longer remains correct, exhibiting high-frequency estimation errors that are fed to the closed-loop system. In order to avoid these destabilizing effects we propose a lemma based on the next dynamic property.

Let us consider the transfer function $G_\Gamma(s)$ between the motor angle and the motor-beam coupling torque (the other measured variable). Then we state that this transfer function exhibits the interlacing property of its poles and zeros on the imaginary axis if it verifies that

$$G_\Gamma(s) = c\frac{s^2(s^2+\varpi_1^2)\cdots(s^2+\varpi_i^2)\cdots(s^2+\varpi_{\bar n}^2)}{(s^2+\omega_0^2)(s^2+\omega_1^2)\cdots(s^2+\omega_i^2)\cdots(s^2+\omega_{\bar n}^2)},\tag{15.29}$$

$\omega_{i-1} < \varpi_i < \omega_i$, $1 \leqslant i \leqslant \bar n$ being $\bar n$ the number of vibration modes of the flexible link considered, and $c > 0$.

The simplest flexible robot is our arm with simplified dynamics described in Section 15.3. In this case it is found that

$$G_\Gamma(s) = G_b(s)C(s) = \frac{cs^2}{s^2+\omega_0^2},\tag{15.30}$$

from combining (15.5) and (15.16). This transfer function corresponds to (15.29) in the case of a single vibration mode ($\bar n = 1$).

This property is also verified by uniform single-link flexible manipulators with distributed mass and a payload at the tip. This is illustrated next.

The governing equation of a flexible link (Euler–Bernoulli equation) can be normalized by defining $t_b = \sqrt{\rho L^4/(EI)}$, where ρ is the mass per unit length, L the link length, E the Young's modulus, I the second moment of area about the bending axis, and introducing the non-dimensional time $t_n = t/t_b$. Consequently the tip payload is also normalized with respect to the beam mass $m_n = m/\rho L$. Transfer functions $G_\Gamma(s)$ are obtained for the normalized beam for different m_n ratios, and the poles and zeros associated with the first six modes are calculated. We assume that modeling six modes is enough to study the spillover effects in most flexible manipulators. Figure 15.13, shows the values of these poles and zeros for mass ratios ranging from negligible link mass ($m_n^{-1} = 0$) to the case of a link mass 10 times larger than the tip payload ($m_n^{-1} = 10$). This figure shows that the interlacing property mentioned earlier is verified by any uniform beam at least in the specified range of variation of the link mass.

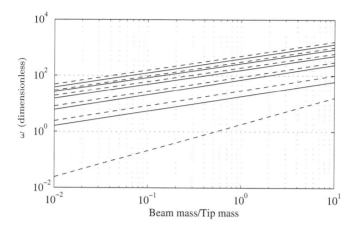

Figure 15.13 Verification of the interlacing property in $G_\Gamma(s)$ in a distributed mass flexible link: the *dashed lines* are ω_i and the *solid lines* ϖ_i

The next lemma proves that, if this property is verified, a family of fractional-order controllers, to which the controller proposed in the previous section belongs, is robust to unmodeled higher modes (spillover).

Lemma 15.1. Assume that our flexible arm verifies the interlacing property (15.29), and that the inner-loop dynamics is negligible. Then any outer-loop controller with a structure

$$R_e(s) = K_p + K_\alpha s^\alpha, \quad K_\alpha > 0, \quad 0 \leqslant \alpha \leqslant 1, \tag{15.31}$$

which uses as feedback signal a tip position estimator of a form similar to (15.17)

$$\theta_t^e = \theta_m - \frac{\Gamma_{\text{coup}}}{\hat{c}}, \tag{15.32}$$

where parameter $\hat{c}$ verifies

$$\hat{c} \geqslant c, \tag{15.33}$$

and c is the value given in (15.29), which is equivalent to the value c in (15.1), keeps stable the closed-loop system.

Proof. Figure 15.14 shows a block diagram of the control scheme. Operating this block diagram (and assuming that $M(s) = 1$) we obtain the equivalent transfer function

$$H_\Gamma(s) = \frac{\Gamma_{\text{coup}}(s)}{\Theta_t^r(s)} = \frac{G_\Gamma(s)}{1 + G_\Gamma(s)(-1 + R_e^{-1}(s))/\hat{c}}. \tag{15.34}$$

If $G_\Gamma(s)$ verifies the above interlacing property, the alternation between poles and zeros of (15.29) produces a Nyquist plot of $G_\Gamma(j\omega)$ of the form shown in Figure 15.15 (a). It exhibits as many half-turns in the infinity as vibration modes has the transfer function (as many as terms $(s^2 + \omega_i^2)$ are in the denominator of this transfer function). This plot shows that the closed-loop system associated with $G_\Gamma(s)$ is marginally stable. It is clear that if the product $(-1 + R_e^{-1}(s))/\hat{c}$ subtracts phase to the system, it is equivalent to approximately rotating the before Nyquist plot clockwise increasing the phase margin, as shows Figure 15.15 (b). Moreover, in order to guarantee that the Nyquist plot does not embrace the point $(-1, j0)$, it must be verified that

$$\lim_{\omega \to \infty} \frac{1}{\hat{c}} G_\Gamma(s)(-1 + R_e^{-1}(s)) \geqslant 1. \tag{15.35}$$

Assuming that $R_e(s)$ is of the form (15.31), and taking into account (15.29), it easily follows that condition (15.35) becomes $\hat{c} \geqslant c$, and (15.33) is proven.

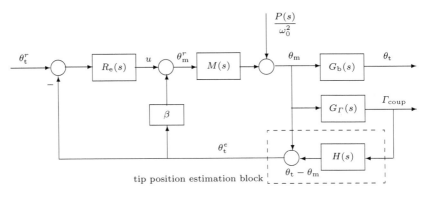

Figure 15.14 Detailed block diagram of the control system including the dynamics of the tip position estimator

Moreover if the controller is of the form (15.31), after some operations we have that

$$\xi(j\omega) = \frac{1}{\hat{c}} G_\Gamma(s)(-1 + R_e^{-1}(s))$$

$$= \frac{K_p(1-K_p) + K_\alpha \omega^\alpha \cos(\pi\alpha/2)(1-2K_p) - K_\alpha^2 \omega^2 - jK_\alpha \omega^\alpha \sin(\pi\alpha/2)}{\hat{c}(K_p^2 + 2K_\alpha K_p \omega^\alpha \cos(\pi\alpha/2) + K_\alpha^2 \omega^{2\alpha})}.$$

$$\tag{15.36}$$

The imaginary component of this equation is negative $\forall \omega \geqslant 0$ provided that $0 \leqslant \alpha \leqslant 1$ and $K_\alpha > 0$. Then $\angle\xi(j\omega) \leqslant 0$, $\forall \omega \geqslant 0$, and it subtracts phase from $G_\Gamma(j\omega)$ at all frequencies, as the Nyquist stability condition requires.   □

Remark 15.2. Conditions (15.31) and (15.33) make the closed-loop system stable for any single-link flexible arm that fulfils the interlacing property (15.29), independently of the number of high-frequency modes considered.

□

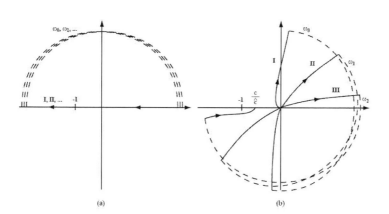

(a)   (b)

Figure 15.15 Nyquist plots of: (a) $G_\Gamma(s)$ and (b) $G_\Gamma(s)(-1 + R_e^{-1}(s))/\hat{c}$

15.6.2 Effect of the Non-negligible Inner-loop Dynamics

In the practical robot studied here, the dynamics of the inner-loop is not negligible, being given by the transfer function (15.11) with $\gamma = 0.022$. Notice that $M(s)$ is independent of the value of the tip payload as its effects on the motor dynamics are removed by the compensation term V_{coup} in (15.9) based on the measurement of the motor-beam coupling torque. The introduction of $M(s)$ implies that the response of the controlled system will be affected by this dynamics, since the simplifying-loop does not result in a double integrator anymore.

Besides, it is important to note that step inputs are not very appropriate for robotic systems, the use of smoother references being more suitable to avoid surpassing the physical limitations of the robot, such as the maximum torque allowed to the links before reaching the elastic limit or the maximum feasible control signal value, (V for a DC motor-amplifier set). For that reason, a fourth-order polynomial reference θ_t^r has been used in our case.

It has been observed that with the introduction of $M(s)$ the settling time of the response gets longer. To reduce it, a proportional part K_p is introduced in the controller to make the output converge faster to its reference. However, it must be taken into account that the introduction of this constant affects the frequency response of the system, changing the specifications. Therefore, there must be a trade off between the fulfilment of the frequency specifications and the settling time required, resulting in $K_p = 0.25$ in our case. Then, the final controller is

$$R_e(s) = 0.25 + 0.02s^{0.85}. \tag{15.37}$$

Only a slight modification of the frequency specifications is obtained with this controller, resulting in $\omega_{cg} = 6.6\,\mathrm{rad/sec}$ and $\varphi_m = 70°$. Recall that this controller keeps the structure (15.31), which is robust to spillover effects, as was shown in the previous section.

15.6.3 Fractional-order Controller Implementation

No physical devices are available to perform the fractional-order derivatives, so approximations are needed to implement fractional-order controllers. These approximate implementations of FOC have been studied in Chapter 12. In this particular case, an indirect discretization method is used. That is, first a finite-dimensional continuous approximation is obtained, and second the resulting s-transfer function is discretized.

It must be taken into account that the fractional-order derivative $s^{0.85}$ has been implemented as $s^{0.85} = ss^{-0.15} = s/s^{0.15}$, that is, an integer-order derivative multiplied by a fractional-order integrator. This way, the resulting open-loop system in the ideal case would be $R_e(s)\omega_0^2/s^2 = \omega_0^2 s^{-0.15}/s$, guaranteeing the cancelation of the steady-state-position error due to the effect of the pure (integer-order) integral part. Therefore, only the fractional-order part $R_d(s) = s^{-0.15}$ has been approximated.

To obtain a finite-dimensional continuous approximation of the fractional-order integrator, a frequency domain identification technique is used, provided by the MATLAB function invfreqs(); see Section 12.3. An integer-order transfer function that fits the frequency response of the fractional-order integrator R_d in the range $\omega \in (10^{-2}, 10^2)$ is obtained. Later, the discretization of this continuous approximation is made by using the Tustin rule with prewarp-frequency ω_{cg} and sampling period $T = 0.002\,\mathrm{sec}$, obtaining a fifth-order digital IIR filter

$$R_d(z) = \frac{-0.1124z^{-5} + 0.7740z^{-4} - 2.0182z^{-3} + 2.5363z^{-2} - 1.5523z^{-1} + 0.3725}{-0.4332z^{-5} + 2.6488z^{-4} - 6.3441z^{-3} + 7.4747z^{-2} - 4.3462z^{-1} + 1}.$$

Therefore, the resulting total fractional-order controller is a sixth-order digital IIR filter given by

$$R_e(z) = 0.25 + 0.02 \left(\frac{1 - z^{-1}}{T} \right) R_d(z), \tag{15.38}$$

where the integer-order derivative is implemented by the standard finite differences formula.

15.7 Experimental Results

The control strategy proposed here, with the use of the outer-loop controller in (15.38), has been tested in the experimental platform of the picture in Figure 15.16, whose dynamics corresponds to the one described previously for the single-link flexible manipulator. In this section, the experimental results obtained are presented. The robustness of the system has been tested by changing the payload at the tip. Motor and tip position records are shown. Simulated control signals are plotted together with the experimental motor control signals for comparison purposes. Note that the simulations neglect Coulomb friction, whilst in the experimental platform a compensation term, +0.3 V for positive motor velocities and −0.25 V for negative ones, has been added to the control signal. These compensation values have been found by a trial and error process.

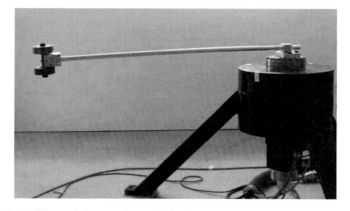

Figure 15.16 Photo of the experimental platform

The time responses of the system for payload changes are shown in Figure 15.17. Bigger masses than 3.2 kg have not been considered since they could cause the beam to reach its elastic limit and, hence, they will be neither simulated nor experimented with. For the nominal mass an overshoot $M_p = 0\%$ is obtained. As far as the robustness is concerned, a slight change in the overshoot of the response appears when the payload changes, due to the effect of the non-negligible inner-loop dynamics. However, only a 0.59% variation in the overshoot is obtained for the different masses.

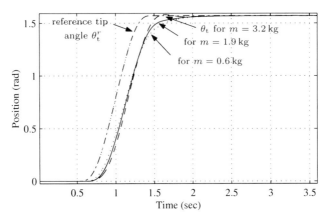

Figure 15.17 Time responses of the system with controller $R_e(s)$, considering non-negligible inner-loop dynamics

Figure 15.18 (a) shows the measurements of the tip angle θ_t and the motor angle θ_m, while 15.18 (b) shows the motor voltage V, both figures corresponding to a mass $m = 3.2\,\mathrm{kg}$. A relay type control appears in the transient and steady states due to Coulomb friction compensation. For this reason the experimental voltage signal obtained presents quick oscillations and is not zero in the steady state.

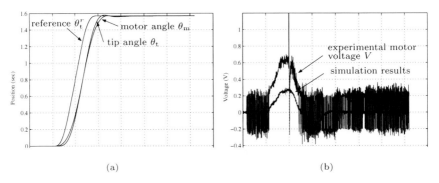

      (a)   (b)

Figure 15.18 Experimental results obtained using controller $R_e(s)$ for $m = 3.2\,\mathrm{kg}$: (a) experimental tip angle θ_t and motor angle θ_m, and (b) a comparison between the experimental and simulated motor voltage V

Figure 15.19 shows the measurements of the tip and motor angles and motor voltage obtained for the nominal mass $m = 1.9\,\mathrm{kg}$. And finally, Figure 15.20 shows the results when $m = 0.6\,\mathrm{kg}$.

Through Figures 15.18 (b), 15.19 (b), and 15.20 (b), it can be observed that the peak of the control signal keeps lower than the saturation limit (which is

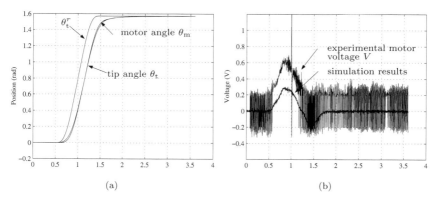

Figure 15.19 Experimental results obtained using controller $R_e(s)$ for $m = 1.9\,\text{kg}$: (a) experimental tip angle θ_t and motor angle θ_m, and (b) a comparison between the experimental and simulated motor voltage V

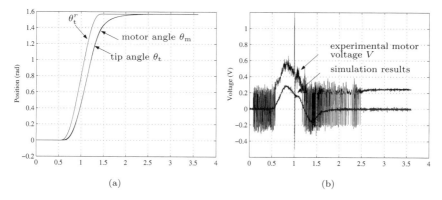

Figure 15.20 Experimental results obtained using controller $R_e(s)$ for $m = 0.6\,\text{kg}$: (a) experimental tip angle θ_t and motor angle θ_m, and (b) a comparison between the experimental and simulated motor voltage V

3 V) and remains almost constant in the presence of payload changes, with a value around 0.65 V, making this control strategy very suitable for avoiding motor saturation problems.

Another important aspect to note is that the fractional-order differentiator part of the controller, $s^{0.85}$, is implemented by $s^{0.85} = s/s^{0.15}$. That is, the fractional-order integrator part acts like a low-pass filter of the signal that enters the derivative operator and reduces the noise introduced through the control loop. Therefore, with the fractional-order controller, the system is not only more robust to payload changes, but also to noise presence. In fact, the control signals obtained with controller (15.28) when repeating the above experiments show a much noisier aspect than those exhibited here.

15.8 Summary

A fractional-order controller has been developed in this chapter to control single-link lightweight flexible arms in the presence of payload changes. The overall controller consists of three nested control loops. Once the Coulomb friction and the motor-beam coupling torque have been compensated, the inner-loop has been designed to give a fast motor response, by using a PD controller with high gains. The middle simplifying-loop reduced the system transfer function to a double integrator. The fractional-order derivative controller was used to shape the outer-loop into the form of a fractional-order integrator. The result is an open-loop constant phase system whose closed-loop responses to a step command exhibit constant overshoot, independent of variations in the load, unlike with a PD controller. It is also noted that the obtained fractional-order controller has a much simpler structure, and can be much more easily designed, than any robust controller designed to achieve similar specifications with any robust control methodology ($\mathcal{H}_\infty$, QFT, *etc.*), which imply costly optimization procedures and the definition of specifications by auxiliary transfer functions.

The fractional-order controller has been tested in an experimental platform by using approximate discrete implementations. From the results obtained it can be concluded that an interesting feature of the fractional-order control scheme is that the overshoot is independent of the tip mass. This allows a constant safety zone to be delimited for any given placement task of the arm, independent of the load being carried, thereby making it easier to plan collision avoidance. It must be noted that with the fractional-order controller the control signal is less noisy than with a standard PD controller, since the fractional-order integrator acts like a low-pass filter and reduces the effects of the noise introduced in the control loop. Besides, with this control strategy, changes in the payload impose only slight variations in the maximum value of the control signal, avoiding possible saturation issues.

Chapter 16
Automatic Control of a Hydraulic Canal

Let us now turn our attention to the problem of the automation of water transportation processes, which are often implemented by means of open hydraulic canals. We will show how simple fractional-order PI controllers like those described in Chapter 5 can substantially improve the robustness of standard PI or PID controllers. Hydraulic canals are a typical example of dynamical systems with important delays and whose parameters may vary over a large range. Fractional-order controllers are designed that improve phase and/or gain margins — which are classical indices that measure closed-loop process robustness — while keeping the desired closed-loop behavior of the canal with the nominal dynamics. Moreover, it is shown that, for canals with significant and variable time delays, fractional-order controllers behave better than standard controllers when all of them are combined with the Smith predictor.

16.1 Background and Motivations

Nowadays water is becoming a precious, rare, and scarce resource in many countries of the world. Then there is a growing interest in the application of advanced management methods to prevent wastage of this vital resource. Irrigation systems are the major water users, with a world's average of 71% of water use [239]. The most important objective of irrigation systems is to provide the demanded quantity of water to the different users at specified instants, and to guarantee the safety of the infrastructure [240].

Water losses in irrigation canals are large. Many irrigation systems are still being managed manually, leading to low efficiency in terms of delivered water vs water taken from the resource [241]. It is widely accepted that these losses can be substantially reduced by employing automatic control systems [242, 243]. In particular, automatic control leads to more efficient water management in irrigation systems which are based on open main canals subject to large losses [244]. The main objectives of these automatic control systems are: 1. improving water efficiency and distribution, 2. reducing water losses, and 3. supplying water users in due time.

New canals often have an automatic control system of water distribution. Additionally different old canals are being modernized with control equipments that help the canal operators to manage water distribution better. Therefore, canal automation has become a significant research area [245].

A typical main irrigation canal consists of several pools separated by gates that are used for regulating the water distribution from one pool to the next one. Figure 16.1 shows a scheme of a main irrigation canal with gates. In automatically regulated canals, the controlled variables are the water levels $y_i(t)$, the manipulated variables are the gate positions $u_i(t)$, and the fundamental perturbation variables are the unknown offtake discharges $q_i(t)$. In this figure $Q_i(t)$ is the discharge through the transversal section of the canal. If the water levels are measured near the end of the pool, the control system is called distant downstream control. Downstream control is considered as being superior to upstream control (water level measured immediately after the control gate) because it increases the efficiency of water use and improves the reliability and flexibility of the system [246]. The choice between downstream and upstream control is to a certain extent dictated by the design of the canal, and it may not be a variable the control engineer can play with [247]. It is not necessary to know the water level variations along the whole pool in order to control the water levels in the canals. It is enough to measure it at some specific points that depend on the specific canal control method to be used. The previous considerations make it possible to approximate the main irrigation canal dynamics by linear models with concentrated parameters and a time delay for control purposes [244, 247].

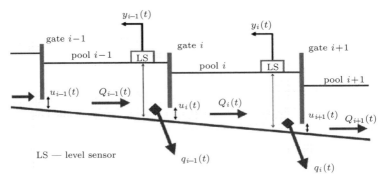

Figure 16.1 Scheme of a main irrigation canal with gates

Research pursued by some authors show that the parameters that characterize the dynamic behavior of different irrigation canals may exhibit wide variations when the water discharge regimes change in the operation range [248–250]. These canals are known as canals with time-varying dynamical parameters [251]. Thus any controller to be designed for this class of irrigation canal has to be robust to time variations in some parameters of the dynamical behavior [252].

Several strategies have been proposed for canal control. The most popular one is based on the classical PI controller [253–256]. However, many studies have shown that simple PI controllers do not seem to be well suited to solve the problem of effective water distribution control in irrigation canals characterized by significant time-varying dynamical parameters [249, 256, 257].

Then it would be desirable to provide a simple way to tune PI controllers according to a minimum performance, guaranteed for a set of hydraulic conditions. This is called the robust performance design problem [258]. The first robust tuning method of PI controller for an irrigation canal was proposed [248]. The designed controller appears to be efficient and robust, but the method lacks some flexibility in the design, since the controller parameters cannot be changed according to performance or robustness criteria. Litrico and coworkers developed in [258] a method to tune a robust distant downstream PI controller for an irrigation canal pool based on gain and phase margins requirements.

In this chapter we will explore the possibility of using quite simple fractional-order PI controllers in order to improve the robustness of the classical PI controller. According to the basic idea of enlarging phase and/or gain margins, we will show that these margins can be conveniently increased by using fractional-order controllers. We will develop a simple method for designing fractional-order robust controllers, and experimental results obtained in a laboratory hydraulic canal will be reported. More details about these contents can be found in [259, 260]. We also develop a fractional-order PI controller embedded in a Smith predictor structure to regulate canals with significant and variable time delays, according to the proposals in [261, 262].

This chapter is divided into two parts.

The first part begins with the description of the experimental canal. Then a linearized model of the dynamics of the canal is obtained, together with the range of variation of the model parameters. Afterward, the method to design robust fractional-order PI controllers is developed, and its robustness features are compared with those of an equivalent PI controller in the frequency domain. Later robustness features of the proposed controller in the time domain are compared with those of the equivalent PI controller and, as a consequence of this, a modification is proposed for the fractional-order controller in order to improve its steady-state behavior. Then experimental results are reported for all these controllers. Next the fractional-order PI controllers are compared with equivalent PID controllers, both in the time and frequency domains. Experimental responses of the fractional-order controllers are also compared with the best equivalent PID controller (the most robust one) and, finally, some conclusions are drawn.

The second part of the chapter begins by introducing the problem of real irrigation main canals, where the time delay is significant. Then the methodology developed in the first part to design controllers using frequency specifications is extended to the case of using a Smith predictor: a section is devoted to design standard controllers and another to design fractional-order ones. In the following section the dynamic model of a real canal pool is obtained. Afterward, several controllers (of integer and fractional-orders) are compared using this model. Later some conclusions are drawn, and the chapter ends with a section that summarizes the main results.

16.2 Description of the Laboratory Hydraulic Canal

The laboratory hydraulic canal where the fractional-order controllers will be tested is located in the Fluids Mechanics Laboratory of the Castilla-La Mancha University in Ciudad Real (Spain). It is a variable slope rectangular canal 5m long, 8cm wide, and the height of the walls is 25cm. All experiments were developed with a very small slope (very close to zero slope). The canal has two motorized adjustable slide undershoot gates (upstream and downstream) of 8cm width that allow its division into pools of different lengths. It has been arranged as three pools of different length separated by two submerged flow gates, the canal main pool being 4m long. However, considering its small dimensions, the canal is fundamentally operated as a single main pool approximately 4.7m long (the second submerged gate is kept open and pools two and three have been united) and in a downstream end operation method. The canal has a relatively small upstream pool. The nominal operation depth is 50mm and depth relative to nominal depth is $\pm$10mm. Figure 16.2 shows this experimental prototype canal, where the upstream canal pool can be seen at the right end and the canal main pool at the center of this figure.

A schematic representation of the control system of this experimental laboratory hydraulic canal is sketched in Figure 16.3. The water flows in a closed circuit from the upstream reservoir to the downstream storage reservoir in order to economize water. The water return to the upstream reservoir is

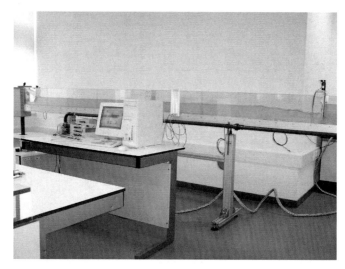

Figure 16.2 Experimental prototype canal of Fluids Mechanics Laboratory of the Castilla-La Mancha University (Spain)

maintained by two electric parallel variable speed pumps that can operate in an independent way (as redundant hardware) or simultaneously, depending on the water inflow needed to supply the canal. The total canal inflow is adjustable from 0 to $14\,\mathrm{m^3/h}$ ($\approx 3.88\,\mathrm{L/sec}$). Therefore, the canal does not have water losses except evaporation effects which are not significant in this case.

Two piezoelectric pressure sensors (PS), located in the canal external bottom, are used to monitor and control the upstream and downstream end water levels. The motorized slide undershoot gates are equipped with DC motors and gate position sensors (GPS). The canal is also equipped with an electromagnetic flowmeter (EMF) installed in the water return tube which measures the canal water inflow pumped by the electric pumps from the storage downstream reservoir toward the upstream reservoir. This flowmeter facilitates the supervision of the pumping operations. The canal uses a Pentium PC as canal control station. A SCADA application is installed in this PC to ensure the canal automatic control and supervision. Signals from different installed sensors and downstream gate positions, upstream and downstream end canal water levels, and canal water inflow are available through a profibus connection protocol. The developed SCADA application allows the implementation of different control strategies like fractional-order, standard and advanced, as well as different canal operation methods (upstream, downstream, downstream end, mixed, Bival, *etc.*), and set-points change of primary and secondary control loops.

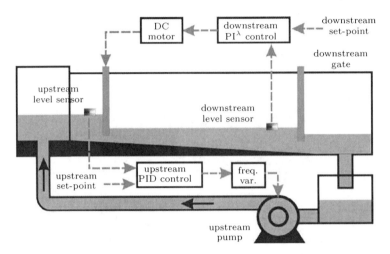

Figure 16.3 Overall control system of experimental prototype canal

Control of this hydraulic canal is very complicated because the maneuvers of opening and closing the upstream gate produce very large changes in the upstream water levels, due to the small dimensions of the upstream pool. In order to solve this problem, a secondary control loop that controls the upstream water level and keeps it in a fixed reference was implemented. The upstream water level is controlled by a PID controller, which acts on two variable speed pumps. The primary control loop carries out the control of the downstream end water level in the main pool of the hydraulic canal, by means of a fractional-order controller or a classical controller. All hydraulic canal controllers (control strategies) are installed inside the canal control station and are managed by the SCADA application. Figure 16.3 also shows a scheme of the overall control system of this canal. The primary control loop can be seen in the upper part of this figure, and the secondary standard control loop can be seen in its lower part.

The primary control loop has a second-order band-pass filter in order to reduce measurement noises whose transfer function is

$$F(s) = \left(\frac{8}{s+8}\right)^2. \tag{16.1}$$

This filter has a gain crossover frequency of 8 rad/sec which is high enough not to modify the dynamics of the canal (the next section will show that the largest time constant is around 10 sec and the smallest around 1 sec, having associated gain crossover frequencies of 0.1 rad/sec and 1 rad/sec respectively, both of them being much smaller than the crossover frequency of the filter). In turn the filter crossover frequency is much smaller than the noise frequencies, allowing for effective noise attenuation. This filter is applied to the signal obtained from the water level sensor. Moreover a saturation filter is also inserted between the controller and the DC motor in order to ensure that the control signal will not be out of the control range (0~50 mm).

16.3 Control-oriented Hydraulic Canal Dynamic Model

The dynamics of water flowing in irrigation open canals is modeled by using the so-called *Saint-Venant equations*, which are nonlinear hyperbolic partial differential equations, and are given by [263]

$$\frac{\partial A_h}{\partial t} + \frac{\partial Q_h}{\partial x_h} = q_h,$$

$$\frac{\partial Q_h}{\partial t} + \frac{\partial Q_h{}^2/A_h}{\partial x_h} + gA_h\frac{\partial y_h}{\partial x_h} = -gA_h S_f + K_h q_h V_h, \tag{16.2}$$

where, $A_h(x_h, t)$ is the canal cross section area; $Q_h(x_h, t)$ is the discharge through section A_h; $q_h(x_h, t)$ is the lateral discharge; $V(x_h, t)$ is the mean velocity in section A_h; $y_h(x_h, t)$ is the absolute water surface elevation; x_h is the distance along the canal; g is the gravity acceleration; t is the time variable; K_h is the weighting coefficient, $K_h = 0$ if $q_h > 0$ and $K_h = 1$ if $q_h < 0$; $S_f(x_h, t)$ is the friction slope. Nowadays, different methods exist for the solution of the Saint-Venant equations, but all of them exhibit large mathematical complexities [264]. Moreover, these equations are very difficult to use directly for controller design [244]. Often, the Saint-Venant equations are linearized around a set-point, and equivalent first-order systems plus a delay are used to model the canal dynamic behavior [247]. Sometimes second-order systems plus a delay are also used [250]. As was mentioned in Section 16.1, these linearized models have the strong drawback that their parameters may experience large changes when the canal operation regime varies, as a consequence of having been obtained from a highly nonlinear dynamic system. Then any controller to be designed for an irrigation canal has to be robust to variations in the parameters of such a linearized model.

The previous model of a single canal pool must be completed with the equation that describes the interaction between consecutive pools, and the influence of the gate opening signal $u(t)$. This is given by the equation of the discharge through a submerged flow gate [263]

$$Q_h(t) = C_d L \sqrt{2g} u(t) \sqrt{y_{up}(t) - y_{dn}(t)}, \tag{16.3}$$

where C_d is the gate discharge coefficient; L is the gate width; $y_{up}(t)$, $y_{dn}(t)$ is the upstream and downstream water levels, respectively.

The discharge variations through an upstream gate originate changes in the canal pool flow propagation, which is determined as a function of the wave velocity $C_h(t)$ and the already defined mean flow velocity $V_h(t)$ [265, 266]. If the measure of the water level variations is carried out at the downstream end of the canal pool, variations of the flow propagation imply variations in the canal pool time delay in the range (τ_{min}, τ_{max}). In this case, the time delay canal pool variations may be determined by [266]

$$\tau(t) = \frac{l_h}{C_h(t) + V_h(t)}, \tag{16.4}$$

where l_h is the canal pool length.

Taking into account all the above considerations, we use in this chapter a linearized dynamic model for the main canal pool consisting of a second-order system plus a delay, where all the parameters of the model (including the delay) may change with time:

$$T_1(t)T_2(t)\frac{d^2\Delta y(t)}{dt^2} + (T_1(t) + T_2(t))\frac{d\Delta y(t)}{dt} + \Delta y(t) = K(t)\Delta u(t - \tau(t)), \tag{16.5}$$

where $\Delta u(t)$ is the incremental gate opening and $\Delta y(t)$ is the incremental downstream water level.

As we consider that all these parameters only change when the flow regime changes — the linearization point changes — main canal dynamics can be described by the following transfer function:

$$G(s) = \frac{\Delta Y(s)}{\Delta U(s)} = \frac{K}{(1 + T_1 s)(1 + T_2 s)} e^{-\tau s}, \tag{16.6}$$

whose parameters are regarded as constant during a maneuver. Then the control problem of an irrigation main canal can be stated as the robust control of system (16.6) whose parameters may take values in specified ranges. This implies that the control-oriented dynamic model must include a set of nominal plant parameters and the range of variation of each of these parameters.

Figure 16.4 shows a scheme of the experiment carried out to identify the dynamics of our canal. A set of step like opening maneuvers are applied to the upstream gate of the main canal pool $u(t)$, and the downstream water level $y(t)$ is measured. A wide range of gate opening operations is used in order to characterize the nominal model and its range of variation.

Figure 16.5 shows the experimental results obtained with one of the gate maneuvers sequences carried out. This figure plots 1. the sequence of gate openings, 2. the downstream water level of the first pool (the water level immediately before the control gate), i.e., the value $y_{up}(t)$ of (16.3), measured by a piezoelectric pressure sensor, and 3. the second pool downstream water level $y(t)$ signal measured by a piezoelectric pressure sensor. It can

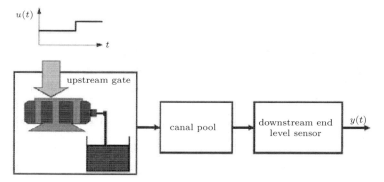

Figure 16.4 Step test of the canal pool

be observed that the PID controller of the secondary control loop keeps approximately constant the downstream water level of the first pool at a set-point of 60 mm. At the beginning of each gate maneuver some interaction can be noticed between the first and second pools: an instantaneous variation of this pool downstream water level with respect to the set-point is produced which is quickly removed by the controller. Experiments showed that the effects of this interaction on the main pool dynamics (second pool) are of secondary order and have little influence on the identification process.

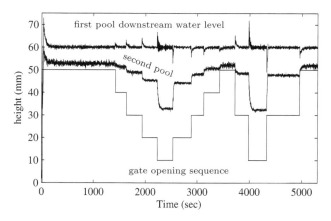

Figure 16.5 Canal responses to a step sequence test applied to a pool

Figure 16.6 shows the fittings attained with linear dynamic models of the form (16.6). The solid line shows the downstream water level signal of the second pool after having been filtered with (16.1). The dashed line of this figure shows the best fittings achieved with model (16.6), having run a standard least squares algorithm to identify its parameters. Fitting of model (16.6) has been carried out for each maneuver yielding different sets of parameter values depending on the initial and final control gate openings $u(t)$. Then the dashed line represents the simulated responses of the different models fitted to the different experimental transient responses, obtained for the different gate opening maneuvers. This figure shows quite good agreement between experimental and simulated responses, demonstrating that model (16.6) adequately represents the dynamics of the canal.

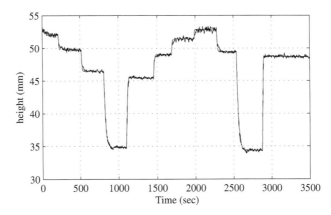

Figure 16.6 Identified models fittings of the experimental responses (*solid line*) for the second pool downstream water-level filtered signal (*dashed line*, not easily distinguishable, or the *smoother line*)

We define a nominal operating point given by $Q_{\text{nom}}(t) \approx 7\text{m}^3/\text{h}(1.94\text{L/sec})$, a first pool downstream water level (or immediately upstream from the control gate) $y_{\text{up}}(t) = 65\,\text{mm}$, and a second pool downstream water level $y(t) = 55\,\text{mm}$. In this case, the least squares identification procedure applied to the downstream water level response of the main pool when a step command is applied in the gate opening yields the following nominal values of the model (16.6) (hereinafter denoted as nominal model): $K_0 = 0.6$, $T_{10} = 10\,\text{sec}$, $T_{20} = 1\,\text{sec}$, and $\tau_0 = 2.6\,\text{sec}$.

However, when the discharge regime changes through the upstream gate in the operation range $(Q_{\text{min}}, Q_{\text{max}})$, where $Q_{\text{min}} = 1\,\text{m}^3/\text{h} \approx 0.27\,\text{L/sec}$ and $Q_{\text{max}} = 14\,\text{m}^3/\text{h} \approx 3.88\,\text{L/sec}$, the dynamical parameters of our canal prototype undergo wide variations. Carrying on the least square fitting identification procedure on several experiments like that shown in Figure 16.5 for different discharge regimes yields the following ranges of variation for the parameters of model (16.6):

$$0.2 \leqslant K \leqslant 1.1, \quad 7.9 \leqslant T_1 \leqslant 12.4, \quad 0.3 \leqslant T_2 \leqslant 1.6, \quad \tau = 2.6. \qquad (16.7)$$

We consider that T_1 is the dominant time constant (the larger time constant associated with the dynamics of the canal pool), while T_2 is the smaller time constant, which represents the motor + gate dynamics and the canal secondary dynamics, which is much faster than the canal pool dominant dynamics and can be regarded as almost invariant with regard to discharge regimes. This hypothesis is clearly supported by the obtained experimental results (16.7).

Moreover, as a result of the small size of our canal prototype, its time delay is approximately constant (and small). In real canals this value would be much larger and would undergo noticeable variations.

Figure 16.7 shows the Bode plots of the nominal plant and the extreme cases in the operation range $(Q_{\min}, Q_{\max})$. In these and the following plots the x axis expresses decimal logarithms of the frequency expressed in radians per second (decades). The effective control of hydraulic canals whose dynamic behavior is characterized by means of mathematical models with time-varying parameters requires the implementation of controllers that are robust to these parameter variations, and to all the frequency responses included between the limits drawn in Figure 16.7.

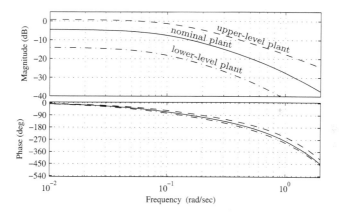

Figure 16.7 Bode plots of the nominal and extreme plants in the operation range $(Q_{\min}, Q_{\max})$

Finally, we mention that our hydraulic canal has no offtake discharges $q_h(t)$, so a disturbance model is not needed for the controller design.

16.4 Design and Experimental Studies of Fractional-order PI Controllers

16.4.1 Fractional-order PI Controller Design

Considering that the PI control strategy is the most commonly used in real hydraulic canal control systems because it can be tuned properly more easily than a PID [253, 267], in this section a fractional-order control strategy based on the generalization of a PI controller is proposed to control our laboratory hydraulic canal. Our controller may exhibit enhanced stability and robustness performance to hydraulic canal parameters variations compared to the standard (classical) control strategies (PI or PID). The extension of the derivation and integration orders from integer to non-integer numbers provides more flexible tuning strategies and, according to Chapter 5, an easier way of achieving control requirements with respect to classical control strategies. The proposed fractional-order PI controller, hereinafter denoted an FPI controller, is of the form

$$u_{\mathrm{FPI}}(t) = K_{\mathrm{p}} \left[\frac{1}{T_{\mathrm{i}}} \mathscr{D}^{-\lambda} e(t) + \mathscr{D}^{1-\lambda} e(t) \right], \quad 0 \leqslant \lambda \leqslant 1, \qquad (16.8)$$

whose transfer function is:

$$C_{\mathrm{FPI}}(s) = K_{\mathrm{p}} s^{1-\lambda} + \frac{K_{\mathrm{i}}}{s^{\lambda}} = \frac{K_{\mathrm{p}} s + K_{\mathrm{i}}}{s^{\lambda}}, \qquad (16.9)$$

where $K_{\mathrm{i}} = K_{\mathrm{p}}/T_{\mathrm{i}}$. Notice that (16.8), (16.9) becomes a PI controller when $\lambda = 1$, and a PD controller when $\lambda = 0$. We chose this particular controller structure because we wanted to have a PD controller (with its nice features of providing large positive phase, and faster and more damped time responses), but modified (by including the fractional-order denominator) in order to achieve closed-loop zero permanent error to a step command, and reduce the amplification of high-frequency noises. With a standard PI controller these last two features can be achieved but paying the price of a constant phase lag of $90°$. Our control structure allows achieving these features with a reduced constant phase lag of $90\lambda°$, which allows better transient dynamics than with a PI. Regarding high-frequency noise amplification, the magnitude Bode plot of a PD controller has a slope at high frequencies of $20\,\mathrm{dB/dec}$ while the slope of our FPI controller is smaller: $20(1 - \lambda)\,\mathrm{dB/dec}$. In this aspect the PI controller is the best one as its slope at high frequencies is $0\,\mathrm{dB/dec}$. Three parameters can be tuned in this controller: K_{p}, K_{i}, and λ. They are one more than in the case of the standard PI controller. The fractional-order can be used to fulfil additional specifications of the controlled system. The block diagram of the fractional-order control system of our experimental laboratory hydraulic canal is shown in Figure 16.8. This block diagram also includes a disturbance $D(s)$ caused by error in flow settings, emergency pump shutoff, changes in offtake backwater effects, offtake gate clogging, *etc.*

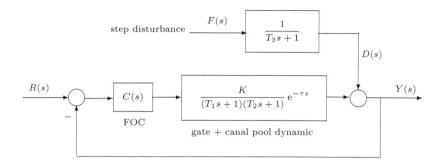

Figure 16.8 Block diagram of the laboratory hydraulic canal control system with fractional-order controller

We design a controller (for the nominal plant) that verifies the typical control system design frequency specifications: 1. a desired phase margin φ_m, which guarantees desired nominal damping and robustness to changes in the delay; 2. a desired gain crossover frequency ω_{cg}, which guarantees a desired nominal speed of response, and 3. zero steady-state error. The parameters of controller (16.9) that fulfil specifications 1–3 can be calculated by the following procedure. According to the Final-Value Theorem [22], condition 3 implies that $\lambda > 0$ must be verified. Conditions 1 and 2 can be expressed in a compact form by the complex equation

$$G(j\omega_{cg})C_{FPI}(j\omega_{cg}) = -e^{j\varphi_m}, \tag{16.10}$$

which involves two real equations. If the form of controller (16.9) is substituted in (16.10) and operating it thus follows that

$$K_i + j\omega_{cg}K_p = -\frac{(j\omega_{cg})^\lambda e^{j\varphi_m}}{G(j\omega_{cg})}. \tag{16.11}$$

If one takes into account that $(j\omega_{cg})^\lambda = \omega_{cg}^\lambda e^{j(\pi/2)\lambda}$, the controller gains are easily determined from

$$K_i = -\omega_{cg}^\lambda \Re\left(\frac{e^{j[(\pi/2)\lambda+\varphi_m]}}{G(j\omega_{cg})}\right), \quad K_p = -\omega_{cg}^{\lambda-1}\Im\left(\frac{e^{j[(\pi/2)\lambda+\varphi_m]}}{G(j\omega_{cg})}\right), \tag{16.12}$$

where $\Re(\cdot)$ and $\Im(\cdot)$ take the real and imaginary parts of a complex number respectively. Then a different controller that achieves conditions 1–3 is obtained for any value of λ. We use this free parameter to increase the robustness of the system. The model obtained from identification in Section

16.3 showed that the larger parameter variations are produced in the process static gain K. *Therefore, the parameter λ has been designed in such a way that the closed-loop system robustness, in the sense of stability, to gain changes is maximum, i.e., the gain margin obtains its maximum value.* This procedure is applied to the laboratory hydraulic canal. Having taken into account the nominal values of the hydraulic canal model parameters (nominal model) defined in Section 16.3, the following design specifications were chosen:

1. Desired settling time. The open-loop nominal canal settling time is approximately $t_s^o = 3T_{10} + \tau_0 = 32.6\,\text{sec}$ (see Section 16.3). The designer wishes to make the closed-loop system response about twice as fast. Thus a settling time $t_s^c = 15\,\text{sec}$, was chosen, which implies an equivalent closed-loop time constant $T_{10}^c = t_s^c/3 = 5\,\text{sec}$ and, therefore, a gain crossover frequency $\omega_{cg} = 1/T_{10}^c = 0.2\,\text{rad/sec}$ (this is a standard result in Control Theory which relates the time constant with the frequency response bandwidth [22].

2. A phase margin $\varphi_m = 60°$, which is a fairly standard value for this specification. A well known result for second-order systems is that the damping ratio ζ is related to the phase margin through the equation $\zeta \approx \varphi_m/100$ for values $0 \leqslant \zeta \leqslant 0.6$. By using $M_p = e^{-\pi\zeta/\sqrt{1-\zeta^2}}$, which relates the overshooting M_p and the damping ratio for second-order systems, one finds that this phase margin corresponds to a value of $M_p \approx 10\%$. These equations can often be used as reasonable approximations for higher order systems [22]. Our open-loop transfer function $G(s)C_{\text{FPI}}(s)$ is more complex than a second-order system. But the delay of our canal is relatively small and the secondary time constant T_2 is much smaller than the main time constant T_1. Consequently, the overshooting value given by these equations can be regarded as an accurate estimation.

According to the dynamic model (16.7) obtained in Section 16.3, the gain K is the canal parameter that undergoes the largest changes: the ratio between the maximum and minimum gains, given in that equation, is about 5 (T_2 may experience variations, which are as large as the gain but this is not taken into account in this stability analysis as its influence on the open- and closed-loop dynamics is very small). Thus, gain change is the most critical feature to be taken into account when designing a robust controller for the writers' canal.

Gain margin (the inverse of the magnitude of the frequency response at the frequency at which the phase is $-180°$) expresses how much the plant gain can be increased before the closed-loop system becomes unstable. It would therefore be desirable to maximize the gain margin in order to allow large gain changes without destabilizing the closed-loop system. Figure 16.9 shows the gain margins attained with the nominal model of the laboratory canal controlled by the proposed FPI regulator, as a function of the fractional order parameter λ. Figure 16.9 shows a maximum gain margin of 3.1 obtained for a value $\lambda = 0.37$. As the gain margin of the standard PI is 2.4 (see Figure 16.9 when $\lambda = 1$), the improvement achieved is 29 %. Thus the designed controller, obtained from (16.12) with that λ value, is

$$C_{\text{FPI}}(s) = 2\frac{1 + 1.6s}{s^{0.37}}. \qquad (16.13)$$

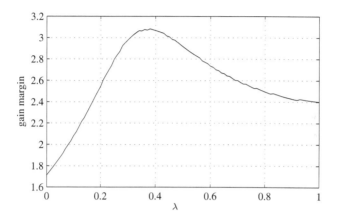

Figure 16.9 Gain margin in function of the fractional order parameter λ

For comparison purposes, the standard PI controller has also been designed. By using (16.12) with $\lambda = 1$, the following is obtained:

$$C_{\text{PI}}(s) = 0.2\frac{1 + 18.1s}{s}. \qquad (16.14)$$

The frequencies (ω_{cp}) corresponding to the gain margins with these controllers are, respectively, $\omega_{\text{cp}} = 0.54\,\text{rad/sec}$ (16.13) and $\omega_{\text{cp}} = 0.47\,\text{rad/sec}$ (16.14).

Figure 16.10 shows the magnitude and phase plots of the Bode plots of the open-loop systems, with the designed FPI and the standard PI controllers, respectively. The phase of the FPI controller is always less negative than the phase of the PI (with the exception of the frequency design point ω_{cg}, where both must logically coincide). Moreover, the magnitude with both controllers is quite similar in the frequency range from ω_{cg} to 1.2 rad/sec. This last frequency is much larger than frequencies ω_{cp}, which correspond to both the designed controllers. In fact, Figure 16.10 shows that the difference of magnitudes is less than 1 dB in the aforementioned frequency range.

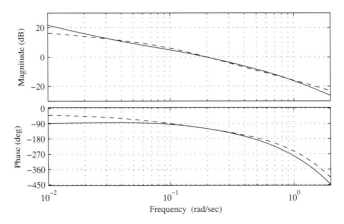

Figure 16.10 Bode plots for the nominal plant, with solid line for PI and dash line for FPI controller

These Bode plots show that:

- If the magnitude of both controllers is similar and the phase is less negative with the FPI than with the PI, then ω_{cp} with the FPI is greater than with the PI, and the gain margin is consequently also greater, thus increasing the robustness to plant gain changes.
- The fact that the FPI exhibits a less negative phase than the PI also implies more robustness (in the sense of stability) to changes in the main time constant of the plant. This last assertion can be justified as follows: if T_1 increases then the variation of the term $1/(1+j\omega T_1)$ causes attenuation in the magnitude of $G(j\omega)C_{FPI}(j\omega)$ and a decrease in its phase. These changes increase as T_1 increases. The system becomes unstable when the magnitude of $G(j\omega)C_{FPI}(j\omega)$ is $0\,dB$ and the phase $-180°$ (marginally stable condition). As the magnitude of the open-loop frequency response is similar for both controllers in a wide frequency range, and both remain similar in that range under changes in T_1, the value of T_1 that makes the closed-loop system marginally stable depends on the phase response of the open-loop transfer functions. A frequency response with smaller phase implies a smaller value for the T_1 that makes the closed-loop system marginally stable. As the fractional-order controller exhibits larger phase than the PI controller in all the frequency range, one will obtain a larger stability limit value for T_1 with the fractional-order controller than with the PI. A similar reasoning can be applied in the case of decrease in T_1. The arguments presented above are valid only under the assumption that frequencies ω_{cp} remain within the frequency range where the magnitudes of the frequency responses remain similar.

Otherwise, the robustness of the FPI and PI controllers to changes in the time delay (assuming that all the other plant parameters remain at their nominal values) is the same as this only depends on the phase margin, which has been designed to be the same for both controllers under nominal conditions. If time delay were the only parameter to change, the stability robustness to these changes would be given by $\omega_{cg}\Delta\tau = \varphi_m$, where $\Delta\tau$ is the maximum increment allowed from the nominal time delay τ_0 ($\Delta\tau = \tau - \tau_0$) and the phase margin is given in radians. Given the general control scheme shown in Figure 16.8, any controller $C(s)$ designed to have a given phase margin would therefore exhibit the same robustness to changes in the time delay, as is the case of the two controllers used here.

A further question is that of the stability robustness to simultaneous changes in all the plant parameters (K, T_1, and T_2 can take values in all the range (16.7)). Figure 16.11 shows the Nichols chart of the nominal open-loop dynamics, and the curves enclosing all the possible open-loop dynamics given by ranges (16.7), for the PI controller (*i.e.*, frequency responses of the open-loop transfer functions for any combination of parameters belonging to the intervals defined in (16.7) will be within the mentioned two extreme curves). This figure shows that the closed-loop system with the PI controller may become very undamped for some combinations of parameters related to the highest plant gain cases (the band defined in the Nichols chart includes the point $(-180, 0)$). Figure 16.12 shows the Nichols charts of the nominal open-loop dynamics, and the curves enclosing all the possible open-loop dynamics for the FPI controller. This last figure shows that the closed-loop system with the FPI controller remains more damped than the PI, even for the worst combination of the three aforementioned parameters (the band defined in the Nichols chart passes to the right of the point $(-180, 0)$, further away than with the PI). Then, the proposed fractional-order PI controller increases the robustness of the standard PI controller and improves the dynamical behavior while keeping a similar temporal response of the closed-loop system in the nominal plant case.

One final comment is that if one changed the design specifications (and the controller design problem) in the sense of allowing for the same robustness in the PI and the FPI controllers, then one would obtain an FPI controller which would attain a faster response with about the same damping as the PI for all the range of variation of canal parameters.

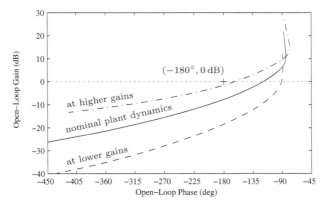

Figure 16.11 Nichols chart of the system with the PI controller

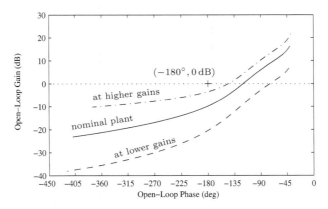

Figure 16.12 Nichols chart of the system with the FPI controller

16.4.2 Time Response of the Fractional-order PI Controller

In this section the temporal responses of both the PI and FPI controllers are studied by carrying out some simulations. Temporal responses of the closed-loop systems sketched in Figure 16.8 to step commands of amplitude 1 are obtained for the nominal plant, and two cases of the plant with extreme dynamics. Moreover, the ability to reject perturbations of these controllers is also compared. Considered perturbations are represented in Figure 16.8, where the input disturbance $f(t)$ is a step of amplitude -1, and the time constant T_3 of the associated block is assumed to be a third of the nominal value of T_1: $T_3 = 3.33$ sec (it is assumed that an offtake discharge is produced somewhere between the upstream gate and the downstream sensor, in a location closer to the downstream end of the pool than to the upstream end).

Plant dynamics are simulated with a simulation step of 0.01 sec. However controllers are simulated assuming a sampling period $T = 0.1$ sec and a zero-order hold with this period too, in order to reproduce the control system hardware used in the experimental setup.

Figure 16.13 shows the closed-loop responses with the PI controller to a unit step command for three cases: the nominal plant ($K_0 = 0.6$ m/m, $T_{10} = 10$ sec, $T_{20} = 1$ sec, $\tau_0 = 2.6$ sec), an extreme case with minimum gain ($K = 0.2$ m/m, $T_1 = 12.4$ sec, $T_2 = 1.6$ sec, $\tau_0 = 2.6$ sec), and an extreme case with maximum gain ($K = 1.1$, $T_1 = 7.9$ sec, $T_2 = 0.3$ sec, and $\tau_0 = 2.6$ sec).

Figure 16.14 shows the closed-loop responses with the FPI controller to the unit step command for the same three cases as above. Figure 16.15 shows the responses of both PI and FPI controllers in the cases of the three mentioned plant dynamics, when the offtake discharge disturbance described above is produced at instant 0 sec.

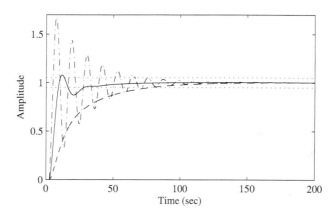

Figure 16.13 Temporal response of the system to a step command with the PI controller, *solid line* for nominal gain, *dashed line* and *dashed dotted line* for minimum and maximum gains, and *dotted lines* for ±5% of reference variations

Comparing Figures 16.13 and 16.14 it is observed that the fractional-order controller provides more damped responses than the PI while having similar rise times. However, the settling time of the closed-loop response is much larger when using FPI than PI controllers. Similar conclusions can be obtained from Figure 16.15 when comparing the ability of both controllers for removing the effects of offtake discharge disturbances

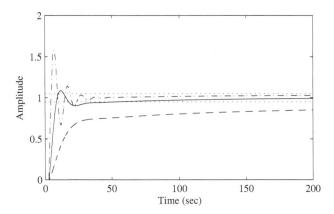

Figure 16.14 Temporal response of the system to a step command with the FPI controller, *solid line* for nominal gain, *dashed line* and *dashed dotted line* for minimum and maximum gains, and *dotted lines* for ±5% of reference variations

The reason why the fractional-order controller exhibits a much larger settling time than the PI controller is because the order of the integral term of the fractional-order controller is $\alpha = 0.37$. The Final-Value Theorem states that the fractional-order controller exhibits null steady-state error to a step reference if $\alpha > 0$ [22]. But in the case of our fractional-order controller, where $\alpha \ll 1$, the output converges to its final-value more slowly than if a PI controller is used.

Next a modification of the fractional-order controller is proposed in such a way that the settling time is decreased without changing the dynamics and robustness properties of this controller. In order to make the system reach its steady state faster, the open-loop frequency response is shaped in the sense of modifying its characteristics at low frequencies by increasing the type of the system, but leaving the frequency characteristics unchanged at medium and high frequencies. This can be achieved by multiplying the fractional-order controller by a term of the form $(s+\eta)/s$. If the parameter η is small enough, then this term improves the steady-state behavior, increases the type of the system, while leaving the dynamics unchanged (this is a standard procedure for designing PI controllers). After some numerical simulations we conclude that the best value of η for the referenced system is 0.02. Then the modified fractional-order controller, hereinafter denoted as FPI-PI controller, is given by

$$C_{\text{FPI-PI}}(s) = 2 \left(\frac{1+1.6s}{s^{0.37}} \right) \left(\frac{s+0.02}{s} \right). \qquad (16.15)$$

Note that $\eta \ll \omega_{\text{cg}}$. Then frequency specifications attained by the previous FPI are not therefore modified by this term.

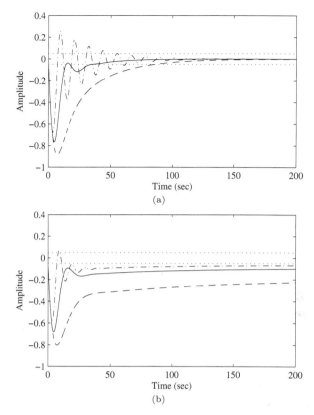

Figure 16.15 Temporal response of the system to an offtake discharge disturbance with the PI and FPI controllers, with nominal in *solid line*, while the *other two lines* for minimum and maximum gains: (a) PI controller and (b) fractional-order PI controller

Figure 16.16 shows the responses of the closed-loop system to a unit step command with this new controller in the cases of the three plant dynamics mentioned before.

Table 16.1 shows the settling times (defined as the time required by the system response to enter the band of ±5% of the desired steady-state value without exiting later) of the PI and FPI-PI controllers for these three plant dynamics. It can be observed that both controllers behave very similarly in the case of the nominal plant, while the fractional-order controller reaches in significantly less time the band of ±5 % of the desired state than the PI in the cases of non-nominal dynamics. It is also mentioned that the settling time attained by both controllers with the nominal plant (28.4 sec is quite different from the design value defined in the specifications of Section 16.4.1. This is because the used relationship between this specification and

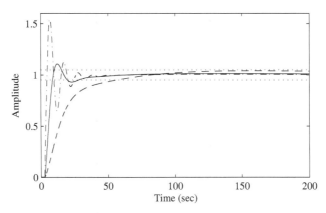

Figure 16.16 Temporal response of the system with the FPI-PI controller, with nominal in *solid line*, while the *other two lines* for minimum and maximum gains

the gain crossover frequency ($\omega_{cg} = 3/t_s^c$) is exact only for certain second-order systems. It is often used as an approximate equation, but in the case of plants with delays the errors may become noticeable. Therefore if a faster closed-loop system was wished, a larger design value for ω_{cg} must be chosen.

Table 16.1 Settling times of the PI and FPI-PI controllers

Controller	Settling time (sec)		
	Minimum gain plant	Nominal plant	Maximum gain plant
PI	79.4	28.4	71.6
FPI-PI	56.1	28.4	25.1

Figure 16.17 draws the Nichols charts of the nominal open-loop dynamics, and the curves enclosing all the possible open-loop dynamics for the FPI and FPI-PI controllers. It shows that frequency responses of both controllers are very similar with the exception of at very low frequencies (the region of larger gains of this figure).

16.4.3 Experiments with the Fractional-order PI Controller

In order to show the feasibility and robust performance of the proposed FPI controllers, real-time experiments were carried out in the laboratory hydraulic canal. The fractional-order control algorithm has been implemented in a

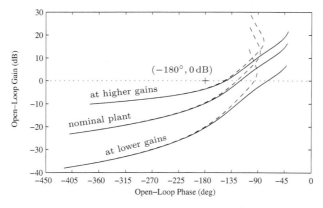

Figure 16.17 Nichols chart of the system with the FPI (*solid lines*) and the FPI-PI controller (*dashed lines*)

SCADA application with a home-tailored fractional-order control LabView library. Fractional-order operators have been approximated by FIR filters based on the Grünwald–Letnikov's definition of the discretized fractional-order operator as expressed in (2.14), combined with the short memory principle (see Chapter 2). A sampling period $T = 0.1$ sec has been used for all the controllers implemented in the SCADA application (fractional-order or standard). Taking into account the variation ranges of the dominant time constant T_1 and the time delay τ of the laboratory hydraulic canal (the dominant time constant can reach values of up to $T_1 = 12.4$ sec, whereas the time delay experiences only small variations), the transient of this canal to a step input may last up to 50 sec. Then fractional-order operators were implemented with a memory of 50 sec, which takes into account most of the transient history in the discretized fractional-order operator. If the sampling period was $T = 0.1$ sec, then $N = 500$ was chosen for the truncation value applied to (2.14). Therefore controller (16.13) was implemented as

$$u(t) = 2\mathscr{D}^{-0.37}e(t) + 3.2\mathscr{D}^{0.63}e(t), \tag{16.16}$$

by using the approximation

$$\mathscr{D}^{\alpha}e(t) = T^{-\alpha} \sum_{j=0}^{\min(\hat{j},N)} (-1)^j \binom{\alpha}{j} e(t - jT), \tag{16.17}$$

with this N value, being $\hat{j}$ such that $\hat{j}T \leqslant t < (\hat{j} + 1)T$. Figure 16.18 shows the frequency responses of the ideal fractional-order controller (16.13) and its discretized implementation using (16.16) and (16.17), with $T = 0.1$ sec and $N = 500$. This figure shows that both responses are quite similar in the range of one decade over and down the designed gain crossover frequency ω_{cg}, which is the interval of interest for the closed-loop behavior (also includes the frequency of the phase crossover frequency ω_{cp}). In the surroundings of ω_{cg} the magnitude difference is smaller than 1 dB, and the phase difference is smaller than $5°$. Moreover these differences diminish as the frequency increases. Then it is assumed that the proposed implementation is accurate enough to reproduce the fractional-order behavior of the FPI controller.

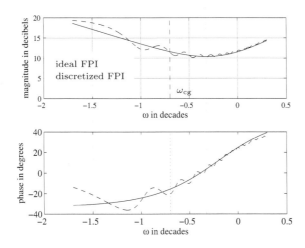

Figure 16.18 Frequency responses of fractional-order controllers: continuous and its implementation for the real time application with $T = 0.1\,\text{sec}$

The robustness performances of controllers (16.13) and (16.14) are evaluated in comparison with changes in the laboratory hydraulic canal dynamical parameters. Both controllers were designed to present the same phase margin φ_m. This means that both controllers with the nominal plant should exhibit the same robustness (from the point of view of stability) to changes in the process time delay τ. Both controllers were designed in order to present the same gain crossover frequency ω_{cg} with the nominal plant. They should therefore exhibit approximately the same settling time. However, the gain margins are different, which implies different robustness (again from the point of view of stability) to changes in the laboratory hydraulic canal static gain K. Moreover, the fact that the FPI controller has a less negative phase than the PI controller while exhibiting a similar magnitude in a significant range of frequencies makes the former more robust to changes in the main time constant. In Figures 16.19 and 16.20 the real-time closed-loop responses of the PI and FPI control systems are, respectively, shown for a working regime which does not correspond to the nominal operation regime (nominal parameters). In this case, the downstream water level set-point was increased by 5 mm. If one had the nominal operation regime, both responses would have exhibited similar responses as a result of the equivalent behavior of both controllers in the nominal regime. Figures 16.19 and 16.20 show that the FPI control system exhibits a better dynamic response than that of the PI (more damped and faster). But the settling time of the FPI control system is much larger than with the PI. Figure 16.20 shows that when the system output approaches the desired final-value the speed of convergence abruptly diminishes (it also shows that a small perturbation occurs at approximately 550 sec). This illustrates the phenomenon mentioned in Section 16.4.2 that if $\lambda < 1$ then the control system output converges on its referenced value more slowly than in the case of an integer controller with $\lambda = 1$.

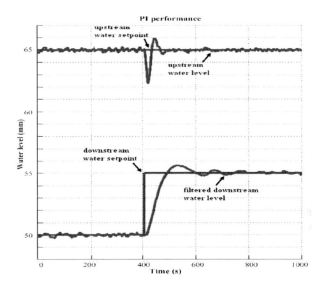

Figure 16.19 Real-time closed-loop responses of the PI control system

Next, the modified FPI controller of (16.15) was implemented by passing the output of (16.16) through the PI term

$$\hat{u}(t) = u(t) + 0.02 \int_0^t u(\tau)d\tau. \qquad (16.18)$$

The integral term of this last equation may be discretized by using any standard method (for example, simple summation of samples of $u(t)$ multiplied by the sampling period). Thus, the overall controller maintains the

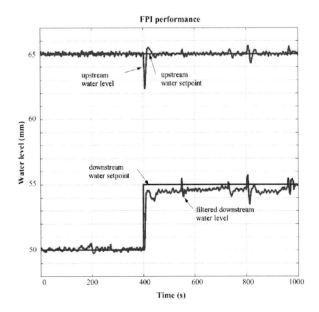

Figure 16.20 Real-time closed-loop responses of the FPI control system

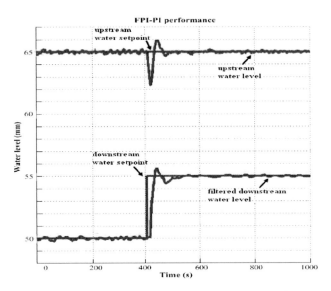

Figure 16.21 Real-time closed-loop responses of the FPI-PI control system

integration effect after discretization, in spite of this effect having been lost in the discretization of the FPI component by using (16.16) and (16.17).

Figure 16.21 shows the real-time closed-loop response of this FPI-PI control system. From Figure 16.21 it may be observed that the settling time has been improved significantly. Figures 16.19–16.21 demonstrate what was stated in Sections 16.4.1 and 16.4.2 by analysis and simulation: that the FPI-PI controller gives a faster response and with less overshooting than the PI controller, while keeping the zero steady-state error, unlike the simple FPI controller. It is mentioned that in all the experiments the step command signal has been passed through a first-order filter of time constant 3 sec in order to smooth the step and prevent the actuator saturation at the first instants of the control action.

With regards to the real-time implementation of the controller, the computation of (16.17) required the implementation of an FIR filter of length $N = 500$. This signifies that implementing control law (16.16) involves approximately 1000 additions and 1000 multiplications (two fractional-order operators have to be calculated), which have to be carried out in $T = 0.1$ sec. These operations can easily be carried out by any industrial computer in this time. Moreover, real canals exhibit slower dynamics than that used here. A fractional-order PI controller has been implemented in a real main irrigation canal pool in the Ebro River, Spain. In this implementation, the sampling period was $T = 60$ sec and the open-loop transient dynamics had a possible duration of up to 15,000 sec. A FIR filter of 500 operations was also used (details can be found in [268]). This number of operations can be drastically reduced by implementing an IIR filter rather than an FIR, as was explained in Chapter 12. In this case an IIR filter of order 3 or 4 is sufficient to reproduce the fractional-order operator behavior in the frequency range of interest for this application.

16.5 Design and Experimental Studies of PID Controllers

16.5.1 Design of a PID Controller

In Section 16.4.3 it was demonstrated that the proposed fractional-order controller is more robust than the standard PI controller, for controllers designed to achieve the same frequency specifications for the nominal dynamics. It can be argued that it has little merit since a PI has two parameters to be tuned while the FPI has three. Then the extra parameter of the FPI can be tuned to improve the robustness, but it is not clear that this controller can perform better than any other standard controller which would also offer three parameters to be tuned. This section is devoted to compare the proposed fractional-order controller with a standard PID controller, which also has three parameters to be tuned:

$$u_{\mathrm{PID}}(t) = K_{\mathrm{p}} \left[\frac{1}{T_{\mathrm{i}}} \mathscr{D}^{-1} e(t) + e(t) + T_{\mathrm{d}} \mathscr{D}^{1} e(t) \right],
\qquad (16.19)$$

whose transfer function is

$$C_{\mathrm{PID}}(s) = K_{\mathrm{p}} + \frac{K_{\mathrm{i}}}{s} + K_{\mathrm{d}} s,
\qquad (16.20)$$

where $K_{\mathrm{i}} = K_{\mathrm{p}}/T_{\mathrm{i}}$ and $K_{\mathrm{d}} = K_{\mathrm{p}} T_{\mathrm{d}}$.

First a method to design the PID controller from frequency specifications is proposed which is an extension of the method developed in Section 16.4.1. Assume that the PID controller has to achieve the same three specifications required in Section 16.4.1 for the PI and FPI controllers. If the form of controller (16.20) is substituted in (16.10) and operating it thus follows that

$$K_{\mathrm{i}} - K_{\mathrm{d}} \omega_{\mathrm{cg}}^{2} + j\omega_{\mathrm{cg}} K_{\mathrm{p}} = -\frac{(j\omega_{\mathrm{cg}}) e^{j\varphi_{m}}}{G(j\omega_{\mathrm{cg}})},
\qquad (16.21)$$

and the controller gains have to verify equations

$$K_{\mathrm{i}} - \omega_{\mathrm{cg}}^{2} K_{\mathrm{d}} = -\omega_{\mathrm{cg}} \Re\left(\frac{j e^{j\varphi_{m}}}{G(j\omega_{\mathrm{cg}})} \right), \quad
K_{\mathrm{p}} = -\Im\left(\frac{j e^{j\varphi_{m}}}{G(j\omega_{\mathrm{cg}})} \right).
\qquad (16.22)$$

Then there are now three parameters with which to tune a controller that must fulfil two specifications: 1 and 2 (specification 3 is always verified provided that the closed-loop system is stable) of Section 16.4.1. A further specification is therefore needed. It is chosen to maximize the robustness to changes in the gain K as the additional specification, accordingly to what was done in Section 16.4.1 for the FPI controller.

The design problem can therefore be reformulated as: *maximize the gain margin M_{g} subject to constraints (16.22) and $K_{\mathrm{i}} \geqslant 0$ (this is an additional closed-loop stability constraint obtained from the Nyquist stability criterion).* Note that this optimization problem is very simple as only one parameter has to be determined to get the maximum. In fact this optimization problem is of similar complexity to that solved in Section 16.4.1 for the FPI controller design.

Figure 16.22 shows the gain margin attained for the family of PID controllers that fulfil (16.22), in function of K_{i}. This plot exhibits a maximum gain margin of $M_g = 3.1$ at $K_{\mathrm{i}} = 0.37$, which corresponds to the controller

$$C_{\mathrm{PID}}(s) = 3.67 + \frac{0.37}{s} + 4.18 s.
\qquad (16.23)$$

This maximum gain margin value is very close to the maximum gain margin attained with the FPI controller.

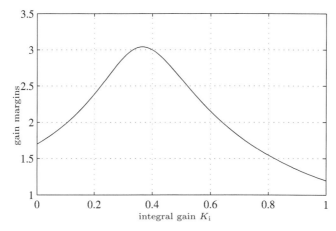

Figure 16.22 Gain margin in function of the K_i gain

Figure 16.23 draws the Nichols charts of the nominal open-loop dynamics, and the curves enclosing all the possible open-loop dynamics for the FPI and PID controllers. It shows that the bands corresponding to both controllers do not enclose the point $(-180, 0)$, and pass similarly closely to that point. Instead, in general, the frequency response of the PID controller has higher magnitude than that of the FPI.

Figure 16.24 plots the Bode plots of the nominal plant with the three controllers: PI, FPI, and PID. From these plots the phase crossover frequency with controller (16.23) is obtained: $\omega_{cp} = 0.645$ rad/sec. Figure 16.24 shows that the PID controller is the one with the largest magnitude both at low and high frequencies (the phase crossover frequency of the PID is also the largest among the three controllers). This means that though the FPI and PID controllers have the same gain margin, exhibiting the same robustness features to plant gain changes, the closed-loop system with the PID controller is the most sensitive to the effects of high-frequency unmodeled dynamics, or high-frequency noises (which are quite likely to appear in water level sensors). For example, a common case is having sensor noise at the Nyquist frequency of the sampling, which in this case would be $\omega_N = \pi/T = 31.4$ rad/sec. Noise attenuation attained by the closed-loop system implemented with the PID controller would be 42 dB while the attenuation using the FPI controller would be 55.4 dB. This means that the FPI controller attenuates this noise 4.7 times more than the PID controller. In any case the controller that would attenuate this noise most is the PI controller as it exhibits a slope of -20 dB/dec at high frequencies while the slope of the FPI is $-20\lambda = -7.4$ dB/dec.

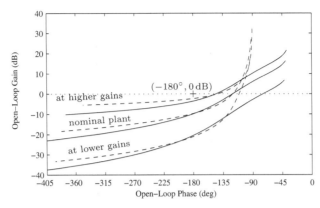

Figure 16.23 Comparison of Nichols chart of the system with the FPI (*solid lines*) and the PID controller (*dashed lines*)

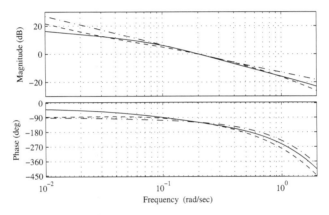

Figure 16.24 Bode plots for the nominal plant with FPI (*solid lines*), PI (*dashed lines*) and PID (*dashed dotted lines*) controllers

16.5.2 Experimental Comparison of the PID Controller and the Fractional-order Controllers

Experiments with these controllers showed that the PID controller (16.23) often became very oscillatory (mainly with low flows) while both FPI controllers (16.13) and (16.15) remained stable in all of the cases. Moreover, in the cases where the PID controller remained stable, its behavior was much more oscillatory than the behaviors of both fractional-order controllers. Figures 16.25–16.27 show the closed-loop responses with the PID, FPI, and FPI-PI controllers, respectively, in a maneuver that implies two steps: the first one involves passing from a very low water level to a low water level (from a very low to a low flow), and the second step involves passing from the previous low water level to a medium water level (the nominal level). These figures illustrate the aforementioned dynamic features. System specifications obtained in this experiment with these three controllers are shown in Table 16.2. Note that these specifications differ from the design ones because the parameters of the plant in the working regime of this experiment are not the nominal ones (though the last steady state is close to the nominal regime).

Table 16.2 System specifications with controllers (16.23), (16.13), and (16.15)

PID			FPI			FPI-PI		
Rise time (sec)	Overshoot (%)	Settling time(sec)	Rise time (sec)	Overshoot (%)	Settling time(sec)	Rise time (sec)	Overshoot (%)	Settling time(sec)
11	50	117	9	19	50	10	6	55

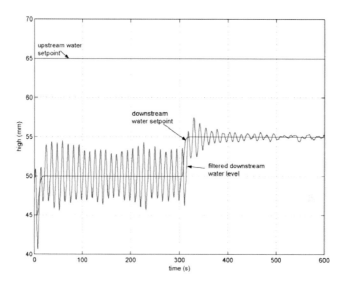

Figure 16.25 Real-time closed-loop responses of the PID control system

Figure 16.28 shows the control signals generated by the three controllers (16.23), (16.13), and (16.15). In these experiments the command signal has also been passed through the same first-order filter of time constant 3 sec as in the PI control experiments, in order to avoid the actuator saturation at the first instants of the control action. Such filtered command signal is shown in the upper part of Figure 16.28.

In the first time interval of the experiment, the PID controller is detuned as a consequence of the values of the canal parameters in this particular

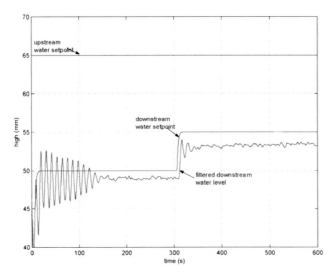

Figure 16.26 Real-time closed-loop responses of the FPI control system

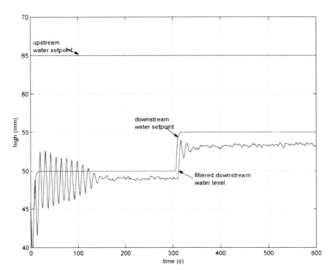

Figure 16.27 Real-time closed-loop responses of the FPI-PI control system

working regime, leading to an oscillatory control signal. In the target working regime the new canal parameters are closer to the nominal parameters leading to a better tuning of the PID parameters and a much better behavior. Both fractional-order controllers remain satisfactorily tuned under these two working regimes, always yielding good closed-loop dynamic performances. Note in Figure 16.28 that signals generated by the fractional-order controllers are smoother and less oscillatory than the signal generated by the PID controller in both maneuvers. Experimentally obtained responses in other working conditions of the hydraulic canal exhibit similar features to those presented here, showing the increased robustness of the FPI-PI controller over the PID. Moreover, the control signal of the FPI-PI controller (16.15) is less oscillatory than that of the FPI controller (16.13).

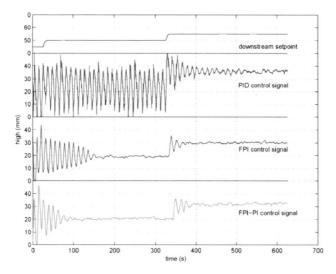

Figure 16.28 Control signals of the three studied controllers

16.5.3 Conclusions on the Robust Control of the Laboratory Hydraulic Canal

The implementation of a simple fractional-order strategy in an experimental laboratory hydraulic canal, which was characterized to present time-varying gain and time constants, but a fixed time delay, has been described. A class of fractional-order PI controllers has been proposed (other possible structures of fractional-order PI controllers could have been nominated), and a straightforward method to design it from frequency specifications has been developed. Moreover, the use of a fractional-order PI controller rather than a PD controller (or a fractional version of this) was motivated by the desire to attain zero steady-state error to step changes in the reference, and attenuate the high frequency noises as far as possible.

The fractional-order controller has been compared with two industrial controllers: a PI and a PID. In order to study equivalent controllers, the three controllers had to achieve the same frequency specifications $(\omega_{\mathrm{cg}}, \varphi_{\mathrm{m}})$ with the nominal plant.

As it was noticed that the canal gain parameter is that which experiences the largest variations, the fractional order of the controller transfer function was designed in order to maximize the robustness of the closed-loop system to changes in a such parameter (thus maintaining the system's stability). This involved maximizing the gain margin of the open-loop system. The PID controller was also designed in order to maximize such robustness (gain margin was also maximized) for comparison purposes. The Nichols charts show that, for all the possible combinations of canal parameter values, the FPI and PID plots pass far away from the point $(-180, 0)$. However the PI band passes very close to such point, showing that this controller is not robust to the highest plant gains.

PID and FPI controllers exhibit similar robustness to gain changes. However, their magnitude Bode diagrams show that the open-loop system with the PID has larger magnitudes than with the FPI. This means that the FPI is less sensitive to high gain noises and unmodeled high-frequency dynamics (*e.g.*, changes in the main time constant T_1).

Therefore, the FPI design has increased the robustness of the closed-loop system not only to changes in the gain but also to changes in the other parameters of the process. The robustness of this fractional-order controller was then studied under realistic conditions in the laboratory hydraulic canal, where canal parameters could vary within the intervals given in Section 16.3, showing better behavior than the two industrial controllers. The FPI controller has also been modified in order to achieve an acceptable settling time by adding a PI factor. By doing this, the design of the steady-state error behavior is decoupled from the design of the frequency characteristics. In fact, parameter λ, which influences both the phase margin and the steady-state error, is designed in the method proposed here to attain only the maximum gain margin.

16.6 Control of Hydraulic Canals with Significant Delays

Real irrigation main canals are systems which are distributed over long distances, with significant time delays and dynamics that change with the operating hydraulic conditions [247, 250, 269, 270].

Moreover, experiments reported by some authors confirm that several irrigation main canal pools (IMCP) exhibit large time-varying time delays (LTVD) when their discharge regimes change in a given operation range

[251, 256, 269, 271–273]. Controllers designed for this class of IMCP must therefore be robust to these time delay variations [241, 252, 256, 274].

Many studies have shown that simple PID controllers appear to be unsuitable for solving the problem of effective water distribution control in an IMCP with LTVD [240, 256, 258, 275–277].

Some authors have proposed the use of the Smith predictor in IMCP control systems to overcome the time delay that characterizes these systems [241, 252, 278]. However, it is well known that small modeling errors can cause instability in Smith predictor based control systems if the controller is not properly designed [279–281].

Various works concerning the application of fractional-order PID controllers to control water distribution in IMCP, which are characterized by large time-varying dynamic parameters, have recently appeared [261, 262, 274, 282]. These papers explore the robustness features of fractional-order controllers combined with the Smith predictor when applied to the effective water distribution control in an IMCP with LTVD. This section and the next ones report some of these results. In particular they are focused on the design of a robust fractional-order PI controller combined with a Smith predictor [261, 262] (SP-FPI controller). We will show that this class of controllers increases the robustness to changes in the process time delay, which is the most determinant parameter in the stability of the closed-loop control system of water distribution in an IMCP [258, 266, 283].

In the following some basics of the Smith predictor based control will be presented.

16.6.1 Standard Control Scheme

Assume it is desired to design a controller for an IMCP with LTVD exhibiting a linearized model of the form (16.5), (16.6).

First a control system is designed for the nominal model $G_0(s)$ with parameters $\{K_0, T_{10}, T_{20}, \tau_0\}$, which must verify the typical design frequency specifications: 1. a desired phase margin (φ_m), which guarantees the desired nominal damping and robustness to changes in time delay; 2. a desired crossover frequency (ω_{cg}), which guarantees the desired nominal speed of response, and 3. zero steady state error to a step command, which implies that the controller must include an integral term. As was stated previously, these three specifications can be attained through the use of a PI controller arranged according to the standard control scheme of Figure 16.8.

The robustness of this controller to changes in the canal pool time delay is given by $\hat{\tau}$, which is the maximum time delay with which the closed-loop control system remains stable. It can be easily calculated in this control scheme:

$$\hat{\tau} = \frac{\varphi_{\mathrm{m}}}{\omega_{\mathrm{cg}}} + \tau_0. \tag{16.24}$$

It must be noted that any controller used in the scheme of Figure 16.8 that fulfils specifications 1 and 2 will exhibit the same time delay stability margin (16.24), independently of its particular form. A different control structure must therefore be used if it were desired to improve such robustness value.

Consider the frequency response of model (16.6) in the nominal case:

$$G_0(j\omega) = G_0'(j\omega)e^{-j\omega\tau_0}, \tag{16.25}$$

where $G_0'(j\omega)$ is the rational part of the model. Design specifications in the frequency domain 1 and 2 are accomplished if the following condition is verified:

$$G_0'(j\omega_{\mathrm{cg}})C(j\omega_{\mathrm{cg}}) = -e^{j(\varphi_{\mathrm{m}}+\tau_0\omega_{\mathrm{cg}})}, \tag{16.26}$$

which is easily obtained from (16.10) substituting there $G(j\omega)$ by $G_0'(j\omega)$ according to (16.25), and $C_{\mathrm{PI}^\lambda}(j\omega_{\mathrm{cg}})$ by $C(j\omega_{\mathrm{cg}})$.

Defining $X = e^{j(\varphi_{\mathrm{m}}+\tau_0\omega_{\mathrm{cg}})}/G_0'(j\omega_{\mathrm{cg}})$, the parameters of a controller with a particular structure are obtained from

$$C(j\omega_{\mathrm{cg}}) = -X. \tag{16.27}$$

The closed-loop transfer function of the standard control system shown in Figure 16.8 becomes, for the nominal plant

$$M_0(s) = \frac{C(s)G_0'(s)e^{-\tau_0 s}}{1 + C(s)G_0'(s)e^{-\tau_0 s}}. \tag{16.28}$$

The time delay term that appears in the denominator of (16.28) prevents the use of many well-known techniques for the analysis and design of linear control systems. Alternative control schemes have consequently been proposed in the last few decades to overcome this problem. Among these, the most widely used is the Smith Predictor scheme [281, 284, 285], which permits more efficient controllers than the traditional scheme of Figure 16.8. Several Smith predictor based control systems are proposed for our IMCP in the following sections.

16.6.2 Smith Predictor Based Control Scheme

The structure of a Smith predictor based control system for our IMCP is shown in Figure 16.29, in which model (16.6) is considered for the canal dynamics. Given a nominal model according to (16.25), and the real canal model $G(j\omega) = G'(j\omega)e^{-j\omega\tau}$, then the closed-loop transfer function of the scheme shown in Figure 16.29 is

$$Y(s) = M(s)R(s) + N(s)D(s),\qquad(16.29)$$

where

$$M(s) = \frac{C(s)G'(s)e^{-\tau s}}{1 + C(s)(G(s) - G_0(s) + G'_0(s))},\qquad(16.30)$$

$$N(s) = \frac{1 + C(s)(G'_0(s) - G_0(s))}{1 + C(s)(G(s) - G_0(s) + G'_0(s))}.\qquad(16.31)$$

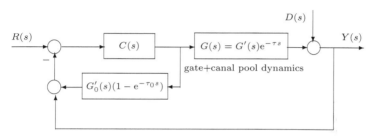

Figure 16.29 Smith predictor based control scheme of an irrigation main canal pool

Note that the time delay stability margin $\hat{\tau}$ now depends on the particular form of the controller $C(s)$, fulfilling specifications 1 and 2 (see the denominators of (16.30) and (16.31)), as opposed to that which occurred in the standard control system of Figure 16.8, in which any controller fulfilling these conditions exhibited the same delay stability margin. A proper design of the controller embedded in a Smith predictor scheme may thus increase such a stability margin. In the nominal case ($G(s) = G_0(s)$) the Smith predictor is tuned, and the time delay term is removed from denominators of expressions (16.30), (16.31) yielding

$$M(s) = \frac{C(s)G'_0(s)e^{-\tau_0 s}}{1 + C(s)G'_0(s)}, \quad N(s) = \frac{1 + C(s)G'_0(s)(1 - e^{-\tau_0 s})}{1 + C(s)G'_0(s)}.\qquad(16.32)$$

According to these equations, the parameters of the $C(s)$ controller may be designed using the time delay-free part of the plant model [281, 286], allowing the use of standard techniques for linear systems. Expression (16.32) can be rewritten as

$$M_0(s) = M'_0(s)e^{-\tau_0 s},\qquad(16.33)$$

where $M'_0(s)$ contains the closed-loop rational dynamics, and the delay is being applied to its output.

16.6.2.1 Design Method for Smith Predictor Based PI Controllers (SP-PI)

In this section we consider that the controller $C(s)$ which is embedded in this control structure is a PI controller which fulfils specifications 1–3. It can be designed by repeating the same process shown in Section 16.6.1, but by taking into account that $G_0'(s)$ must be handled rather than $G_0(s)$ in the characteristic equation of (16.28) (this basically implies that the term $\tau_0 \omega_{cg}$ of the imaginary exponential of (16.26) disappears there and in X of (16.27)). It now yields

$$C_{\text{SP}-\text{PI}}(s) = K_{\text{p}} + \frac{K_{\text{i}}}{s}, \quad K_{\text{p}} = -\Re\left(X_{\text{SP}}\right), \quad K_{\text{i}} = \omega_{cg}\Im\left(X_{\text{SP}}\right), \quad (16.34)$$

where

$$X_{\text{SP}} = \frac{e^{j\varphi_{\text{m}}}}{G_0'(j\omega_{cg})}. \quad (16.35)$$

16.6.2.2 Design Method for Smith Predictor Based PID Controllers (SP-PID)

In order to design a PID controller of the form

$$C_{\text{SP}-\text{PID}}(s) = K_{\text{p}} + K_{\text{d}}s + \frac{K_{\text{i}}}{s}, \quad (16.36)$$

the same process shown in Section 16.6.1 can be repeated, again taking into account that $G_0'(s)$ must be handled rather than $G_0(s)$ in the characteristic equation of (16.28). It now follows that

$$K_{\text{p}} = -\Re\left(X_{\text{SP}}\right), \quad K_{\text{i}} - K_{\text{d}}\omega_{cg}^{2} = \omega_{cg}\Im\left(X_{\text{SP}}\right). \quad (16.37)$$

We now have three parameters with which to tune a controller that must fulfil two specifications: 1 and 2 (specification 3 is always verified provided that the closed-loop system is stable). A further specification is therefore needed. We choose to maximize the robustness to changes in the time delay ($\hat{\tau}$) as the additional specification. The design problem can therefore be reformulated as *maximize $\hat{\tau}$ subject to constraints (16.37) and $K_{\text{i}} > 0$ (this is an additional closed-loop stability constraint obtained from the Nyquist stability criterion)*.

16.7 Fractional-order Control of Hydraulic Canals with Significant Delays

In this section fractional-order PI controllers embedded in a Smith predictor structure are proposed to increase the robustness to time delay changes of IMCP with LTVD. First the simplest fractional-order controller, an I^λ controller, is designed. Later a PI^λ controller is designed. Parameters of both controllers are calculated by using the same methodology of the previous section.

16.7.1 Design Method for Smith Predictor Based I^λ Controllers (SP-FI)

A fractional-order I controller (I^λ) embedded in a Smith predictor based control scheme is described. This controller was studied in [261], and is of the form $u_{\mathrm{FI}}(t) = K_{\mathrm{i}}\mathscr{D}^{-\lambda}e(t)$, $0 < \lambda \leqslant 1$. Its transfer function is

$$C_{\mathrm{SP-FI}}(s) = \frac{K_{\mathrm{i}}}{s^\lambda}. \tag{16.38}$$

Two parameters are to be designed in this controller: K_{i} and λ. They are tuned to fulfil frequency specifications 1 and 2. We should mention that the zero steady state error condition 3 is guaranteed by controller (16.38) provided that $\lambda > 0$, according to the Final Value Theorem [22]. The complex equation given by conditions 1 and 2 now becomes (see (16.35))

$$\frac{K_{\mathrm{i}}}{(\mathrm{j}\omega_{\mathrm{cg}})^\lambda} = -X_{\mathrm{SP}}. \tag{16.39}$$

Taking into account that $(\mathrm{j}\omega_{\mathrm{cg}})^\lambda = \omega_{\mathrm{cg}}^\lambda e^{\mathrm{j}(\pi/2)\lambda}$, and operating in (16.39), this yields the parameters of controller (16.38)

$$\lambda = \frac{2}{\pi}\left(\pi - \angle X_{\mathrm{SP}}\right), \quad K_{\mathrm{i}} = \omega_{\mathrm{cg}}^\lambda |X_{\mathrm{SP}}|. \tag{16.40}$$

16.7.2 Design Method for Smith Predictor Based PI^λ Controllers (SP-FPI)

Here a robust fractional-order PI controller embedded in a Smith predictor based control scheme is designed. It is of the form $u_{\mathrm{PI}^\lambda}(t) = K_{\mathrm{p}}e(t) + K_{\mathrm{i}}\mathscr{D}^{-\lambda}e(t)$, $0 < \lambda \leqslant 1$ which is also denoted as the PI^λ controller. Its transfer function is

$$C_{\mathrm{SP-FPI}}(s) = K_{\mathrm{p}} + \frac{K_{\mathrm{i}}}{s^\lambda}. \tag{16.41}$$

In (16.41), $\lambda = 1$ gives the standard PI controller. Also note that the structure of this controller is different from the fractional-order PI controller used in the laboratory hydraulic canal (16.8), (16.9). Three parameters are to be designed in this controller: K_p, K_i, and λ. Again we have one parameter more than in the case of the SP-PI controller. As occurred with the SP-PID, we design controller (16.41) in order to fulfil frequency specifications 1 and 2, and to maximize the robustness to changes in the time delay (maximize $\hat{\tau}$). As with the SP-FI controller, the zero steady state error condition 3 is guaranteed by the Final Value Theorem. The complex equation given by conditions 1 and 2 now becomes

$$ K_p + \frac{K_i}{(j\omega_{cg})^\lambda} = -X_{SP}. \tag{16.42} $$

Expanding $(j\omega_{cg})^{-\lambda} = \omega_{cg}^{-\lambda} [\cos(\pi\lambda/2) - j\sin(\pi\lambda/2)]$, and operating in (16.42), yields the parameters of controller (16.41):

$$ K_p = -\Re(X_{SP}) - \cot\left(\frac{\pi}{2}\lambda\right)\Im(X_{SP}), \quad K_i = \frac{\omega_{cg}^\lambda}{\sin(\pi\lambda/2)}\Im(X_{SP}). \tag{16.43} $$

16.8 Control-oriented Dynamic Model of an IMCP

The laboratory hydraulic canal of the previous sections cannot be used here because its delay time is small compared with the dominant time constant ($\tau_0 = 2.6$ sec. and $T_{10} = 10$ sec). Moreover its time delay remains quite constant for the different flow regimes.

Then the controllers proposed in the previous two sections will be compared in the regulation problem of a real IMCP with a significant delay. In order to do this, the second pool of the Aragon Imperial Main Canal (AIMC), which pertains to the Ebro Hydrographical Confederation in Zaragoza, Spain, has been considered. It constitutes one of the most important hydrographical confederations in this country, with an approximate area of $86{,}100 \text{ km}^2$. This canal is considered to be an excellent hydraulic work and flows more or less parallel to the right-hand bank of the Ebro River. It supplies drinking water to the city of Zaragoza, guaranteeing 60% of its needs. The surface irrigated by this canal is 26,500 ha. The production of the area under irrigation is mainly oriented towards extensive herbaceous crops, along with fruit cultivation and horticulture. This canal obtains its water from the Ebro river thanks to the elevation of the Pignatelli dam. This dam is 230 m long and 6.5 m high, and its building was completed in 1790. The water passes through the Casa de Compuertas (Gate House) which controls the $30 \text{ m}^3/\text{sec}$ of discharge at its origin, although this value may sometimes be superior as a result of a high flow in the Ebro river. This canal is 108 km long and has a variable depth of between 3 m and 4 m and a trapezoidal cross section. It has ten pools of different lengths which are separated by undershoot flow gates. All the pools of the Aragon Imperial main canal are electrified and equipped with downstream end water level and gate position sensors, motors for the gates' positioning, and systems for data acquisition, processing, and storage. The canal also has a remote supervisory control and data acquisition system (SCADA) with communication by radio and field buses that provide real time data of controlled variables. This system allows the effective implementation of daily water management decisions in the canal, thus permitting water to be supplied according to the irrigation schedule.

Data and results reported in this section were obtained from the second pool in the AIMC, which is known as PK8 and has a complex hydraulic infrastructure. It is a cross structure main canal pool of 11 km in length, a variable depth of between 3.5 m and 3.2 m, a variable width of between 15.0 m and 20.0 m, and a design maximum discharge of $30 \, \text{m}^3/\text{sec}$, in its entire extension. The available measurements are the downstream end water level and the upstream gate position.

Experiments based on the response to a step like input were carried out at PK8 in order to obtain a mathematical model with which to describe its dynamic behavior. In this test the downstream gate was kept in a fixed position, the upstream gate was excited with a step signal, and the downstream end water level was measured with a level sensor. The water level and gate position are given in meters, and were uniformly sampled over a period of $T = 60 \, \text{sec}$. The fully shut gate has a 0 m position, while positive values signify an open gate. The experimental response of this IMCP to a step command is presented in Figure 16.30. Such a response was obtained under its nominal hydraulic operation regime $(Q_{\text{nom}}(t) = 14 \, \text{m}^3/\text{sec})$ and shows that the dynamic behavior of this main canal pool can be described by the second-order transfer function with a time delay (16.6).

When the discharge through the upstream gate corresponds to the nominal hydraulic operation regime, the nominal values of the parameters of model (16.6) are obtained. Then nominal parameters of nominal model $G_0(s)$ are $K_0 = 1.11 \, \text{m/m}$, $T_{10} = 1250 \, \text{sec}$, $T_{20} = 50 \, \text{sec}$, and $\tau_0 = 550 \, \text{sec}$. As may be observed in Figure 16.30, measurements of the water level in this IMCP are not affected by reflecting waves caused by upstream discharge changes, since the downstream end water level sensor is installed inside an off-line stilling well at the downstream end of the pool. Validation results of linear model (16.6) with the parameters' estimated nominal values (nominal model) are also shown in Figure 16.30. This figure shows good agreement between the

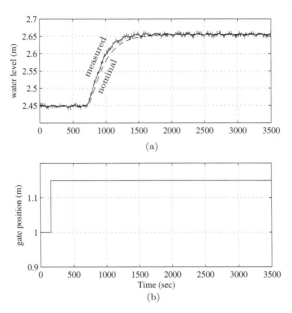

Figure 16.30 Validation result of the linear model (16.6) with estimated nominal values of parameters: (a) PK8 measured water level/nominal model, and (b) PK8 upstream gate position

results obtained from the step test and from the simulation of the nominal linear model (16.6).

Experiments reported in previously cited works [243, 249, 250, 273] on the identification of the dynamic behavior of IMCP have shown that canals similar to this experience large variations in their model parameters when the discharge regimes change across their upstream gates in the operation range. In our specific IMCP we will pay particular attention to changes in the time delay ($\tau_{\min} \leqslant \tau \leqslant \tau_{\max}$), as this is the most determinant parameter in the stability of its closed-loop control system [249, 252]. It is thus assumed that all the other dynamical parameters of (16.6) remain at their nominal values.

By using (16.3) and (16.4) and the experimental temporal responses of PK8, it was determined that if the discharge regime through the PK8 upstream gate varies in its operation range ($Q_{\min}(t) = 10 \, \mathrm{m^3/sec}$, $Q_{\max}(t) = 25 \, \mathrm{m^3/sec}$), then its time delay undergoes an operation variation range ($\tau_{\min} = 500 \, \mathrm{sec}$, $\tau_{\max} = 1450 \, \mathrm{sec}$). Details of this calculation can be found in [262].

16.9 Comparison of Controllers from Simulation Results

In this section the robustness of the previously described control systems to changes in the time delay of our IMCP is compared. Five control laws will then be studied: 1. the standard PI controller, 2. the PI controller with Smith predictor, 3. the PID controller with Smith predictor, 4. the fractional-order I controller with Smith predictor, and 5. the fractional-order PI controller with Smith predictor. All these controllers are designed in order to exhibit the same closed-loop dynamic behavior (the same overshooting and settling time) when the parameters of model (16.6) corresponding to our IMCP take the nominal values obtained in the previous section. Our robustness analysis considers the variation of the time delay obtained at the end of the previous section ($\tau \in [500, 1450]$ sec) as a consequence of the canal operation regime variations. The other parameters remain fixed. The controllers are compared from three points of view: 1. the time delay robustness index, 2. their temporal response in a nominal operation regime, and 3. their control signal amplitude in the nominal operation regime.

In order to compare the controllers, simulated results are reported, obtained using the IMCP model identified in the previous section from real data.

16.9.1 Controller Design Specifications for the Nominal Plant

Crossover frequency and settling time. The open-loop settling time of the nominal plant is approximately $t_{\text{cl}}^o \approx 3T_{10} + \tau_0 = 4300$ sec. We wish to make the closed-loop Smith predictor based control system response about twice as fast. Thus, we choose a crossover frequency of $\omega_{\text{cg}} = 0.0019$ rad/sec which approximately corresponds to a settling time of $t_{\text{cl}}^o \approx 5/\omega_{\text{cg}} = 2630$ sec. This last expression is an approximation obtained from general considerations concerning the inverse relationship existing between the crossover frequency and the settling time [22]. However, for very small values of the integral gain in the PID controller, or small values of λ in the PI$^\lambda$ controller, the convergence to the steady state value of the closed-loop system becomes very slow and the previous relation no longer holds. Therefore, besides the design specification of ω_{cg}, we will also impose the condition that the settling time must always be smaller than 2700 sec.

Phase margin. The same value $\varphi_{\text{m}} = 60°$ is chosen here as in Section 16.4, as this is a quite standard value. In that section it was found that the overshooting corresponding to this phase margin is approximately 10%, which is acceptable for our IMCP application.

16.9.2 Standard PI Controller

Expression (16.27) particularized to a PI controller yields

$$C_{\mathrm{PI}}(s) = K_{\mathrm{p}} + \frac{K_{\mathrm{i}}}{s}, \; K_{\mathrm{p}} = -\Re(X), \; K_{\mathrm{i}} = \omega_{\mathrm{cg}}\Im(X). \tag{16.44}$$

By using the pair of frequency design specifications $(\varphi_{\mathrm{m}}, \omega_{\mathrm{cg}})$ defined above, (16.44), when applied to the nominal plant, yields the PI controller:

$$C_{\mathrm{PI}}(s) = 2.28 - \frac{9.56 \times 10^{-4}}{s}, \tag{16.45}$$

which makes the closed-loop system unstable. The above specifications cannot, therefore, be achieved by a simple PI controller.

16.9.3 Smith Predictor Based PI Controller

The delay is now taken out of the closed-loop. The desired settling time (having taken the delay apart) is therefore $\hat{t}_{\mathrm{cl}}^{o} = t_{\mathrm{cl}}^{o} - \tau_0 = 2080$ sec, yielding a crossover frequency $\omega_{\mathrm{cg}} \approx 5/\hat{t}_{\mathrm{cl}}^{o} = 0.0025$ rad/sec. The phase margin remains the same. If we apply these frequency design specifications to expressions (16.34), (16.35) assuming the nominal plant, the following SP-PI controller is obtained:

$$C_{\mathrm{SP-PI}}(s) = 2.26 + \frac{0.0048}{s}. \tag{16.46}$$

The temporal response of the closed-loop system with controller (16.46) to a unit step command is plotted in Figure 16.31 (a), in which the region of errors lower than $\pm 5\%$ of the desired final value, which defines the settling time, is also drawn (this region will also be plotted in Figure 16.33). This figure shows that the overshooting is 15.2%, the settling time is 2615 sec, and the control system exhibits zero steady state error. Figure 16.31 (b) shows the upstream gate position signal $u(t)$ generated by the control system, whose maximum value is 2.46. Since it is assumed that the time delay is the only variable parameter in the plant, it is verified that $G'(s) = G_0'(s)$, and operating in (16.30) yields

$$M(s) = \frac{C(s)G_0'(s)e^{-\tau s}}{1 + C(s)G_0'(s)(1 + e^{-\tau s} - e^{-\tau_0 s})}. \tag{16.47}$$

From the characteristic equation of (16.47) we find that this control system remains stable for any value of the time delay $\tau < \tau_0$. However, for $\tau > \tau_0$ there is a limit value $\hat{\tau} = 1230\,\text{sec}$ above which the system becomes unstable. Therefore, this controller cannot guarantee stability throughout the entire range of variation of the time delay.

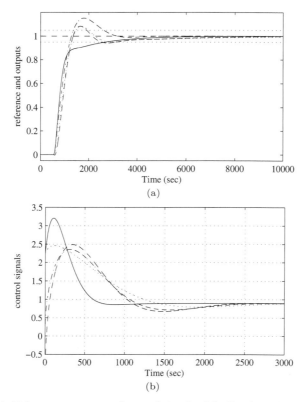

Figure 16.31 Unit step responses and control signals of the Smith predictor based control systems to a unit step command in the nominal plant case, with SP-PID shown in *solid line*, SPI-FPI-M shown in *dashed line*, SP-PI shown in *dotted line*, and SP-FI-M shown in *dashed dotted line*: (a) step responses, and (b) control signals

16.9.4 Smith Predictor Based PID Controller

In accordance with Section 16.6.2.2, we apply (16.37), together with the optimization procedure to maximize $\hat{\tau}$, in order to obtain controller (16.36). Figure 16.32 (a) plots the time delay limit $\hat{\tau}$ with regard to the controller integral gain K_i, and Figure 16.32 (b) plots the settling time t_{cl}^o with regard to the same controller gain. This figure plots the values corresponding to the previous SP-PI controller, and shows that it is possible to obtain an SP-PID controller (also represented in this figure) with the same settling time but a larger time delay limit $\hat{\tau}$, which is

$$C_{\text{SP−PID}}(s) = 2.26 + \frac{0.0018}{s} − 483s. \qquad (16.48)$$

We should also like to note that there is a wide range $0.0033 < K_i < 0.0065$ in which the relationship $\omega_{cg} \approx 5/\hat{t}_{cl}^o$ is approximately verified (see Figure 16.32 (b)) but outside this interval (both for smaller and larger gain values) the settling time grows, making it necessary to impose the additional design constraint of $t_{cl}^o < 2700\,\text{sec}$. In fact the SP-PID controller designed is outside this range and is obtained as a consequence of applying the aforementioned constraint.

The temporal response of the closed-loop system with controller (16.48) to a unit step command is also plotted in Figure 16.31 (a), and shows a settling time of 2618 sec similar to the previous design. Figure 16.31 (b) shows the corresponding control signal $u(t)$ whose maximum value is 3.19. The limit value of the time delay is now $\hat{\tau} = 1410\,\text{sec}$. Robustness has therefore been improved with regard to the previous controller, but it is not yet sufficient to guarantee stability in the entire range $\tau \in [500, 1450]\,\text{sec}$.

16.9.5 Fractional-order I Controller with Smith Predictor

We now apply the previous frequency design specifications to (16.40), assuming the nominal plant. The following SP-FI controller is obtained:

$$C_{\text{SP−FI}}(s) = \frac{0.2}{s^{0.45}}. \qquad (16.49)$$

The temporal response of this control system to a unit step command is plotted in Figure 16.33. This shows that the response provided by the SP-FI controller is more damped than that provided by the SP-PI, but less than that provided by the SP-PID controller. Moreover, the settling time of this SP-FI controller is much larger than in the other two cases. This is due to the fractional order of the integral term of the controller. The Final Value Theorem states that this SP-FI controller exhibits null steady state error if $\lambda > 0$. But the fact of λ being smaller than 1 makes the output converge on its reference value more slowly than in the case of an integer controller with $\lambda = 1$.

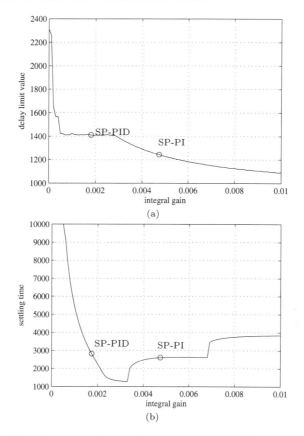

Figure 16.32 Time delay stability limit $\hat{\tau}$ and closed-loop settling time of the SP-PID controller: (a) delay limit value, and (b) settling time

In order to overcome the aforementioned problem, the SP-FI controller can be modified in such a way that the settling time is decreased without changing its dynamic and robustness properties. This can be achieved by shaping the open-loop frequency response so that its characteristics at low frequencies are changed by increasing the type of the control system, while leaving the frequency characteristics at medium and large frequencies unchanged. This can be achieved by multiplying the SP-FI controller (16.49) by a term of the form $(s + \mu)/s$:

$$C_{\mathrm{SP-FI-M}}(s) = \frac{0.2}{s^{0.45}} \left(\frac{s + \mu}{s} \right). \tag{16.50}$$

This technique has been successfully used in previous works [259, 260], and was justified and utilized in Section 16.4.2. If the parameter is sufficiently small, then this term improves the steady state behavior (increases the type of the control system) while leaving the dynamic behaviors unchanged. After some numerical simulations we concluded that the best value of μ for our modified SP-FI-M controller is 0.00007. Note that since $\mu \ll \omega_{cg}$, the frequency specifications attained with the SP-FI-M controller are only slightly modified by this term with regard to those which are desired. The temporal response of this controller is plotted in Figure 16.31 (a) and Figure 16.33, and its control signal in Figure 16.31 (b). These figures show that the settling time of this controller is 1980 sec, the overshot is 8.3%, and the maximum value of the control signal is 2.36. We have therefore shown that this controller, which exhibits approximately the same frequency specifications as before — these were modified to $\varphi_m = 58.4°$, $\omega_{cg} = 0.0025 \, \text{rad/sec}$ — has less settling time than with the SP-PI and SP-PID controllers, and less overshooting than with the SP-PI, but uses a control signal of less amplitude. This suggests that the SP-FI-M controller "manages" the control effort better than the classical controllers. The characteristic equation of (16.47) now states that this control system remains stable for time delay values under $\hat{\tau} = 1408 \, \text{sec}$.

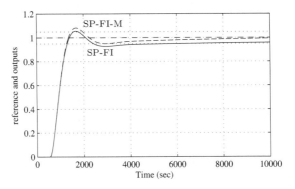

Figure 16.33 Responses of the Smith predictor based control system with the FI controller with and without the term $(s + \mu)/s$

16.9.6 Fractional-order PI Controller with Smith Predictor

According to Section 16.7.2, we should apply (16.43) together with the optimization procedure to maximize $\hat{\tau}$, in order to obtain controller (16.41). Figure 16.34 plots the maximum allowed time delay $\hat{\tau}$ with regard to λ, calculated for controller (16.41). Since $\hat{\tau}(\lambda)$ is a strictly decreasing function, the optimum is given by the boundary condition $t_{cl}^o < 2700 \, \text{sec}$. As was previously mentioned, small values of the fractional-order integral term make the output converge very slowly to its final value, and we propose adding an integral term of the form $(s + \mu)/s$ to the controller to solve this. In turn, such a term decreases the robustness (the limit value $\hat{\tau}$). Therefore, for each controller (16.41) (for a given λ), the parameter μ must be optimized in the sense of becoming the minimum value (in order to influence the value $\hat{\tau}$ as little as possible), which makes the closed-loop system verify condition $t_{cl}^{o} < 2700 \, \text{sec}$. Figure 16.35 plots the optimal values of μ with regard to λ. It shows that for values of $\lambda > 0.55$ such a term is not necessary. Figure 16.34 also plots the maximum allowed time delay $\hat{\tau}$ for the SP-FPI controller plus the optimized integral term (given by Figure 16.35). From now on this will be denoted as the SP-FPI-M controller. We should mention that the magnitudes of controller gains K_p, K_i grow very quickly as λ decreases. In Figure 16.34 we therefore choose the controller which corresponds to the maximum value of λ (in order to obtain moderate controller gains), which verifies $\hat{\tau} = 1450 \, \text{sec}$, and this controller is thus:

$$C_{\text{SP-FPI-M}}(s) = \left(-3.71 + \frac{1.89}{s^{0.2}}\right)\left(\frac{s + 1.52 \times 10^{-4}}{s}\right). \tag{16.51}$$

The temporal response of the closed-loop system with controller (16.51) to a unit step command is also plotted in Figure 16.31 (a), and shows a settling time of 1871 sec which is the smallest among all the controllers, and an overshooting of 8.3 %. Figure 16.31 (b) shows the corresponding control signal $u(t)$ whose maximum value is 2.5. As the limit value of the time delay is $\hat{\tau} = 1450 \, \text{sec}$, the robustness condition in the time delay range $\tau \in [500, 1450] \, \text{sec}$ is now fulfilled.

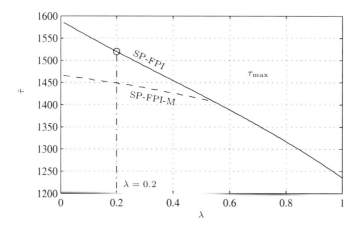

Figure 16.34 Time delay stability limit $\hat{\tau}$ with regard to λ of the SP-FPI controllers with and without the term $(s + \mu)/s$

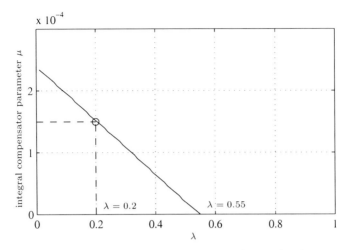

Figure 16.35 Values of the optimized μ parameter with regard to λ

16.9.7 Conclusions on the Robust Control of the IMCP with LTVD

Simulations of several Smith predictor based controllers applied to the model of a real canal pool have been carried out: PI, PID, fractional-order integral FI, and fractional-order PI. For the same given frequency specifications, these simulations have shown that:

- Fractional-order controllers (SP-FI-M and SP-FPI-M) behave better than standard controllers (SP-PI and SP-PID) in the nominal regime: there is much less settling time, and less overshooting.
- The maximum amplitude of the control signal is smaller with the fractional-order controllers than with the standard ones, thus the former better prevent the saturation of the actuators. This issue, combined with the first issue, allows us to conclude that the proposed fractional-order controllers manage the control effort more efficiently than their standard counterparts (SP-FI-M vs SP-PI) and (SP-FPI-M vs SP-PID).
- A computer costly optimization procedure allows us to obtain an SP-PID controller that significantly improves the robustness value attained by the SP-PI: 14.6% better. But similar results can be achieved by using the proposed SP-FI-M controller, which is obtained by applying very simple formulas (16.40).

- The use of an optimization procedure to design the proposed SP-FPI-M controller allows us to increase the limit delay value further. In our particular canal pool, this allows us to attain the stability margin needed to operate the canal safely in all its possible operation regimes.

The design procedure of the SP-FPI-M controller is as laborious as that of the standard SP-PID controller, and more laborious than that of the PI controller, but only requires optimization with regard to one parameter λ, or in some cases as many as two parameters λ, and μ. However, it yields controllers whose robustness is similar to those which can be obtained from standard robust control techniques, such as $\mathcal{H}_\infty$, which requires a much more costly optimization process in which many control parameters have to be tuned.

16.10   Summary

The design of simple robust fractional-order controllers for hydraulic canals, characterized to present time-varying dynamical parameters, has been described.

First, a simple fractional-order control strategy was designed and implemented in an experimental laboratory hydraulic canal, which was characterized to present time-varying gain and time constants, but a fixed time delay. A class of fractional-order PI controllers has been proposed, and a straightforward method to design it from frequency specifications has been developed. It was shown that this controller outperformed standard PI and PID controllers – all of them designed to have the same closed-loop behavior than the fractional-order controller at nominal conditions – in its robustness features.

Second, another simple FPI control strategy combined with a Smith predictor was designed and simulated using an experimentally obtained model of an irrigation main canal pool. This plant was characterized to present basically a large time-varying time delay, the other dynamic parameters remaining approximately constant. The fractional-order PI controller designed here had a different structure from the previous one, but the design methodology employed was the same as that used in the laboratory hydraulic canal. It was shown that this controller allowed to attain better robustness features than standard PI and PID controllers embedded in a Smith predictor – all of them designed to have the same closed-loop behavior as the fractional-order controller at nominal conditions. Moreover it was shown that the fractional-order I controller attained similar robustness characteristics as the

PID controller, both of them combined with the Smith predictor, while being much easier to design.

The comparison of real-time and simulated responses of FPI, and standard PI and PID controllers (both in the standard control scheme and the Smith predictor scheme) proved the effectiveness of the proposed fractional-order control strategies in terms of performance, and supported the robustness results obtained in the frequency response theoretical analysis. Moreover, these results suggest that the best robustness features are not strictly associated with having more controller parameters to tune (FPI and PID controllers both have three parameters to tune), but also the structure of the controller makes an impact on the robustness. In this chapter we have tried to prove that, in dynamical systems with a delay, fractional-order controllers like those developed here, possess a structure that allows for more robustness to combinations of changes of plant parameters. It should be pointed out that the interest in such fractional-order controllers in this application is justified by the fact that dynamical parameters of irrigation main canals may change drastically as a function of their operating hydraulic regimes.

It should be mentioned that a main canal usually has multiple pools. Then modern canal control systems may be more complex than a PI, sometimes including feedforward terms to compensate for the interactions between consecutive pools. These control systems often include a series of simple PI controllers as the lowest control level which are coordinated by other more complex controllers at upper levels [255]. Substituting these PI controllers for the fractional-order controllers proposed here at such a low level of control may improve the global control system as local control robustness is increased and thus makes an impact on the overall control system performance.

Chapter 17
Mechatronics

In this chapter, the tuning and auto-tuning methods described in Part III of the book will be applied to the control of a real mechatronic laboratory platform consisting of position and velocity servos. This type of devices are very commonly used in industrial environments and many other processes have the same type of transfer functions modeling their dynamics. For this reason, this application is rather practical and representative of a class of industrial processes.

The experimental platform and the implementation of the control strategy are described in the following sections.

17.1 The Experimental Platform

The connection scheme in Figure 17.1 shows the different elements of the experimental platform:

- Data acquisition board AD 512, by Humusoft, running on MATLAB 5.3 and using the real-time toolbox "Real-Time Windows Target." This board was previously used for the implementation of robust fractional-order controllers [287].
- A computer Pentium II, 350MHz, 64M RAM, which supports the data acquisition board and where the programs run for the implementation of the method proposed.
- A servomotor 33-002 by Feedbak, consisted of 1. a mechanical unit 33-100, which constitutes the servo, strictly speaking, 2. an analog unit 33-110, which connects to the mechanical unit through a 34-way ribbon cable which carries all power supplies and signals enabling the normal circuit interconnections to be made on the analog unit, and 3. a power supply 01-100 for the system. The mechanical unit has a brake the position of

which changes the gain of the system, that is, the brake acts like a load to the motor.

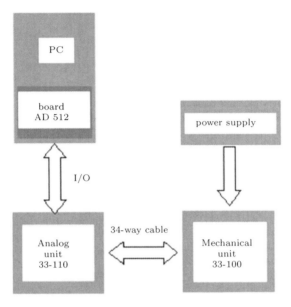

Figure 17.1 Connection scheme of the experimental mechatronic platform

In Figure 17.2 a photo of the system described here is displayed.

Figure 17.2 Photo of the experimental mechatronic platform

For the implementation of the control loop in MATLAB, the scheme in Figure 17.3 has been used, where:

- **Board**. Refers to the data acquisition board described above. The type AD 512 is selected and its libraries are loaded.
- **Input/Output**. These blocks refer to the analog input and output that will be used for the control loop. The sampling period is set to $T = 0.01$ sec. The output signal of the servo is connected to the input of the loop (*Input*), and the controller output (*Output*) is connected to the input of the servo, closing the loop this way.
- **Adaptor**. It has to be taken into account that a gain scaling must be done for the input and output signals, since the board amplifies both the signals to and from the servo. These adaptor blocks ensure a unity gain for the inputs and outputs.
- **Controller** $C(z)$. The controller is implemented by using a discrete transfer function $C(z)$, obtained as will be described in the following section.
- **ZOHs**. Since a discrete version of the controller is used and the inputs and outputs are continuous signals, two zero-order holds (ZOH block) are used for the continuous-discrete and discrete-continuous signal conversions.
- **Step**. Refers to the step reference signal.
- **Scope**. To show the signals generated through the process.

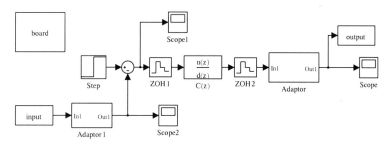

Figure 17.3 Simulink scheme for the implementation of the fractional-order controller

17.2 Experimental Tuning of the Control Platform

In this section, the design of the controller is done by using an .m file (MATLAB file) in which the basis of the tuning method proposed in Chapter 8 will be taken into account. Let us show the experimental results obtained when controlling the position servo and the velocity servo.

17.2.1 Position Servo

The experimentally obtained transfer function of the position servo is

$$G_1(s) = \frac{1.4}{s(0.7s + 1)} e^{-0.05s}. \tag{17.1}$$

The design specifications are: phase margin $\varphi_m = 80°$; gain crossover frequency $\omega_{cg} = 2.2\,\mathrm{rad/sec}$; velocity error constant $K_v = 0.82$, that is, $k' = K_c x^\alpha = 1$. As can be observed, a lead compensator is needed to fulfil these requirements, giving

$$C_1(s) = \left(\frac{2.0161s + 1}{0.0015s + 1} \right)^{0.7020},$$

with $k' = 1$, $x = 7.4012 \times 10^{-4}$, $\lambda = 2.0161$, and $\alpha = 0.7020$. The Bode plots of this compensator are shown in Figure 17.4. At the gain crossover frequency $\omega_{cg} = 2.2176\,\mathrm{rad/sec}$, the compensator has a magnitude of 9.2803 dB and a phase of 54.1961°. At that frequency the magnitude of the open-loop system is 0 dB, and the phase is $-100°$, fulfilling the frequency specifications.

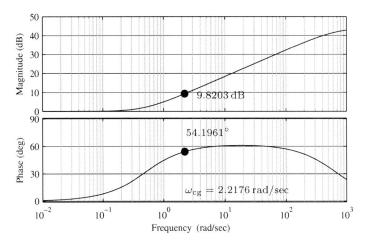

Figure 17.4 Bode plots of controller $C_1(s)$

Once the compensator is calculated, its implementation is carried out. An indirect digital method is used for the discretization of the controller, with MATLAB function `invfreqs()` and Tustin method with prewarping (see Chapter 12). This way, an integer-order transfer function is obtained which fits the frequency response of the fractional-order controller in the range $\omega \in (10^{-2}, 10^2)$, with five poles and zeros. Later, the discretization of this continuous approximation is made by using Tustin rule with prewarping, with a sampling period $T = 0.01$ sec and prewarp frequency ω_{cg}, resulting controller $C_1(z)$. This controller is a 5th-order digital IIR filter, given by

$$C_1(z) = \frac{1.976z^{-5} - 4.807z^{-4} - 34.46z^{-3} + 130.25z^{-2} - 149.2z^{-1} + 56.2538}{-0.0093z^{-5} - 0.121z^{-4} + 0.431z^{-3} + 0.284z^{-2} - 1.575z^{-1} + 1}.$$

The experimental step response of the closed-loop system is shown in Figure 17.5. This response has been compared with that obtained in simulation. It can be observed that the experimentation fits perfectly well the simulation results.

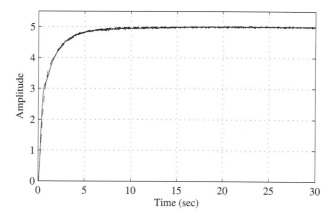

Figure 17.5 Step response of the position servo system with controller $C_1(s)$, *solid line* for experimental data, and *dashed smooth line* for simulation results

17.2.2 Velocity Servo

The experimental transfer function of the velocity servo to be controlled is

$$G_2(s) = \frac{1.4}{0.7s + 1}e^{-0.05s}. \tag{17.2}$$

In order to cancel the steady-state error, an integrator has been added to the compensator. Therefore, the final controller will consist of an integrator plus a fractional-order compensator. The compensator will be designed taking into account the contribution in magnitude and phase of the integrator. The frequency specifications to fulfil are $\omega_{cg} = 4\,\text{rad/sec}$, $\varphi_m = 80°$, and $k' = 1$.

The resulting controller is

$$C_2(s) = \frac{1}{s}\left(\frac{3.1394s + 1}{0.0047s + 1}\right)^{0.85},$$

with $k' = 1$, $x = 0.0015$, $\lambda = 3.1394$, and $\alpha = 0.85$. Figure 17.6 shows the Bode plots of $C_2(s)$. At the frequency ω_{cg} the controller has a magnitude of 6.6871 dB and a phase of $-18.3488°$. The magnitude of the open-loop system at this frequency is 0 dB, and the phase $-100°$. Once again the specifications are fulfilled.

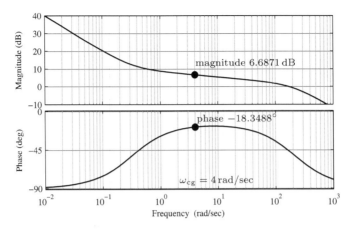

Figure 17.6 Bode plots of controller $C_2(s)$

Following the indirect discretization method explained in Chapter 12, with a frequency range $\omega \in (10^{-2}, 10^2)$ and sampling period $T = 0.01$ sec, the resulting discrete controller, a third-order IIR digital filter, is

$$C_2(z) = \frac{0.01}{2} \frac{1 + z^{-1}}{1 - z^{-1}} \frac{75.9289z^{-2} - 210.8592z^{-1} + 135.5044}{0.0084z^{-2} - 0.6858z^{-1} + 1}.$$

Figure 17.7 shows the experimental step response of the system with controller $C_2(z)$, comparing it with the response obtained from the simulation results.

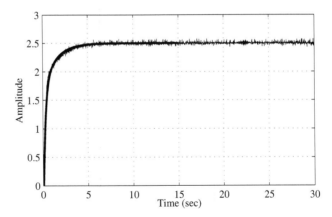

Figure 17.7 Step response of the position servo system with controller $C_2(s)$, *solid line* for experimental data, and emphdashed smooth line for simulation results

The robustness constraint regarding plant gain variations has not been tested in these experiments. For this purpose, the auto-tuning method proposed in Chapter 9 will be implemented and tested in the following section for the control of this mechatronic platform.

17.3 Experimental Auto-tuning on the Mechatronic Platform

The same experimental platform presented in Section 17.1 will be used to test the robustness performance of the controlled system when using the auto-tuning method described in Chapter 9.

The Simulink block diagram used for the implementation of the relay test is the one in Figure 17.8. The blocks are already described in Section 17.1. The relay is implemented by using Simulink block "relay," with amplitude δ and hysteresis $\epsilon = 0$. Equally, the delay is also a Simulink block. *Impulse* refers to the impulse signal to initialize the relay test.

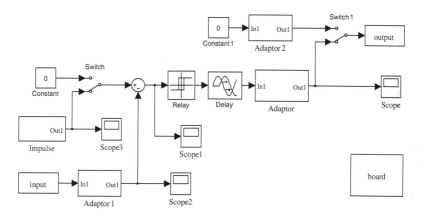

Figure 17.8 Scheme for the implementation of the relay test using MATLAB

The control loop is implemented as previously explained in Section 17.1.

Specifications of gain crossover frequency, phase margin, and robustness to plant gain variations are given. In this case, the desired gain crossover frequency is $\omega_{cg} = 2.3$ rad/sec. The relay has an output amplitude of $\delta = 6$, without hysteresis, $\epsilon = 0$. The two initial values (θ_{-1} and θ_0) of the delay used to reach the frequency specified are 0.1 sec and 0.04 sec, respectively. After several iterations the output signal shown in Figure 17.9 is obtained.

The value of the delay θ_a obtained for the selection of the frequency specified is $\theta_a = 0.2326$ sec, and the corresponding frequency is $\omega_u = 2.2789$ rad/sec. The amplitude and period of this oscillatory signal are $a = 1.8701$ and $T_u = 2.7571$ sec, respectively. Therefore, the magnitude and phase of the plant estimated through the relay experiment at the frequency $\omega_u = 2.2789$ rad/sec are $|G(j\omega_u)|_{dB} = -12.2239$ dB and $\arg(G(j\omega_u)) = -149.6328°$, respectively. Measuring experimentally the frequency response of the system in order to validate these values, a magnitude of -11.8556 dB and a phase of $-150.2001°$ are obtained. So, only a slight error is committed in the

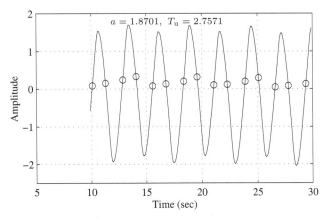

Figure 17.9 Output signal of the relay test

estimation. Next, a fractional-order $PI^\lambda D^\mu$ controller is designed with the proposed tuning method to obtain a phase margin $\varphi_m = 60°$ at the gain crossover frequency $\omega_u = 2.2789$ rad/sec. The gain of the controller will be fixed to 1, that is, $k' = K_c x^\alpha = 1$.

The first step is the design of the fractional-order PI^λ part, in (9.2). For that purpose, the slope of the phase of the plant v is estimated by (9.7). The slope obtained in this case is $v = -0.2568$ sec . With the value of the slope and applying the criterion described for the fractional-order PI^λ controller (see (9.9) and (9.11)), the controller that cancels the slope of the phase curve of the plant is

$$PI^\lambda(s) = \left(\frac{0.4348s + 1}{s}\right)^{0.8468}. \tag{17.3}$$

At the frequency ω_u this fractional-order PI^λ controller has a magnitude of -3.5429 dB, a phase of $-38.3291°$ and a phase slope of 0.2568. Therefore, the estimated system $G_{\text{flat}}(s)$ has a magnitude of -15.7668 dB and a phase of $-187.9619°$. These values can be easily obtained through the values of the magnitude and phase of the plant estimated by the relay test at the frequency ω_u and the magnitude and phase of the controller PI^λ at the same frequency. Next, the controller $PD^\mu(s)$ is designed to fulfil the specifications of phase margin and gain crossover frequency required for the controlled system. Following the iterative process described previously, the resulting controller is given by

$$PD^\mu(s) = \left(\frac{4.0350s + 1}{0.0039s + 1}\right)^{0.8160}. \tag{17.4}$$

At the frequency $\omega_u = 2.2789$ rad/sec the controller $PD^\mu(s)$ has a magnitude of 15.7668 dB and a phase of $67.9619°$.

Then, the resulting total controller $C(s)$ is the following one:

$$C(s) = \left(\frac{0.4348s + 1}{s}\right)^{0.8468} \left(\frac{4.0350s + 1}{0.0039s + 1}\right)^{0.8160}. \tag{17.5}$$

The Bode plots of $C(s)$ are shown in Figure 17.10. The magnitude and phase of this controller at the frequency ω_u are 12.2239 dB and 29.6328°, respectively. Therefore, the open-loop system $F(s)$ has a phase margin of 60° and a magnitude of 0 dB at the gain crossover frequency $\omega_u = 2.2789$ rad/sec, fulfilling the design specifications.

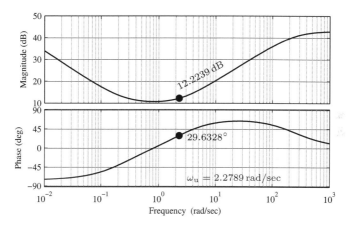

Figure 17.10 Bode plots of the fractional-order controller $C(s)$

For the implementation of the resulting fractional-order controller $C(s)$, the frequency domain identification technique using MATLAB function `invfreqs()` is applied again (Chapter 12). An integer-order transfer function is obtained which fits the frequency response of the fractional-order controller in the range $\omega \in (10^{-2}, 10^2)$, with three poles/zeros for the PI^λ part and three poles/zeros for the PD^μ part. Later, the discretization of this continuous approximation is made by using the Tustin rule with prewarping, with a sampling period $T = 0.01$ sec and prewarp frequency ω_{cg}. With this controller the phase of the open-loop system $F(s)$ is the flattest possible, ensuring the maximum robustness to variations in the gain of the plant, as can be seen in the step responses of the controlled system for $k = K_{nom}$ (nominal gain), $k = 2K_{nom}$, and $k = 0.5K_{nom}$ (Figure 17.11). The gain variations are provoked by changing the position of the motor brake. Figure 17.12 shows the control laws of the system for the different gains. It can be observed that for this gain range this control strategy is very suitable, since the peak of the control laws is much lower than 10 V, the saturation voltage of the motor.

Comparing the step responses with those obtained (in simulation) with the
PID controller $C_{zn}(s) = 22.1010\,(1 + 1/(0.55s) + 0.1375s)$ designed by the
second method of Ziegler–Nichols (Figure 17.13), the better performance of
the system with the fractional-order controller $C(s)$ can be observed.

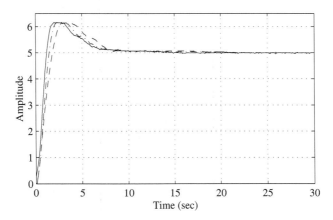

Figure 17.11 Step responses of the system with controller $C(s)$, with *solid line*, *dashed
line* and *dashed dotted line* for $k = 2k_{\mathrm{nom}}$, $k = 0.5k_{\mathrm{nom}}$, and $k = k_{\mathrm{nom}}$, respectively

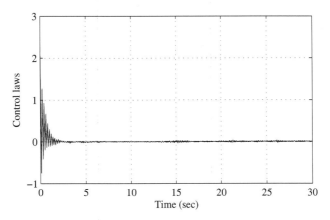

Figure 17.12 Experimental control laws of the controlled system with controller $C(s)$

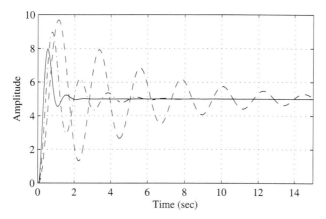

Figure 17.13 Step responses of the system with controller $C_{zn}(s)$, with *solid lines*, *dashed lines* and *dashed dotted lines* for $k = 0.5k_{nom}$, $k = k_{nom}$, and $k = 2k_{nom}$, respectively

17.4 Summary

From these experimental results, we can conclude that the tuning and auto-tuning methods proposed are very effective and easy to implement. The relations among the parameters of the controller are direct and simple, and a constraint regarding robustness to plant gain variations is fulfilled together with specifications of gain crossover frequency and phase margin.

Chapter 18
Fractional-order Control Strategies for Power Electronic Buck Converters

This chapter presents several alternative methods for the control of power electronic buck converters applying fractional-order control (FOC). For achieving this goal, the controller design will be carried out by two strategies. On the one hand, the design of a linear controller for the DC/DC buck converter will be considered. In that sense, the Bode's ideal loop transfer function presented in Chapter 2 will be used as reference system. On the other hand, the fractional calculus is proposed in order to determine the switching surface applying a fractional sliding mode control (F_RSMC) scheme to the control of such devices. In that sense, switching surfaces based on fractional-order PID and PI structures are defined. An experimental prototype has been developed and the experimental and simulation results confirm the validity of the proposed control strategies.

18.1 Introduction

Switched mode DC/DC power converters are used in a wide variety of applications, including power supplies for personal computers, DC motor drives, active filters, *etc*. *Pulse-width modulation* (PWM) sets the basis for the regulation of switched mode converters. The operation of these devices is often based on the control of the output voltage of a passive filter. A basic DC/DC converter circuit known as the *buck converter* is illustrated in Figure 18.1. The buck converter consists of a switch network that reduces the DC component of voltage and a low-pass filter that removes the high-frequency switching harmonics. Several control strategies, both linear and nonlinear, have been used for the control of DC/DC converters, such as PI, dead beat, sliding mode control, *etc*. [288]

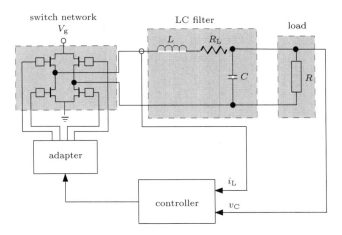

Figure 18.1 Buck converter

On the one hand, one of the goals with the control of the buck converter is to achieve more robust control systems despite input voltage changes or load perturbations. The input voltage variations in a buck converter can be viewed as changes in the process gain. One of the main advantages of fractional-order controllers is the possibility of obtaining an open-loop transfer function in the form of a fractional-order integrator, that is, the mentioned Bode's ideal loop transfer function, providing a controlled system robust to changes in the process gain. In this way, here we propose a method for the control of power electronic converters by using a linear controller based on FOC and Bode's ideal function [18,90] and then to obtain a discrete equivalent that allows its practical implementation.

On the other hand, since power electronic converters inherently include switching devices which exhibit a discontinuous behavior, the DC/DC buck converter can be modeled as a bilinear system, which achieves a different linear topology for every state of the control signal, u. From this point of view, the DC/DC converter can be considered as a variable structure system (VSS) since its structure is periodically changed by the action of the controlled switches. Sliding mode control (SMC) for VSS [141] offers an alternative way to implement a control action which exploits the inherent variable structure of DC/DC converters. In particular, the converter switches are driven as a function of the instantaneous values of the state variables in such a way as to force the system trajectory to follow a suitable selected surface on the phase space called the sliding surface, s. The use of techniques based on switching surfaces in the control of such devices have been referenced since the earlier 1980s [289–291]. SMC is well known for its robustness to disturbances and parameters variations. This chapter also deals with the design of switching surfaces for a buck converter using alternative techniques based on FOC. In this sense, a fractional-order form of linear compensation PID and PI networks will be used in order to obtain fractional sliding surfaces of the form $\text{PI}^\lambda \text{D}^\mu$ (see Section 18.4.2.1).

The rest of the chapter is organized as follows. Section 18.2 describes the processes of modeling and linearization of the plant and a real system model is obtained. Section 18.3 deals with the design of linear controllers for the buck converter based on the fractional-order control and the Bode's ideal loop transfer function. Section 18.4 deals with the design of switching surfaces for a buck converter using alternative techniques based on FOC. Section 18.5 shows simulation and experimental results and Section 18.6 states some conclusions.

18.2 Model of the Buck Converter

To design the control system of a converter, it is necessary to model the converter dynamic behavior. Unfortunately, modeling of converter dynamic behavior is hampered by the nonlinear time-varying nature of the switching and PWM process. The formulation in the form of a bilinear system of a DC/DC buck converter defined on $\mathbb{R}^n$ is

$$\dot{x} = Ax + BV_g u, \tag{18.1}$$

where $x \in \mathbb{R}^n$ is the state vector; $A \in \mathbb{R}^{n \times n}$ and $B \in \mathbb{R}^{n \times m}$ are matrices with constant real entries; u is a scalar control ($m = 1$) taking values from the discrete set $u = \{0, 1\}$. The state-space model of the buck converter is

$$\begin{aligned}
\dot{i}_L &= -\frac{1}{L} v_c + \frac{V_g}{L} u, \\
\dot{v}_c &= \frac{1}{C} i_L - \frac{1}{RC} v_c,
\end{aligned} \tag{18.2}$$

and its state-space model in phase canonical form is

$$\begin{bmatrix} \dot{v}_c \\ \ddot{v}_c \end{bmatrix} = \begin{bmatrix} 0 & 1 \\ -\dfrac{1}{LC} & -\dfrac{1}{RC} \end{bmatrix} \begin{bmatrix} v_c \\ \dot{v}_c \end{bmatrix} + \begin{bmatrix} 0 \\ \dfrac{1}{LC} \end{bmatrix} V_g u. \tag{18.3}$$

In general, in order to obtain positive output voltages only two structures are necessary (Figure 18.2). These structures correspond to different state-space models. The desired output voltage is obtained by changing these structures temporarily. A state-space model for every structure yields to

$$\begin{aligned}
\dot{x} &= A_0 x + B_0 V_g, \quad \text{for } u = 0, \\
\dot{x} &= A_1 x + B_1 V_g, \quad \text{for } u = 1.
\end{aligned} \tag{18.4}$$

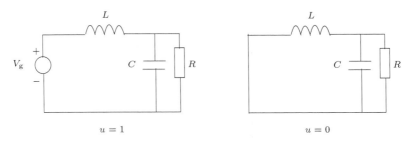

Figure 18.2 DC-DC buck converter topologies

The methods of design proposed in Section 18.3 are based on frequency domain techniques. Therefore, a linearized model of the plant is necessary in order to apply the design process.

18.2.1 Discrete Plant Model

In order to obtain a valid model of the DC/DC converter, this system is considered as composed of two subsystems: LC filter and a PWM actuator. A time-invariant linear LC filter may be represented by a state-space model:

$$
\begin{aligned}
\dot{\boldsymbol{x}}(t) &= \boldsymbol{A}\boldsymbol{x}(t) + \boldsymbol{B}v_{\mathrm{s}}(t), \\
y(t) &= \boldsymbol{C}\boldsymbol{x}(t),
\end{aligned}
\tag{18.5}
$$

where $v_{\mathrm{s}}(t)$ is the input of the filter and it is provided by a PWM actuator where the control input is the duty cycle, d. In the case of an "ON-OFF-ON" PWM actuator of symmetric pulse with respect to $T/2$ (being T the sampling period), its model is

$$
v_{\mathrm{s}}(t) = \begin{cases} 0, & \text{for } kT \leqslant t < kT + t_1(k), \\ V_{\mathrm{g}}, & \text{for } kT + t_1(k) \leqslant t < kT + t_2(k), \\ 0, & \text{for } kT + t_2(k) \leqslant t < (k+1)T, \end{cases}
\tag{18.6}
$$

where $t_1(k) = T(1 - d(k))/2$ and $t_2(k) = T(1 + d(k))/2$.

An equivalent discrete model of the joint system PWM actuator-LC filter can be obtained from the solution of (18.5)

$$
\boldsymbol{x}(t) = e^{\boldsymbol{A}(t-t_0)}\boldsymbol{x}(t_0) + \int_{t_0}^{t} e^{\boldsymbol{A}(t-\tau)}\boldsymbol{B}v_{\mathrm{g}}(\tau)\mathrm{d}\tau.
\tag{18.7}
$$

Particularizing for $t_0 = kT$, $t = (k+1)T$, and $v_s(t)$ given by (18.6), the discretized model is given by the equation

$$x(k+1) = e^{AT}x(k) + e^{AT}\left[e^{-At_1(k)} - e^{-At_2(k)}\right]A^{-1}BV_g,$$
$$y(k) = Cx(k). \tag{18.8}$$

The matrix A is diagonalizable, and its Jordan canonical form is

$$A = V\Lambda V^{-1}, \quad \Lambda = \text{diag}(\lambda_1, \lambda_2, \cdots, \lambda_n). \tag{18.9}$$

Then, (18.8) becomes

$$x(k+1) = e^{AT}x(k) + V\Psi(k)V^{-1}BV_g, \tag{18.10}$$

where $\Psi(k)$ is a diagonal control matrix of the form

$$\Psi(k) = e^{\Lambda T/2}\left[e^{\Lambda Td(k)/2} - e^{-\Lambda Td(k)/2}\right], \tag{18.11}$$

which defines the relation between Ψ and $d(k)$ in the case of a symmetric "ON-OFF-ON" actuator. Equation 18.10 defines a nonlinear relationship between the control variable $d(k)$ and the state $x(k+1)$.

The control design methods proposed in Section 18.3 are based on a frequency domain approach. Consequently, the first step in the control design process is to obtain a linear model of the DC/DC converter. Several methods have been proposed to carry out this linearization: model averaging [292], first-order truncation of a Taylor series expansion [293], and methods based on the functional minimization [294]. The last one is used in this case, which allows the use of lower switching frequencies.

18.2.2 Linearized Plant Model

The method used for obtaining the linearized discrete model of the converter is developed in [294] and is based on the approximation of the diagonal matrix $\Psi(k)$ by

$$\Psi(k) \approx \hat{\Gamma}v(d(k)), \tag{18.12}$$

where $\hat{\Gamma}$ is a constant diagonal complex matrix of the same dimension of Ψ, and v is a real function of the physical control variable $d(k)$.

Applying this method, the linearized model becomes

$$x(k+1) = e^{AT}x(k) + V\text{diag}\left(\hat{\Gamma}\right)V^{-1}BV_g\bar{v}(k), \tag{18.13}$$

where $\bar{v}(k)$ is a fictitious control signal related to the real control signal $d(k)$. The inverse of $\hat{v}(d)$ can be tabulated and used in the control algorithm in order to obtain the duty cycle $d(k)$ to be applied. In the case of an "ON-

OFF-ON" actuator the equation to use is

$$d(k) = \bar{v}^{-1}\left(\bar{v}\left(k\right)\right).\qquad(18.14)$$

18.2.3 Real System Model for Design

The selected parameter values for the converter shown in Figure 18.2 are $L = 3.24\,\text{mH}$, $R_L = 0\,\Omega$, $C = 48\,\mu\text{F}$, $R = 117\,\Omega$, and $V_g = 20\,\text{V}$, for a switching frequency of $2\,\text{kHz}$ ($T = 0.5\,\text{ms}$). This choice depends on the application of the converter. This one is used as a voltage compensator for canceling wave subharmonic perturbations up to $2\,\text{ms}$ and so a settling time less than 1ms is required. This fact limits the values of L and C. The switching frequency is limited by the switching devices used. With these parameter values, the obtained linear discrete model is

$$\boldsymbol{x}\left(k+1\right) = \begin{bmatrix} 0.3070 & -0.1100 \\ 7.4237 & 0.2585 \end{bmatrix} \boldsymbol{x}(k) + \begin{bmatrix} 0.1798 \\ 1.0571 \end{bmatrix} \times 10^3 V_g \bar{v}(k),\qquad(18.15)$$

where $\bar{v}(k)$ is the fictitious control signal and the duty ratio $d(k)$ is obtained from (18.14).

Then, the discrete-time transfer function of the converter is

$$G(z) = k\frac{c_1 z^{-1} + c_2 z^{-2}}{d_0 + d_1 z^{-1} + d_2 z^{-2}},\qquad(18.16)$$

being $k = 10^3$, $c_1 = 1.0571$, $c_2 = 1.01$, $d_0 = 1$, $d_1 = -0.5655$, $d_2 = 0.8958$.

As the model system is discrete and the design method used is based on the continuous frequency domain, the discrete model must be converted into a pseudo-continuous system using the bilinear w-transformation. So, the obtained pseudo-continuous system is

$$\begin{aligned} G(w) &= k\frac{1 - wT/2}{1 + wT/2}\mathscr{T}_w \left[\frac{c_1 + c_2 z^{-1}}{d_0 + d_1 z^{-1} + d_2 z^{-2}}\right] \\ &= k\frac{2/T - w}{2/T + w}\mathscr{T}_w \left[\frac{c_1 + c_2 z^{-1}}{d_0 + d_1 z^{-1} + d_2 z^{-2}}\right], \end{aligned}\qquad(18.17)$$

where $\mathscr{T}_w$ represents the w-transformation.

As can be observed, the previous system (18.17) corresponds to a non-minimum-phase system, whose transfer function, $G(w)$, can be written as the composition of a minimum-phase function, $G_{\mathrm{mp}}(w)$, with an all-pass filter $A(w)$ [295]:

$$G(w) = A(w)G_{\mathrm{mp}}(w),\qquad(18.18)$$

where

$$G_{\mathrm{mp}}(w) = k\frac{(w + p)\left(b_1 w + b_2\right)}{a_1 w^2 + a_2 w + a_3},\qquad(18.19)$$

and

$$A(w) = \frac{-w + p}{w + p}, \tag{18.20}$$

being $k = 10^3$, $p = 4 \times 10^3$, $b_1 = 471$, $b_2 = 8.2684 \times 10^7$, $a_1 = 2.4613 \times 10^4$, $a_2 = 8.336 \times 10^6$, $a_3 = 2.12848 \times 10^{11}$.

18.3 Linear Fractional-order Control

The final objective of the control of a DC/DC buck converter is to achieve a constant output voltage despite load variations and input voltage (V_g) disturbances. The gain factor, k, in (18.16) depends directly on V_g and so, the input voltage variations can be viewed as changes in the process gain and so Bode's ideal function can be used as reference system. Since the methods of design proposed in this section are based on frequency domain techniques, a linearized model of the plant is necessary in order to apply the design process.

By using the w-transform, the pseudo-continuous system obtained is a non-minimum-phase system, whose phase lag must be canceled. To take into account this phase lag two methods are proposed. One of them is based on considering the non-minimum-phase term like a delay and to use a controller that combines a fractional-order integrator and a Smith predictor structure in order to achieve the working specifications. The other one deals with the design of the controller taking into account the phase lag at the frequency of interest for determining the order of the Bode's ideal loop transfer function. Finally, in order to avoid the handling of non-minimum-phase systems due to the use of the w transform, the design problem is approached considering directly the discrete linearized model of the DC/DC converter and applying the discrete version of the Bode's ideal function as reference system.

18.3.1 Controller Design Based on the Smith Predictor Structure

This method of design is applied in two steps. First, the minimum-phase subsystem $G_{mp}(w)$ will be compensated. Next, the non-minimum-phase subsystem $A(w)$ will be considered.

18.3.1.1 Minimum-phase Subsystem Compensation

For this purpose, the Bode's ideal loop transfer function will be taken as reference system. Considering the transfer function of the minimum-phase subsystem (18.19) and a desired open-loop transfer function for the compensated minimum-phase subsystem, the parameters K_c and λ will be selected to obtain a specified phase margin φ_m and crossover frequency ω_c. So, the transfer function of the compensator $D_o(w)$ can be obtained as

$$D_o(w) = \frac{K_c}{w^\lambda G_{mp}(s)} = \frac{K_c\left(a_1 w^2 + a_2 w + a_3\right)}{w^\lambda k\left(w + p\right)\left(b_1 w + b_2\right)}, \qquad (18.21)$$

being the design equations

$$K_c = \omega_c^\lambda, \quad \lambda = \frac{2}{\pi}(\pi - \varphi_m). \qquad (18.22)$$

Taking as working specifications for the design: $\varphi_m = 64°$; $\omega_c = 1.36 \times 10^3$ rad/sec, the parameters of the compensator are $K_c = 1.1 \times 10^4$, $\lambda = 1.29$. Taking into account the original non-minimum-phase system $G(w)$, the open-loop transfer function of the compensated system responds to the equation

$$G(w)D_o(w) = \frac{K_c}{w^\lambda}\frac{(-w + p)}{(w + p)}, \qquad (18.23)$$

which consists of the Bode's ideal loop transfer function plus an all-pass filter. The phase margin has fallen considerably until 26.21°. This is due to the phase lag introduced by the all-pass filter.

18.3.1.2 Non-minimum-phase Subsystem Compensation Using the Smith Predictor Structure

The previous analysis reveals that the non-minimum-phase effects must be considered in the controller design process. In this section, the non-minimum-phase term will be compensated by using the Smith predictor structure shown in Figure 18.3.

This structure was proposed by Smith to deal with delay systems. In this work, this structure is applied to a non-minimum-phase system with the aim of designing the controller taking the minimum-phase subsystem and then to implement the resultant controller in the form shown in Figure 18.3. In this control scheme, the non-minimum-phase subsystem, $A(w)$, is used instead of the delay block of the Smith Predictor structure.

By using this control structure, the total controller $D_1(w)$, marked with a dotted line in Figure 18.3, will be

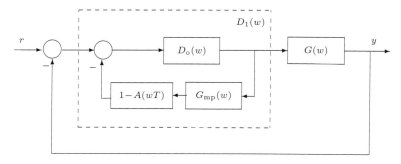

Figure 18.3 Smith predictor structure

$$D_1(w) = \frac{D_o(w)}{1 + D_o(w)G_{mp}(w) - D_o(w)G(w)}. \tag{18.24}$$

Consequently, the global transfer function of the closed-loop compensated system will be

$$F(w) = \frac{K_c}{w^\lambda + K_c} A(w). \tag{18.25}$$

18.3.2 Controller Design Using Phase-lag Compensation

The second proposed design technique takes into account the additional phase-lag introduced by the all-pass filter. This additional phase delay can be evaluated at the frequency of interest w_c being

$$\angle (A(jv))\Big|_{v=\omega_c} = -37.55°$$

resulting in a new phase margin for design:

$$\varphi'_m = \varphi_m - \angle (A(jv))\Big|_{v=\omega_c} = 64 + 37.55 = 101.55°.$$

The open-loop transfer function of the compensated system in this case will be:

$$D_2(w)G(w) = \frac{1}{G_{mp}(w)} \frac{K_c}{w^\lambda} G(w). \tag{18.26}$$

The controller $D_2(w)$ will be obtained from the minimum-phase subsystem, considering the phase lag of $A(w)$. From (18.22) controller parameters are obtained as $K_c = 532$, $\lambda = 0.87$.

Figure 18.4 shows the resulting open-loop Bode plots. As can be observed, the phase margin achieved with both controllers, $D_1(w)$ and $D_2(w)$, are 63.65° and 64.01° at the indicated gain crossover frequencies of 1.2 ×

10^3 rad/sec and 1.36×10^3 rad/sec, respectively. Therefore, the design specifications are very approximately fulfilled.

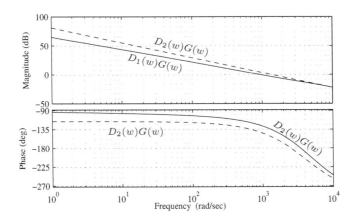

Figure 18.4 Bode plots of the compensated pseudo-continuous system

18.3.3 Controller Design Based on the Discrete Linearized Model

In this section a new design alternative for the controller is proposed from the discrete linearized model of the converter. For this purpose a discrete version of the Bode's ideal loop transfer function will be taken as reference system. So, the open-loop transfer function of the desired compensated system will become

$$G(z)D(z) = \frac{K_c}{[\Delta(z)]^\lambda}, \tag{18.27}$$

where $\Delta(z)$ denotes the discrete equivalent of the Laplace operator s. The parameters K_c and λ will be selected to obtain the specified phase margin φ_m and crossover frequency ω_c. The transfer function of the compensator $D(z)$ is

$$D(z) = \frac{K_c}{[\Delta(z)]^\lambda} \frac{1}{G(z)}, \tag{18.28}$$

where $G(z)$ corresponds to (18.16) and the design equations are (18.22).

In the frequency range of interest, we suppose that the frequency responses of the following equations coincide:

$$\left. \frac{K_c}{s^\lambda} \right|_{s=j\omega,\ \omega\in[10^0,5\times10^3\ \mathrm{rad/sec}]} \approx \left. \frac{K_c}{[\Delta(z)]^\lambda} \right|_{z=e^{j\omega}}.$$

Taking as working specifications for the design $\varphi_m = 64°$, $\omega_c = 1.36 \times 10^3$ rad/sec, the parameters of the compensator are $K_c = 1.1 \times 10^4$, $\lambda = 1.29$. Figure 18.5 shows the open-loop Bode plots of the compensated system using the discrete linearized model. The phase margin achieved is $63.92°$ at the gain crossover frequency of 1.31×10^3 rad/sec. Therefore, the design specifications are very approximately fulfilled. This fact validates the design hypothesis.

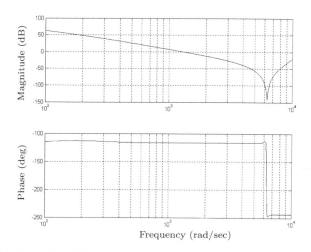

Figure 18.5 Bode plot of the compensated discrete linearized model $G(z)$

18.4 Fractional Sliding Mode Control (F$_R$SMC)

Sliding mode control is well known for its robustness to disturbances and parameter variations. The sliding mode design approach consists of two steps. The first one involves the design of a switching function, $S = 0$, so that the sliding motion satisfies the design specifications. The second one is concerned with the selection of a control law which will enforce the sliding mode, therefore existence and reachability conditions are satisfied. These conditions are obtained from geometrical considerations: the deviation from the switching surface S and its time derivative $\dot{S}$ should be of opposite signs in the vicinity of a switching surface $S = 0$:

$$\lim_{S \to 0^+} \dot{S} < 0, \quad \text{and,} \quad \lim_{S \to 0^-} \dot{S} > 0. \tag{18.29}$$

These conditions are usually written more conveniently as $S\dot{S} < 0$.

This section deals with the design of switching surfaces for a buck converter using alternative techniques based on FOC. Two alternative switching surfaces are presented in order to achieve a good agreement between both transitory and stationary responses. First, sliding surfaces based on linear compensation networks PID or PI are presented. Then, the fractional-order form of these networks, $PI^\lambda D^\mu$ or PI^λ, are used in order to obtain the sliding surfaces. The integral component of this networks makes zero the steady-state error. The control law used in this section is based on the PWM technique applied by computer combined with the generation of trajectories from the different structures of the bilinear model [296].

18.4.1 Sliding Surfaces Through PID and PI Structures

First, this section presents a class of linear sliding surface, based on the canonical form of the converter, using a PID structure to achieve a zero steady-state error.

By using the state-space model in phase canonical form (18.3), the open-loop dynamics of v_c can be expressed as

$$\ddot{v}_c + \frac{1}{RC}\dot{v}_c + \frac{1}{LC}v_c = \frac{V_g}{LC}u. \tag{18.30}$$

From (18.30) a candidate sliding surface for a buck converter can be obtained of the form [297]

$$S = K_p\left(v_r - v_c\right) + K_i\int\left(v_r - v_c\right)dt + K_d\frac{d\left(v_r - v_c\right)}{dt}, \tag{18.31}$$

where v_r is the reference voltage and K_p, K_i, and K_d are the design parameters to be determined. The control system block diagram is shown in Figure 18.6. Since the measurement variables are i_L and v_c, and the design method used here is based on the phase canonical form (the state vector is $x = \begin{bmatrix} v_c & \dot{v}_c \end{bmatrix}^T$), a state-space transformation block is needed.

In sliding motion ($S = 0$, $\dot{S} = 0$), the closed-loop dynamics of v_c becomes

$$\ddot{v}_c + \frac{K_p}{K_d}\dot{v}_c + \frac{K_i}{K_d}v_c = \frac{K_i}{K_d}u. \tag{18.32}$$

The steady-state solution of (18.32) shows asymptotic stability if the following conditions are fulfiled:

$$\frac{K_p}{K_d} > 0, \quad \frac{K_i}{K_d} > 0.$$

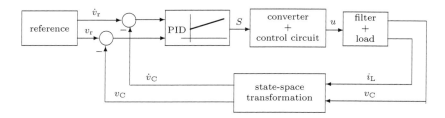

Figure 18.6 Block diagram of the control system

The existence of the sliding mode requires that inequality $S\dot{S} < 0$ is fulfilled. From (18.31) the equivalent control law u_{eq} is obtained. By doing $\dot{S} = 0$ and applying the equivalent control definition, u_{eq} is obtained as

$$u_{eq} = \frac{v_c}{V_g} + \frac{K_i LC}{K_d V_g}(v_r - v_c) - \frac{L}{K_d V_g}\left(K_p - \frac{K_d}{RC}\right)\left(i_L - \frac{v_c}{R}\right), \qquad (18.33)$$

where

$$i_L = C\dot{v}_c + \frac{v_c}{R}. \qquad (18.34)$$

The equivalent control, u_{eq}, can be interpreted as the continuous control law that maintain $\dot{S} = 0$ if the dynamics were exactly known and, therefore, this value is not computed for obtaining the control variable u. From (18.33) and the constraint $|u_{eq}| \leqslant 1$ and considering the equilibria conditions in stationary regime, $v_c = v_r$ and $i_L = 0$, the conditions that limit the existence region of the design parameters are obtained as

$$0 < \frac{K_i}{K_d} < \frac{V_g}{LCv_r},$$
$$0 < \frac{K_p}{K_d} < \frac{1}{RC} + \frac{R}{L}\left(\frac{V_g}{v_r} - 1\right). \qquad (18.35)$$

Another candidate sliding surface, using a PI structure, can be obtained of the form [297]

$$S = i_r - i_L, \qquad (18.36)$$

where

$$i_r = K_p(v_r - v_c) + K_i \int (v_r - v_c)\,d\tau,$$

and v_r is the reference voltage, K_p and K_i are the parameters to be determined. The choice of this structure is due to the fact that the closed-loop converter behaves in sliding mode as a linear first-order system. Previous sliding surface guarantees a null steady-state error because of the integral term. The proportional term allows to get a satisfactory transient response.

The sensed state-space variables are the inductor current i_L and the capacitor voltage v_c. Now the state-space transformation that appears in Figure 18.6 is not necessary and the state-vector is $\boldsymbol{x} = \begin{bmatrix} i_L & v_c \end{bmatrix}^T$. From (18.36) and letting $\dot{S} = 0$, the equivalent control law u_{eq} in this case becomes

$$u_{eq} = \frac{L}{V_g} \left[K_i \left(v_r - v_c \right) - \frac{K_p}{C} i_L + \left(\frac{K_p}{RC} + \frac{1}{L} \right) v_c \right]. \tag{18.37}$$

From (18.37) and using the constraint $|u_{eq}| \leqslant 1$, and considering the aforementioned equilibria conditions, the conditions that limit the existence region of the design parameters are obtained as

$$
\begin{aligned}
0 &< K_i < \frac{V_g}{L v_r}, \\
0 &< K_p < \frac{V_g - v_r}{v_r} \frac{RC}{L}.
\end{aligned}
\tag{18.38}
$$

From (18.33) and (18.37), the existence region in steady state takes the form $0 < v_c < V_g$, which corroborates the step-down behavior of the buck converter. Finally, the control law is chosen in order to satisfy the reachability condition $S\dot{S} < 0$.

18.4.2 Fractional Sliding Surfaces

This section develops two extensions of the aforementioned sliding surfaces which apply fractional calculus to the design of such surfaces. The goal of this procedure is to get a hybrid system that combines the advantages in terms of robustness of the fractional-order control and the sliding mode control. First, the design of switching surfaces by the application of fractional-order PID compensation networks is carried out. The use of $PI^\lambda D^\mu$ allows one to choose, besides the parameters of the classical PID (K_p, K_i, and K_d), the orders of integration λ and derivation μ. Next, the sliding surfaces are obtained using the power model of the converter (taking as state-vector $\begin{bmatrix} i_L & v_c \end{bmatrix}^T$) and applying a fractional-order PI^λ compensation network.

18.4.2.1 $PI^\lambda D^\mu$ Sliding Surfaces

Taking into account the integro-differential equation which defines $PI^\lambda D^\mu$ control action

$$u(t) = K_p e(t) + K_i \mathscr{D}^{-\lambda} e(t) + K_d \mathscr{D}^\mu e(t), \tag{18.39}$$

where, K_p, K_i, K_d, λ, and μ are the design parameters to be determined (and, for simplicity in notation, $\mathscr{D}^{(*)} \equiv {}_o\mathscr{D}_t^{(*)}$), a candidate fractional sliding surface with generalized fractional-order PI$^\lambda$D$^\mu$ structure can be obtained of the form

$$S = K_p (v_r - v_c) + K_i \mathscr{D}^{-\lambda} (v_r - v_c) + K_d \mathscr{D}^\mu (v_r - v_c), \qquad (18.40)$$

with $0 < \mu < 1, 0 < \lambda < 2$. The first derivative of the sliding surface will be

$$\dot{S} = K_p \mathscr{D} (v_r - v_c) + K_i \mathscr{D} \mathscr{D}^{-\lambda} (v_r - v_c) + K_d \mathscr{D} \mathscr{D}^\mu (v_r - v_c). \qquad (18.41)$$

By using definitions and properties of the fractional-order derivatives and integrals [3, 298], where these properties have been used for obtaining an alternative state-space representation of fractional-order systems, we can express the former condition as

$$\dot{S} = K_p (\dot{v}_r - \dot{v}_c) + K_i \mathscr{D}^{1-\lambda} (v_r - v_c) + K_d \mathscr{D}^{\mu-1} (\ddot{v}_r - \ddot{v}_c). \qquad (18.42)$$

By substituting (18.2) into (18.42), with $v_r =$constant ($\dot{v}_r = \ddot{v}_r = 0$), the condition becomes

$$\mathscr{D}^{\mu-1} u_{eq} = -\frac{K_p L}{K_d V_g} i_L + \frac{K_p L}{K_d V_g R} v_c + \frac{K_i LC}{K_d V_g} \mathscr{D}^{1-\lambda} (v_r - v_c)$$
$$+ \frac{1}{V_g} \mathscr{D}^{\mu-1} v_c + \frac{L}{RCV_g} \mathscr{D}^{\mu-1} i_L - \frac{L}{R^2 CV_g} \mathscr{D}^{\mu-1} v_c. \qquad (18.43)$$

So, the equation for equivalent control is

$$u_{eq} = \frac{v_c}{V_g} + \frac{K_i LC}{K_d V_g} \mathscr{D}^{2-\mu-\lambda} (v_r - v_c) + \frac{L}{RCV_g} \left(i_L - \frac{v_c}{R} \right) - \frac{K_p L}{K_d V_g} \mathscr{D}^{1-\mu} \left(i_L - \frac{v_c}{R} \right)$$
$$= \frac{v_c}{V_g} - \frac{K_i LC}{K_d V_g} \mathscr{D}^{1-\mu-\lambda} \dot{v}_c + \frac{L}{RCV_g} \left(i_L - \frac{v_c}{R} \right) - \frac{K_p L}{K_d V_g} \mathscr{D}^{1-\mu} \left(i_L - \frac{v_c}{R} \right),$$
$$\qquad (18.44)$$

that is, equivalent control is a function of the states and their fractional-order derivatives and integrals. If $\lambda = \mu = 1$ (PID structure for sliding surface), (18.33) is obtained.

From (18.44), with restriction $|u_{eq}| \leqslant 1$, and for the equilibria conditions: 1. $v_c = v_r$, $i_L = 0$, and 2. $v_c = 0$, $i_L = 0$, the reachability conditions can be obtained:

1. $v_c = v_r$, $i_L = 0$,

$$0 < \frac{K_p}{K_d} < \frac{1}{\mathscr{D}^{1-\mu} v_r} \left(\frac{RV_g}{L} - \frac{Rv_r}{L} + \frac{v_r}{RC} \right). \qquad (18.45)$$

2. $v_c = 0$, $i_L = 0$,

$$0 < \frac{K_i}{K_d} < \frac{V_g}{LC \mathscr{D}^{2-\mu-\lambda} v_r}. \qquad (18.46)$$

If $\mu = \lambda = 1$, these conditions reduce to (18.35).

The block diagram of the control system in this case is the same as that shown in Figure 18.6 changing the block of the PID network by a $PI^{\lambda}D^{\mu}$ one.

18.4.2.2 PI^{λ} Sliding Surfaces

Next, a PI fractional sliding surface based on the power model of the converter is generated. The proposed sliding surface for achieving this goal is given by

$$S = i_r - i_L, \tag{18.47}$$

being

$$i_r = K_p (v_r - v_c) + K_i \mathscr{D}^{-\lambda} (v_r - v_c), \tag{18.48}$$

where K_p, K_i and λ are the design parameters to be determined.

From (18.47) and (18.48), the sliding surface with generalized fractional-order PI^{λ} structure becomes

$$S = K_p (v_r - v_c) + K_i \mathscr{D}^{-\lambda} (v_r - v_c) - i_L, \tag{18.49}$$

where $0 < \lambda < 2$. Following the procedure of the previous section, the equation obtained for equivalent control is:

$$u_{eq} = \frac{L}{V_g} \left[K_i \mathscr{D}^{1-\lambda} (v_r - v_c) + \left(\frac{K_p}{RC} + \frac{1}{L} \right) v_c - \frac{K_p}{C} i_L \right], \tag{18.50}$$

that is, equivalent control is a function of the states and their fractional-order derivatives and integrals. If $\lambda = 1$ (PI structure for sliding surface), (18.37) is obtained.

From (18.50), with restriction $|u_{eq}| \leqslant 1$, and for the aforementioned equilibria conditions, reachability conditions are obtained:

1. $v_c = v_r$, $i_L = 0$,

$$0 < K_p < \frac{RC}{L} \frac{V_g - v_c}{v_c}. \tag{18.51}$$

2. $v_c = 0$, $i_L = 0$,

$$0 < K_i < \frac{V_e}{L \mathscr{D}^{1-\lambda} v_r}. \tag{18.52}$$

If $\lambda = 1$, these conditions reduce to (18.38).

The control block diagram corresponds to that displayed in Figure 18.6 without the state-space transformation and changing the PID network by a PI^{λ} one.

18.5 Simulation and Experimental Results

18.5.1 Simulation Results

In order to show the performance of the proposed methods, both linear and sliding mode control, simulation and experimental results for all the controllers, with the specifications listed before, are shown here. In the case of linear control, the simulated system corresponds to the block diagram of Figure 18.7 where: the block "table" performs the conversion between fictitious control signal and duty ratio; the block PWM provides to the filter a voltage V_g during the interval $d(k)$; the block "filter" is the LC filter plus the load resistance.

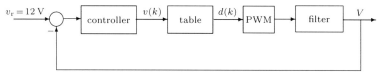

Figure 18.7 Block diagram for simulation

In order to implement the calculated linear regulators in a digital processor, the first step is to find the realizable discrete equivalents of the previous controllers. This is achieved using inverse w-transformation, and, in the case of fractional-order integrators, they are approximated by the continued-fraction expansion (CFE) and the Tustin's rule [165], for obtaining discrete equivalents of order 5. Using the previous approximation, (18.22) does not satisfy the causality principle and so a modification is introduced. Since the order λ is bigger than 1, $\Delta(z)$ is considered as composed of a pure integrator and a fractional-order integrator of order $\lambda' = \lambda - 1$. The fractional-order integrator is approximated by the expressed method and the pure integrator using the forward rectangular rule. Figure 18.8 displays the open-loop Bode plots of the compensated system with the described approximations. These results show that the design specification, phase margin φ_m and gain crossover frequency ω_c, achieved in the design process are kept.

Next, the simulated step responses are presented. Figure 18.9 shows the simulated step responses obtained with the described linear controllers. An overshoot can be observed in the time response of the controlled system using a controller based on the discrete version of the Bode's ideal loop transfer function. This overshoot is due to $\lambda > 1$, and can be removed by changing the design specifications.

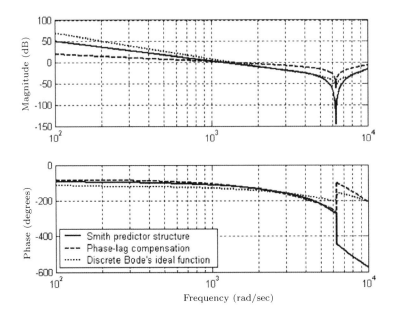

Figure 18.8 Bode plots of the compensated system

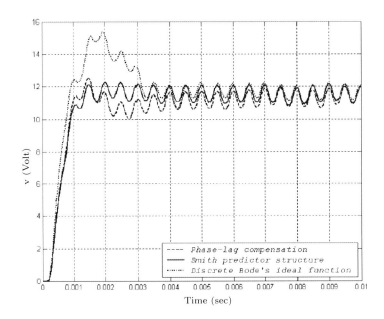

Figure 18.9 Simulated step responses for the designed controllers

For obtaining the simulation results of the controlled system when a fractional structure is used to obtain sliding surfaces, the first step is to find the discrete equivalents of $\mathscr{D}^{-\lambda}(v_r - v_c)$ and $\mathscr{D}^{\mu}(v_r - v_c)$ using the continued fraction expansion method and the Tustin's rule aforementioned, for obtaining discrete equivalents of order 5. Figure 18.10 shows the simulated results obtained with the described controllers. These results are very similar to each other and they agreed with the expected ones from the used methodology to carry out the design of the controllers. The compensation parameters for the sliding mode controllers are displayed in Table 18.1. As can be observed, the fractional orders are close to one, but it must be taken into account that these parameters are mainly introduced to show the possibility of using fractional-order surfaces without obtaining the optimal values. New works are going to deal with the optimizing process.

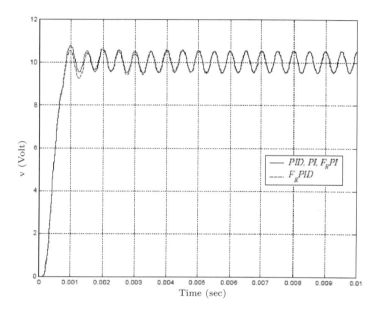

Figure 18.10 Simulation results of the controlled converter for the defined sliding surfaces

Table 18.1 SMC controller parameters

Network	K_p	K_i	K_d	λ	μ
PID	1	20	0.25×10^{-3}	–	–
PI	0.2	15	–	–	–
$PI^\lambda D^\mu$	100	20	0.125×10^{-4}	1.00	0.90
PI^λ	0.2	25	–	1.06	–

18.5.2 Experimental Results

In order to show the feasibility of the proposed methods, a real prototype of the buck converter has been built and experimental results are reported and discussed. The prototype of the converter system is shown in Figure 18.11.

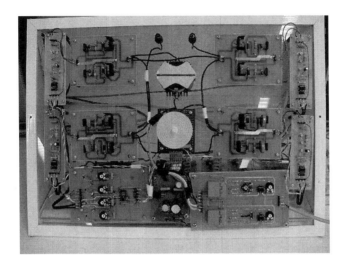

Figure 18.11 Prototype of the converter system

Figure 18.12 shows the block diagram of the controller. The controller algorithms have been implemented in a Pentium 166 MHz machine. The process interface has been carried out with a multifunction data acquisition card PCL 818 and a multifunction Counter/Timer card PCL 836, which provides three PWM channels.

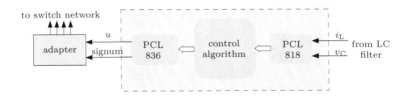

Figure 18.12 Controller block diagram

Figure 18.13 (a-c) shows the experimental responses obtained with the linear controllers. Figure 18.13 (d-g) shows the experimental responses obtained with the sliding mode controllers.

Since there is a good agreement between simulated and experimental results in all the cases, the following comparison is only done considering simulation results.

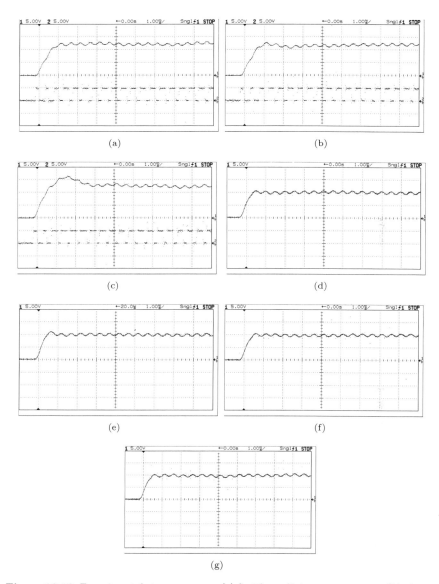

Figure 18.13 Experimental step responses: (a) Smith predictor structure case, (b) phase-lag compensation case, (c) design based on the discrete model case, (d) sliding surface through a PID structure, (e) sliding surface through a PI structure, (f) sliding surface through a PI^λ structure, and (g) sliding surface through a $PI^\lambda D^\mu$ structure

18.5.3 Robustness Comments on Simulation Results

In order to make a comparison between the different controllers, the disturbance rejection characteristics of the converter, namely audio-susceptibility and closed-loop output impedance, have to be considered. There are other methods for achieving this goal, *e.g.*, applying an input voltage feedforward to the PWM or using a peak current mode control, where two control loops are necessary. However with the proposed controllers the requirements are fulfilled with a single control loop. In order to show the robustness and low sensitivity to plant parameters variations of these controllers, a series inductance $L = 3\,\text{mH}$ is considered in the load circuit of the converter. The simulation results in the case of sliding mode controllers are shown in Figure 18.14. As can be observed, there are no significant variations in the step responses with reference to the results shown in Figure 18.10.

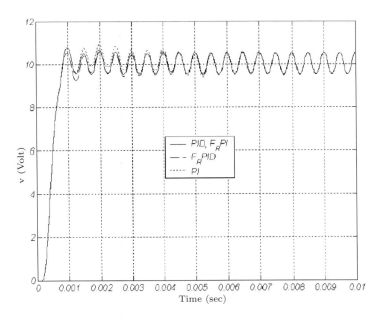

Figure 18.14 Simulation results of the sliding mode controlled converter considering a series inductance $L = 3\,\text{mH}$

Figures 18.15 and 18.16 show the responses to step input voltage changes when a linear controller and a sliding mode controller are used, respectively. These results are compared with those obtained with two control algorithms usually used for controlling converters. The design and implementation of these controllers of reference can be seen in [299]. The best performance is

achieved with the controller based on the discrete version of the Bode's ideal
function.

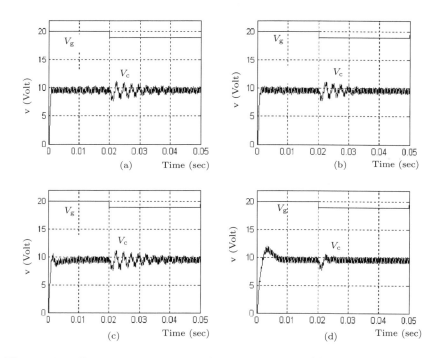

Figure 18.15 Response to step input voltage changes, with: (a) dead beat control, (b)
Smith predictor structure, (c) phase-lag compensation, (d) discrete version of Bode's ideal
function

Figures 18.17 and 18.18 show the response to input voltage ripple when a
linear controller and a sliding mode controller are used, respectively. As can
be observed, the best performance in presence of ripple in the source voltage
V_g is achieved with the PI^λ structure. Therefore, this control scheme shows
much better disturbance rejection properties.

18.6 Summary

In this chapter several alternative methods for the control of power electronic
buck converters applying fractional-order control have been presented. On the
one hand, the design of a linear controller for the DC/DC buck converter is
considered where Bode's ideal loop transfer function presented in Chapter 2
is used as the reference system. On the other hand, the fractional calculus

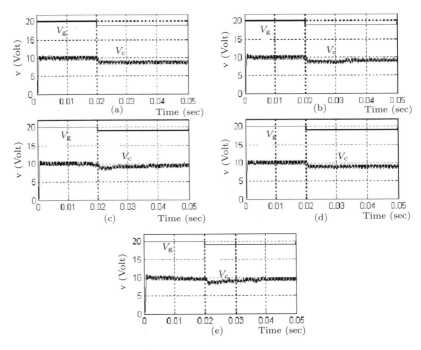

Figure 18.16 Response to step input voltage changes: (a) weighted mean of the state errors, (b) PID structure, (c) PI structure, (d) $PI^\lambda D^\mu$ structure, and (e) PI^λ structure

is proposed in order to determine the switching surface applying a fractional sliding mode control scheme to the control of such devices. In this case, switching surfaces based on fractional-order PID and PI structures are defined. An experimental prototype has been developed and the experimental and simulation results confirm the validity of the proposed control strategies. The practical implementation of the obtained controllers is feasible, as has been demonstrated, to offer good results. New research is being carried out in order to obtain an analytical justification of the robustness properties observed in the behavior of the PI^λ structure, as well as in the definition of new fractional switching surfaces and in the application of the designed controllers to other topologies (as boost and buck-boost converters) and more complex systems (as motor drives). Boost and buck-boost converters show a right-half-plane zero in their control-to-output transfer function, which makes their control more complex. Several of the proposed control strategies for the buck converter can be directly applied to the control of these converters.

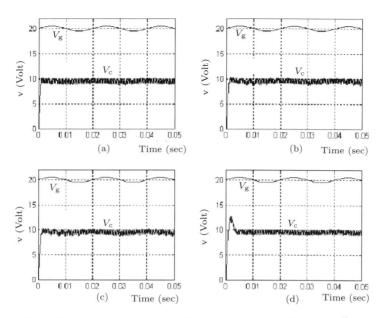

Figure 18.17 Response to step input voltage ripple, with: (a) dead beat control, (b) Smith predictor structure, (c) phase-lag compensation, and (d) discrete version of Bode's ideal function

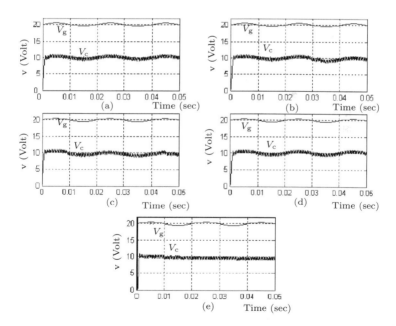

Figure 18.18 Response to input voltage ripple, with: (a) weighted mean of the state errors, (b) PID structure, (c) PI structure (d) $PI^\lambda D^\mu$ structure, and (e) PI^λ structure

Appendix
Laplace Transforms Involving
Fractional and Irrational Operations

As the cases of integer-order systems, Laplace transform and its inverse are very important. In this appendix, the definition is given first. Then some of the essential special functions are described. Finally, an inverse Laplace transform table involving fractional and irrational-order operators is given.

A.1 Laplace Transforms

For a time-domain function $f(t)$, its Laplace transform, in s-domain, is defined as

$$\mathscr{L}[f(t)] = \int_0^\infty f(t)e^{-st}dt = F(s), \tag{A.53}$$

where $\mathscr{L}[f(t)]$ is the notation of Laplace transform.

If the Laplace transform of a signal $f(t)$ is $F(s)$, the inverse Laplace transform of $F(s)$ is defined as

$$f(t) = \mathscr{L}^{-1}[F(s)] = \frac{1}{j2\pi} \int_{\sigma-j\infty}^{\sigma+j\infty} F(s)e^{st}ds, \tag{A.54}$$

where σ is greater than the real part of all the poles of function $F(s)$.

A.2 Special Functions for Laplace Transform

Since the evaluation for some fractional-order is difficult, special functions may be needed. Here some of the special functions are introduced and listed in Table A.1.

A.3 Laplace Transform Tables

An inverse Laplace transform table involving fractional and irrational operators is collected in Table A.2 [86, 300].

Table A.1 Some special functions

Special functions	Definition
Mittag-Leffler	$\mathscr{E}_{\alpha,\beta}^{\gamma}(z) = \sum_{k=0}^{\infty} \dfrac{(\gamma)_k}{\Gamma(\alpha k + \beta)} \dfrac{z^k}{k!}, \ \mathscr{E}_{\alpha,\beta}(z) = \mathscr{E}_{\alpha,\beta}^{1}(z), \ \mathscr{E}_{\alpha}(z) = \mathscr{E}_{\alpha,1}(z)$
Dawson function	$\mathrm{daw}(t) = e^{-t^2} \displaystyle\int_0^t e^{\tau^2} \, d\tau$
erf function	$\mathrm{erf}(t) = \dfrac{2}{\sqrt{\pi}} \displaystyle\int_0^t e^{-\tau^2} \, d\tau$
erfc function	$\mathrm{erfc}(t) = \dfrac{2}{\sqrt{\pi}} \displaystyle\int_t^{\infty} e^{-\tau^2} \, d\tau = 1 - \mathrm{erf}(t)$
Hermit polynomial	$\mathscr{H}_n(t) = e^{t^2} \dfrac{d^n}{dt^n} e^{-t^2}$
Bessel function	$\mathscr{J}_{\nu}(t)$ is the solution to $t^2 \ddot{y} + t\dot{y} + (t^2 - \nu^2)y = 0$
Extended Bessel function	$\mathscr{I}_{\nu}(t) = j^{-\nu} \mathscr{J}_{\nu}(jt)$

Table A.2 Inverse Laplace transforms with fractional and irrational operators

$F(s)$	$f(t) = \mathscr{L}^{-1}[F(s)]$	$F(s)$	$f(t) = \mathscr{L}^{-1}[F(s)]$
$\dfrac{s^{\alpha\gamma-\beta}}{(s^{\alpha}+a)^{\gamma}}$	$t^{\beta-1}\mathscr{E}_{\alpha,\beta}^{\gamma}(-at^{\alpha})$	$\dfrac{1}{s^n\sqrt{s}}, n=1,2,\cdots$	$\dfrac{2^n t^{n-\frac{1}{2}}}{1\cdot 3\cdot 5\cdots(2n-1)\sqrt{\pi}}$
$\dfrac{k}{s^2+k^2}\coth\left(\dfrac{\pi s}{2k}\right)$	$\lvert \sin kt \rvert$	$\arctan\dfrac{k}{s}$	$\dfrac{1}{t}\sin kt$
$\log\dfrac{s^2-a^2}{s^2}$	$\dfrac{2}{t}(1-\cosh at)$	$\dfrac{1}{s\sqrt{s}}e^{-k\sqrt{s}}$	$2\sqrt{\dfrac{t}{\pi}}e^{-\frac{1}{4t}k^2} - k\,\mathrm{erfc}\left(\dfrac{k}{2\sqrt{t}}\right)$
$\log\dfrac{s^2+a^2}{s^2}$	$\dfrac{2}{t}(1-\cos at)$	$\dfrac{e^{-k\sqrt{s}}}{\sqrt{s}(a+\sqrt{s})}$	$e^{ak}e^{a^2 t}\mathrm{erfc}\left(a\sqrt{t}+\dfrac{k}{2\sqrt{t}}\right)$
$\dfrac{(1-s)^n}{s^{n+\frac{1}{2}}}$	$\dfrac{n!}{(2n)!\sqrt{\pi t}}\mathscr{H}_{2n}\left(\sqrt{t}\right)$	$\dfrac{1}{\sqrt{s+b}(s+a)}$	$\dfrac{1}{\sqrt{b-a}}e^{-at}\mathrm{erf}\left(\sqrt{(b-a)t}\right)$
$\dfrac{1}{\sqrt{s^2+a^2}}$	$\mathscr{I}_0(at)$	$\dfrac{(1-s)^n}{s^{n+\frac{3}{2}}}$	$-\dfrac{n!}{(2n+1)!\sqrt{\pi}}\mathscr{H}_{2n+1}\left(\sqrt{t}\right)$
$\dfrac{1}{\sqrt{s^2-a^2}}$	$\mathscr{I}_0(at)$	$\dfrac{(a-b)^k}{(\sqrt{s+a}+\sqrt{s+b})^{2k}}$	$\dfrac{k}{t}e^{-\frac{1}{2}(a+b)t}\mathscr{I}_k\left(\dfrac{a-b}{2}t\right), \ k>0$
$\dfrac{\sqrt{s+2a}-\sqrt{s}}{\sqrt{s+2a}+\sqrt{s}}$	$\dfrac{1}{t}e^{-at}\mathscr{I}_1(at)$	$\dfrac{\sqrt{s+2a}-\sqrt{s}}{\sqrt{s+2a}+\sqrt{s}}$	$\dfrac{1}{t}e^{-at}\mathscr{I}_1(at)$
$\dfrac{(\sqrt{s^2+a^2}-s)^{\nu}}{\sqrt{s^2+a^2}}$	$a^{\nu}\mathscr{J}_{\nu}(at), \nu>-1$	$\dfrac{1}{(s^2-a^2)^k}$	$\dfrac{\sqrt{\pi}}{\Gamma(k)}\left(\dfrac{t}{2a}\right)^{k-\frac{1}{2}}\mathscr{I}_{k-\frac{1}{2}}(at)$
$\dfrac{(\sqrt{s^2-a^2}+s)^{\nu}}{\sqrt{s^2-a^2}}$	$a^{\nu}\mathscr{I}_{\nu}(at), \ \nu>-1$	$\dfrac{1}{(\sqrt{s^2+a^2})^k}$	$\dfrac{\sqrt{\pi}}{\Gamma(k)}\left(\dfrac{t}{2a}\right)^{k-\frac{1}{2}}\mathscr{J}_{k-\frac{1}{2}}(at)$
$(\sqrt{s^2+a^2}-s)^k$	$\dfrac{ka^k}{t}\mathscr{J}_k(at), k>0$	$\log\dfrac{s-a}{s-b}$	$\dfrac{1}{t}\left(e^{bt}-e^{at}\right)$
$\dfrac{1}{s+\sqrt{s^2+a^2}}$	$\dfrac{\mathscr{J}_1(at)}{at}$	$\dfrac{1}{\sqrt{s+a}\sqrt{s+b}}$	$e^{-\frac{1}{2}(a+b)t}\mathscr{I}_0\left(\dfrac{a-b}{2}t\right)$

Table A.2 (continued)

$F(s)$	$f(t) = \mathscr{L}^{-1}[F(s)]$	$F(s)$	$f(t) = \mathscr{L}^{-1}[F(s)]$
$\dfrac{1}{(s+\sqrt{s^2+a^2})^N}$	$\dfrac{N\,\mathscr{J}_N(at)}{at}$, $N>0$	$\dfrac{b^2-a^2}{(s-a^2)(\sqrt{s}+b)}$	$e^{a^2t}\Big[b-a\ \mathrm{erf}\left(a\sqrt{t}\right)\Big] - be^{b^2t}\mathrm{erfc}\left(b\sqrt{t}\right)$
$\sqrt{s-a}-\sqrt{s-b}$	$\dfrac{1}{2\sqrt{\pi t^3}}\left(e^{bt}-e^{at}\right)$	$\dfrac{\sqrt{s+2a}-\sqrt{s}}{\sqrt{s}}$	$ae^{-at}\Big[\mathscr{I}_1(at)+\mathscr{I}_0(at)\Big]$
$\dfrac{1}{s}e^{-k/s}$	$\mathscr{J}_0\left(2\sqrt{kt}\right)$	$\dfrac{1}{\sqrt{s}}e^{-k/s}$	$\dfrac{1}{\sqrt{\pi t}}\cos 2\sqrt{kt}$
$\dfrac{1}{\sqrt{s}}e^{k/s}$	$\dfrac{1}{\sqrt{\pi t}}\cosh 2\sqrt{kt}$	$\dfrac{1}{s\sqrt{s}}e^{-k/s}$	$\dfrac{1}{\sqrt{\pi k}}\sin 2\sqrt{kt}$
$\dfrac{1}{s\sqrt{s}}e^{k/s}$	$\dfrac{1}{\sqrt{\pi k}}\sinh 2\sqrt{kt}$	$\dfrac{1}{s^\nu}e^{-k/s}$	$\left(\dfrac{t}{k}\right)^{\frac{1}{2}(\nu-1)}\mathscr{J}_{\nu-1}\left(2\sqrt{kt}\right)$, $\nu>0$
$e^{-k\sqrt{s}}$	$\dfrac{k}{2\sqrt{\pi t^3}}e^{-\frac{1}{4k}k^2}$	$\dfrac{1}{s^\nu}e^{k/s}$	$\left(\dfrac{t}{k}\right)^{\frac{1}{2}(\nu-1)}\mathscr{I}_{\nu-1}\left(2\sqrt{kt}\right)$
$\dfrac{1}{s}e^{-k\sqrt{s}}$	$\mathrm{erfc}\left(\dfrac{k}{2\sqrt{t}}\right)$	$\dfrac{1}{s\sqrt{s}}e^{-\sqrt{s}}$	$2\sqrt{\dfrac{t}{\pi}}e^{-\frac{1}{4t}}-\mathrm{erfc}\left(\dfrac{1}{2\sqrt{t}}\right)$
$\dfrac{1}{\sqrt{s}}e^{-k\sqrt{s}}$	$\dfrac{1}{\sqrt{\pi t}}e^{-\frac{1}{4t}k^2}$	$\dfrac{e^{-\sqrt{s}}}{\sqrt{s}(\sqrt{s}+1)}$	$e^{t+1}\mathrm{erfc}\left(\sqrt{t}+\dfrac{1}{2\sqrt{t}}\right)$
$\dfrac{1}{(s+a)^\alpha}$	$\dfrac{t^{\alpha-1}}{\Gamma(\alpha)}e^{-at}$	$\dfrac{1}{s^\alpha+a}$	$t^{\alpha-1}\mathscr{E}_{\alpha,\alpha}\left(-at^\alpha\right)$
$\dfrac{a}{s(s^\alpha+a)}$	$1-\mathscr{E}_\alpha\left(-at^\alpha\right)$	$\dfrac{s^\alpha}{s(s^\alpha+a)}$	$\mathscr{E}_\alpha\left(-at^\alpha\right)$
$\dfrac{1}{s^\alpha(s-a)}$	$t^\alpha\mathscr{E}_{1,1+\alpha}(at)$	$\dfrac{s^\alpha}{s-a}$	$-t^\alpha\mathscr{E}_{1,1-\alpha}(at)$, $0<\alpha<1$
$\dfrac{1}{\sqrt{s}}$	$\dfrac{1}{\sqrt{\pi t}}$	$\dfrac{1}{s\sqrt{s}}$	$2\sqrt{\dfrac{t}{\pi}}$
$\dfrac{1}{\sqrt{s}(s+1)}$	$\dfrac{2}{\sqrt{\pi}}\mathrm{daw}\left(\sqrt{t}\right)$	$\dfrac{\sqrt{s}}{s+1}$	$\dfrac{1}{\sqrt{\pi t}}-\dfrac{2}{\sqrt{\pi}}\mathrm{daw}\left(\sqrt{t}\right)$
$\dfrac{1}{\sqrt{s}(s+a^2)}$	$\sqrt{t}\,\mathscr{E}_{1,3/2}\left(-a^2t\right)$	$\dfrac{s}{(s-a)\sqrt{s-a}}$	$\dfrac{1}{\sqrt{\pi t}}e^{at}(1+2at)$
$\dfrac{\sqrt{s}}{s+a^2}$	$\dfrac{1}{\sqrt{t}}\mathscr{E}_{1,1/2}\left(-a^2t\right)$	$\dfrac{1}{\sqrt{s}+a}$	$\dfrac{1}{\sqrt{\pi t}}-ae^{a^2t}\mathrm{erfc}\left(a\sqrt{t}\right)$
$\dfrac{1}{s\sqrt{s+1}}$	$\mathrm{erf}\left(\sqrt{t}\right)$	$\dfrac{\sqrt{s}}{s-a^2}$	$\dfrac{1}{\sqrt{\pi t}}+ae^{a^2t}\mathrm{erf}\left(a\sqrt{t}\right)$
$\dfrac{1}{\sqrt{s}(s-a^2)}$	$\dfrac{1}{a}e^{a^2t}\mathrm{erf}\left(a\sqrt{t}\right)$	$\dfrac{1}{\sqrt{s}(s+a^2)}$	$\dfrac{2}{a\sqrt{\pi}}e^{-a^2t}\displaystyle\int_0^{a\sqrt{t}}e^{\tau^2}d\tau$
$\dfrac{1}{\sqrt{s}(\sqrt{s}+a)}$	$e^{a^2t}\mathrm{erfc}\left(a\sqrt{t}\right)$	$\dfrac{s\sqrt{s}}{s+1}$	$2\sqrt{\dfrac{t}{\pi}}-\dfrac{2}{\sqrt{\pi}}\mathrm{daw}\left(\sqrt{t}\right)$
$\dfrac{1}{\sqrt{s+1}}$	$\dfrac{e^{-t}}{\sqrt{\pi t}}$	$\dfrac{1}{\sqrt{s}(s-1)}$	$e^t\mathrm{erf}\left(\sqrt{t}\right)$
$\dfrac{\sqrt{s}}{s-1}$	$\dfrac{1}{\sqrt{\pi t}}+e^t\mathrm{erf}\left(\sqrt{t}\right)$	$\dfrac{k!}{\sqrt{s}\pm\lambda}$	$t^{(k-1)/2}\mathscr{E}_{1/2,1/2}^{(k)}\left(\mp\lambda\sqrt{t}\right)$, $\Re(s)>\lambda^2$
$\dfrac{1}{\varrho^\alpha}$	$\dfrac{t^{\alpha-1}}{\Gamma(\alpha)}$	$\dfrac{s^{\alpha-1}}{s^\alpha\perp\lambda}$	$\mathscr{E}_\alpha\left(\mp\lambda t^\alpha\right)$, $\Re(s)>|\lambda|^{1/\alpha}$

Table A.2 (continued)

$F(s)$	$f(t) = \mathscr{L}^{-1}[F(s)]$
$\dfrac{1}{\sqrt{s(s+a)}(\sqrt{s+a}+\sqrt{s})^{2\nu}}$	$\dfrac{1}{a^{\nu}}e^{-at/2}\mathscr{I}_{\nu}\left(\dfrac{a}{2}t\right),\ k>0$
$\dfrac{\Gamma(k)}{(s+a)^k(s+b)^k}$	$\sqrt{\pi}\left(\dfrac{t}{a-b}\right)^{k-\frac{1}{2}}e^{-\frac{1}{2}(a+b)t}\mathscr{I}_{k-\frac{1}{2}}\left(\dfrac{a-b}{2}t\right)$
$\dfrac{1}{\sqrt{s^2+a^2}(s+\sqrt{s^2+a^2})^N}$	$\dfrac{J_N(at)}{a^N}$
$\dfrac{1}{\sqrt{s^2+a^2}(s+\sqrt{s^2+a^2})}$	$\dfrac{J_1(at)}{a}$
$\dfrac{b^2-a^2}{\sqrt{s}(s-a^2)(\sqrt{s}+b)}$	$e^{a^2t}\left[\dfrac{b}{a}\mathrm{erf}\left(a\sqrt{t}\right)-1\right]+e^{b^2t}\mathrm{erfc}\left(b\sqrt{t}\right)$
$\dfrac{ae^{-k\sqrt{s}}}{s(a+\sqrt{s})}$	$-e^{ak}e^{a^2t}\mathrm{erfc}\left(a\sqrt{t}+\dfrac{k}{2\sqrt{t}}\right)+\mathrm{erfc}\left(\dfrac{k}{2\sqrt{t}}\right)$
$\dfrac{1}{\sqrt{s+a}(s+b)\sqrt{s+b}}$	$te^{-\frac{1}{2}(a+b)t}\left[\mathscr{I}_0\left(\dfrac{a-b}{2}t\right)+\mathscr{I}_1\left(\dfrac{a-b}{2}t\right)\right]$
$\dfrac{e^{-\sqrt{s}}}{\sqrt{s}+1}$	$\dfrac{e^{-\frac{1}{4k}}}{\sqrt{\pi t}}-e^{t+1}\mathrm{erfc}\left(\sqrt{t}+\dfrac{1}{2\sqrt{t}}\right)$
$\dfrac{e^{-\sqrt{s}}}{s(\sqrt{s}+1)}$	$\mathrm{erfc}\left(\dfrac{1}{2\sqrt{t}}\right)-e^{t+1}\mathrm{erfc}\left(\sqrt{t}+\dfrac{1}{2\sqrt{t}}\right)$

References

1. K.J. Åström, R.M. Murray. Feedback Systems: An Introduction for Scientists and Engineers. Princeton University Press, 2008
2. K.S. Miller, B. Ross. An Introduction to the Fractional Calculus and Fractional Differential Equations. New York: John Wiley and Sons, 1993
3. I. Podlubny. Fractional Differential Equations, Mathematics in Science and Engineering, volume 198. San Diego: Academic Press, 1999
4. R.L. Magin. Fractional Calculus in Bioengineering. Begell House, 2006
5. K.B. Oldham, J. Spanier. The Fractional Calculus. Theory and Applications of Differentiation and Integration of Arbitrary Order. New York: Dover, 2006
6. S. Dugowson. Les Différentielles Métaphysiques: Histoire et Philosophie de la Généralisation de l'Ordre de Dérivation. Ph.D. thesis, University of Paris, 1994
7. V. Kiryakova. Generalized Fractional Calculus and Applications. Number 301 in Pitman Research Notes in Mathematics. Essex: Longman Scientific & Technical, 1994
8. R. Gorenflo, F. Mainardi. Fractional calculus: Integral and differential equations of fractional order. In A. Carpintieri, F. Mainardi *eds.*, Fractals and Fractional Calculus in Continuum Mechanics. Springer Verlag, 1997
9. G. Mittag-Leffler. Sur la représentation analytique d'une branche uniforme d'une fonction monogene. Acta Mathematica, 1904, 29:101–181
10. R. Gorenflo, Y. Luchko, S. Rogosin. Mittag-Leffler type functions: notes on growth properties and distribution of zeros. Preprint A-97-04, Freie Universität Berlin, 1997
11. I. Podlubny. Numerical solution of ordinary fractional differential equations by the fractional difference method. In S. Elaydi, I. Gyori, G. Ladas *eds.*, Advances in Difference Equations. Proceedings of the Second International Conference on Difference Equations. CRC Press, 1997
12. S. Westerlund, L. Ekstam. Capacitor theory. IEEE Transactions on Dielectrics and Electrical Insulation, 1994, 1(5):826–839
13. M. Cuadrado, R. Cabanes. Temas de Variable Compleja. Madrid: Servicio de Publicaciones de la ETSIT UPM, 1989
14. D. Matignon. Stability properties for generalized fractional differential systems. D. Matignon, G. Montseny, *eds.*, Proceedings of the Colloquium Fractional Differential Systems: Models, Methods and Applications, 5. Paris, 1998, 145–158
15. S. Manabe. The non-integer integral and its application to control systems. Japanese Institute of Electrical Engineers Journal, 1961, 6(3-4):83–87
16. A. Oustaloup. La Dérivation non Entière. Paris: Hermès, 1995
17. A. Oustaloup, B. Mathieu. La Commande CRONE: du Scalaire au Multivariable. Paris: Hermès, 1999
18. K.J. Åström. Limitations on control system performance. Preprint, 1999

19. R.L. Bagley, P.J. Torvik. On the appearance of the fractional derivative in the behavior of real materials. Journal of Applied Mechanics, 1984, 51:294–298

20. R.L. Bagley, R.A. Calico. Fractional-order state equations for the control of visco-elastic damped structures. J Guidance, Control and Dynamics, 1991, 14(2):304–311

21. A. A. Kilbas, H. M. Srivastava, J. J. Trujillo. Theory and Applications of Fractional Differential Equations. North-Holland, 2006

22. K. Ogata. Modern Control Engineering, 4th edition. Prentice Hall, 2001

23. R.A. Horn, C.R. Johnson. Topics in Matrix Analysis. Cambridge University Press, 1991

24. M. Abramowitz, I.A. Stegun. Handbook of Mathematical Functions. Dover Publications, 1972

25. D. Matignon, B. d'Andréa-Novel. Some resuts on controllability and observability of finite-dimensional fractional differential systems. http://www-sigenstfr/matignon

26. L. Zadeh, C.A. Desoer. Linear Systems Theory. McGraw-Hill, 1963

27. K.B. Oldham, J. Spanier. The Fractional Calculus. New York: Academic Press, 1974

28. P. Ostalczyk. The non-integer difference of the discrete-time function and its application to the control system synthesis. International Journal of Systems Science, 2000, 31(12):1551–1561

29. A. Dzieliński, D. Sierociuk. Stability of discrete fractional order state-space systems. Journal of Vibration and Control, 2008, 14(9-10):1543–1556

30. S. Guermah, S. Djennoune, M. Bettayeb. Controllability and observability of linear discrete-time fractional-order systems. Int J Applied Mathematics and Computer Science, 2008, 18(2):213–222

31. C.L. Phillips, Jr. H. Troy Nagle. Digital Control System. Analysis and Design. New Jersey, USA: Prentice Hall, 1984

32. D. Sierociuk. Fractional Order Discrete State-Space System Simulink Toolkit User Guide. [online] http://www.ee.pw.edu.pl/dsieroci/fsst/fsst.htm

33. A. Dzieliński, D. Sierociuk. Simulation and experimental tools for fractional order control education. Proceedings of 17th World Congress The International Federation of Automatic Control, Seoul, Korea, July 6-11. IFAC WC 2008, 2008, 11654–11659

34. A. Dzieliński, D. Sierociuk. Observer for discrete fractional order-space state. Proceedings of the Second IFAC Workshop on Fractional Derivatives and Its Applications (FDA'06), Porto, Portugal, 2006

35. D.L. Debeljković, M. Aleksendrić, N. Yi-Yong, et al. Lyapunov and non-Lyapunov stability of linear discrete time delay systems. Facta Universitatis, Series: Mechanical Engineering, 2002, 1(9):1147–1160

36. R. Hilfer. Applications of Fractional Calculus in Physics. World Scientific, 2000

37. A. Dzieliński, D. Sierociuk. Reachability, controllability and observability of fractional order discrete state-space systems. Proceedings of the 13th IEEE/IFAC International Conference on Methods and Models in Automation and Robotics (MMAR'07), Szczecin, Poland, 2007

38. L. Dorčak, I. Petráš, I. Koštial, et al. State-space controller design for the fractional-order regulated aystem. Proceedings of the ICCC'2001. Krynica, Poland, 2001, 15–20

39. D. Matignon, B. d'Andréa-Novel. Observer-based controllers for fractional differential systems. 36st IEEE-CSS SIAM Conference on Decision and Control, 1997

40. Y.Q. Chen, H.S. Ahn, I. Podlubny. Robust stability check of fractional order linear time invariant systems with interval uncertainties. Signal Processing, 2006, 86:2611–2618

41. Y.Q. Chen, H.S. Ahn, D.Y. Xue. Robust controllability of interval fractional order linear time invariant systems. Signal Processing, 2006, 86:2794–2802

42. A. Si-Ammour, S. Djennoune, M. Bettayeb. A sliding mode control for linear fractional dystems with input and state delays. Communications in Nonlinear Science and Numerical Simulation, 2009, 14:2310–2318

43. D. Sierociuk, A. Dzieliński. Fractional Kalman filter algorithm for states, parameters and order of fractional system estimation. International Journal of Applied Mathematics and Computer Science, 2006, 16(1):129–140

44. D. Sierociuk. Estimation and control of discrete dynamic fractional order state space systems. Ph.D. thesis, Warsaw University of Technology, 2007. In polish

45. A. Dzieliński, D. Sierociuk. Adaptive feedback control of fractional order discrete state-space systems. Proceedings of International Conference on Computational Intelligence for Modelling Control and Automation, volume 1. CIMCA'2005, 2005, 804–809

46. H. Bode. Relations between attenuation and phase in feedback amplifier design. Bell System Technical Journal, 1940, 19:421–454

47. H. Bode. Network Analysis and Feedback Amplifier Design. Van Nostrand, 1945

48. B.M. Vinagre, C.A. Monje, A.J. Calderón, et al. The fractional integrator as a reference function. Proceedings of the First IFAC Workshop on Fractional Differentiation and Its Applications. ENSEIRB, Bordeaux, France, 2004, 150–155

49. R.S. Barbosa, J.A. Tenreiro, I.M. Ferreira. A fractional calculus perspective of PID tuning. Proceedings of the ASME 2003 Design Engineering Technical Conferences and Computers and Information in Engineering Conference, Chicago, USA, 2003

50. A. Oustaloup. La Commade CRONE: Commande Robuste d'Ordre Non Entier. Hermes, Paris, 1991

51. A. Oustaloup, F. Levron, F. Nanot, et al. Frequency band complex non integer differentiator : characterization and synthesis. IEEE Transactions on Circuits and Systems I: Fundamental Theory and Applications, 2000, 47(1):25–40

52. I. Podlubny. Fractional-order systems and $PI^{\lambda}D^{\mu}$ controllers. IEEE Transactions on Automatic Control, 1999, 44:208–214

53. B.M. Vinagre, I. Podlubny, L. Dorčák, et al. On fractional PID controllers: a frequency domain approach. Proceedings of the IFAC Workshop on Digital Control. Past, Present and Future of PID Control. Terrasa, Spain, 2000, 53–58

54. R. Caponetto, L. Fortuna, D. Porto. Parameter tuning of a non integer order PID controller. 15th International Symposium on Mathematical Theory of Networks and Systems, Notre Dame, Indiana, 2002

55. R. Caponetto, L. Fortuna, D. Porto. A new tuning strategy for a non integer order PID controller. Proceedings of the First IFAC Workshop on Fractional Differentiation and Its Application. ENSEIRB, Bordeaux, France, 2004, 168–173

56. J.F. Leu, S.Y. Tsay, C. Hwang. Design of optimal fractional-order PID controllers. Journal of the Chinese Institute of Chemical Engineers, 2002, 33(2):193–202

57. B.M. Vinagre. Modelado y Control de Sistemas Dinámicos Caracterizados por Ecuaciones Íntegro-Diferenciales de Orden Fraccional. PhD Thesis, Universidad de Educación a Distancia (UNED), Madrid, Spain, 2001

58. B. M. Vinagre, Y. Q. Chen. Lecture Notes on Fractional Calculus Applications in Automatic Control and Robotics. The 41st IEEE CDC2002 Tutorial Workshop # 2. Las Vegas, Nevada, USA, 2002, 1–310. [Online] http://mechatronics.ece.usu.edu/foc/cdc02_tw2_ln.pdf

59. R.S. Barbosa, J.A. Tenreiro Machado. Describing function analysis of systems with impacts and backlash. Nonlinear Dynamics, 2002, 29(1-4):235–250

60. R. Barbosa, J.A. Tenreiro, I.M. Ferreira. Tuning of PID controllers based on Bode's ideal transfer function. Nonlinear Dynamics, 2004, 38:305–321

61. R. Barbosa, J.A. Tenreiro, I.M. Ferreira. PID controller tuning using fractional calculus concepts. Fractional Calculus and Applied Analysis, 2004, 7(2):119–134

62. Y.Q. Chen, K.L. Moore, B.M. Vinagre, et al. Robust PID controller autotuning with a phase shaper. Proceedings of the First IFAC Workshop on Fractional Differentiation and Its Application. ENSEIRB, Bordeaux, France, 2004, 162–167

63. C.A. Monje, A.J. Calderón, B.M. Vinagre. PI vs fractional DI control: first results. Proceedings of the CONTROLO 2002: 5th Portuguese Conference on Automatic Control. Aveiro, Portugal, 2002, 359–364

64. Y.Q. Chen, B.M. Vinagre, C.A. Monje. Une proposition pour la synthèse de correcteurs PI d'ordre non entier. Proceedings of the Action Thématique Les Systèmes à Dérivées Non Entières, LAP-ENSEIRB, Bordeaux, France, 2003

65. C.A. Monje, B.M. Vinagre, V. Feliu, et al. Tuning and auto-tuning of fractional order controllers for industry applications. Control Engineering Practice, 2008, 16(7):798–812

66. R. Malti, M. Aoun, O. Cois, et al. Norm of fractional differential systems. Proceedings of the ASME 2003 Design Engineering Technical Conferences and Computers and Information in Engineering Conference, Chicago, USA, 2003

67. I. Petráš, M. Hypiusova. Design of fractional-order controllers via $\mathcal{H}_\infty$ norm minimisation. Selected Topics in Modelling and Control, 2002, 3:50–54

68. B.S.Y. Sánchez. Fractional-PID control for active reduction of vertical tail buffeting. MSc Thesis, Saint Louis University, St. Louis, USA, 1999

69. A. Oustaloup, B. Mathieu, P. Lanusse. The CRONE control of resonant plants: application to a flexible transmission. European Journal of Control, 1995, 1(2):113–121

70. D. Valério. Fractional order robust control: an application. Student forum, University of Porto, Portugal, 2001

71. P. Lanusse, T. Poinot, O. Cois, et al. Tuning of an active suspension system using a fractional controller and a closed-loop tuning. Proceedings of the 11th International Conference on Advanced Robotics. Coimbra, Portugal, 2003, 258–263

72. A.J. Calderón, B.M. Vinagre, V. Feliu. Linear fractional order control of a DC-DC buck converter. Proceedings of European Control Conference, Cambridge, UK, 2003

73. A.J. Calderón. Control Fraccionario de Convertidores Electrónicos de Potencia tipo Buck. MS Thesis, Escuela de Ingenierías Industriales, Universidad de Extremadura, Badajoz, Spain, 2003

74. V. Pommier, R. Musset, P. Lanusse, et al. Study of two robust control for an hydraulic actuator. ECC 2003: European Control Conference, Cambridge, UK, 2003

75. C.A. Monje, F. Ramos, V. Feliu, et al. Tip position control of a lightweight flexible manipulator using a fractional order controller. IEE Proceeding-D: Control Theory and Applications, 2007, 1(5):1451–1460

76. J.A. Tenreiro, A. Azenha. Fractional-order hybrid control of robot manipulators. Proceedings of the 1998 IEEE International Conference on Systems, Man and Cybernetics: Intelligent Systems for Humans in a Cyberworld, San Diego, California, USA. 1998, 788–793

77. N.M. Fonseca, J.A. Tenreiro. Fractional-order hybrid control of robotic manipulators. Proceedings of the 11th International Conference on Advanced Robotics, Coimbra, Portugal. 2003, 393–398

78. B.M. Vinagre, I. Petráš, P. Merchán, et al. Two digital realizations of fractional controllers: application to temperature control of a solid. Proceedings of the European Control Conference, Porto, Portugal. 2001, 1764–1767

79. I. Petráš, B.M. Vinagre. Practical application of digital fractional-order controller to temperature control. Acta Montanistica Slovaca, 2002, 7(2):131–137

80. I. Petráš, B.M. Vinagre, L. Dorčák, et al. Fractional digital control of a heat solid: experimental results. Proceedings of the International Carpathian Control Conference, Malenovice, Czech Republic. 2002, 365–370

81. J. Sabatier, P. Melchior, A. Oustaloup. Réalisation d'un banc d'essais thermique pour l'enseignement des systemes non entier. Proceedings of the Colloque sur l'Enseignement des Technologies et des Sciences de l'Information et des Systemes, Université Paul Sabatier, Toulouse, France. 2003, 361–364

82. J. G. Zeigler, N. B. Nichols. Optimum settings for automatic controllers. Transactions of ASME, 1942, 64:759–768

83. H.N. Koivo, J.N. Tanttu. Tuning of PID controllers: survey of SISO and MIMO techniques. Proceedings of Intelligent Tuning and Adaptive Control, Singapore, 1991

84. S. Yamamoto, I. Hasimoto. Present status and future needs: the view from Japanese industry. In Arkun and Ray *ed* Proceedings of 4th International Conference on Chemical Process Control, Texas, 1991

85. L. Debnath. A brief historical introduction to fractional calculus. Int J Math Educ Sci Technol, 2004, 35(4):487–501

86. R.L. Magin. Fractional calculus in bioengineering. Critical Reviews^{TM} in Biomedical Engineering, 2004, 32(1-4)

87. D.Y. Xue, Y.Q. Chen. A comparative introduction of four fractional order controllers. Proc. of The 4th IEEE World Congress on Intelligent Control and Automation (WCICA02). Shanghai, China: IEEE, 2002, 3228–3235

88. Y.Q. Chen. Ubiquitous fractional order controls? Proc. of The Second IFAC Symposium on Fractional Derivatives and Applications (IFAC FDA06, Plenary Paper.). Porto, Portugal, 2006, 19–21

89. D.Y. Xue, C.N. Zhao, Y.Q. Chen. Fractional order PID control of a DC-motor with an elastic shaft: a case study. Proceedings of American Control Conference. Minneapolis, Minnesota, USA, 2006, 3182–3187

90. S. Manabe. The non-integer integral and its application to control systems. Japanese Institute of Electrical Engineers Journal, 1960, 80(860):589–597

91. A. Oustaloup. Linear feedback control systems of fractional order between 1 and 2. Proc. of the IEEE Symposium on Circuit and Systems. Chicago, USA, 1981,

92. M. Axtell, E.M. Bise. Fractional calculus applications in control systems. Proc. of the IEEE 1990 Nat. Aerospace and Electronics Conf. New York, USA, 1990, 563–566

93. J.A.T. Machado. Analysis and design of fractional-order digital control systems. J of Systems Analysis-Modeling-Simulation, 1997, 27:107–122

94. J.A.T. Machado (Guest Editor). Special Issue on Fractional Calculus and Applications. Nonlinear Dynamics, 2002, 29:1–385

95. M.D. Ortigueira and J.A.T. Machado (Guest Editors). Special Issue on Fractional Signal Processing and Applications. Signal Processing, 2003, 83(11):2285–2480

96. C. Ricardo, L. Fortuna, D. Porto. A new tuning strategy for a non integer order PID controller. Proc of 1st IFAC Workshop on Fractional Differentiation and its Applications, Bordeaux, France, 2004

97. D. Valério, J.S. da Costa. Tuning-rules for fractional PID controllers. Proceedings of The Second IFAC Symposium on Fractional Differentiation and its Applications (FDA06), Porto, Portugal, 2006

98. C.A. Monje Micharet. Design Methods of Fractional Order Controllers for Industrial Applications. Ph.D. thesis, University of Extremadura, Spain, 2006

99. C.A. Monje Micharet. Auto-tuning of fractional PID controllers. IEEE Control System Society San Diego Chapter Meeting. In Trex Enterprises,. San Diego, California, USA, 2005, Slides available at http://mechatronics.ece.usu.edu/foc/

100. C.N. Zhao. Research on Analyse and Design Methods of Fractional Order System. Ph.D. thesis, Northeastern University, China, 2006

101. K.J. Åström, H. Panagopoulos, T. Hägglund. Design of PI controllers based on non-convex optimization. Automatica, 1998, 34(5):585–601

102. K.J. Åström, H. Panagopoulos, T. Hägglund. Design of PID controllers based on constrained optimization. IEE Proceedings of Control Theory and Application, 2002, 149(1):32–40

103. K.J. Åström, T. Hägglund. PID Controller: Theory, Design and Tuning. Research Triangle Park, NC: Instrument Society of America, 1995

104. K.J. Åström, T. Hägglund. Automatic Tuning of PID Controllers. Reading, MA: Instrumentation Society, 1998

105. K.J. Åström, T. Hägglund. Revisiting the Ziegler-Nichols tuning rules for PI control. Asian Journal of Control, 2002, 4(4):364–380

106. Y.Q. Chen, K.L. Moore. Analytical stability bound for a class of delayed fractional-order dynamic systems. Nonlinear Dynamics, 2002, 29(1-4):191–200

107. B.J. Lurie. Three-parameter tunable tilt-integral-derivative (TID) controller. US Patent US5371670, 1994

108. A. Oustaloup, X. Moreau, M. Nouillant. The CRONE suspension. Control Engineering Practice, 1996, 4(8):1101–1108

109. A. Oustaloup, P. Melchoir, P. Lanusse, et al. The CRONE toolbox for MATLAB. Proc of the 11th IEEE International Symposium on Computer Aided Control System Design, Anchorage, USA, 2000

110. C.A. Monje, B.M. Vinagre, Y.Q. Chen, et al. Proposals for fractional $PI^\lambda D^\mu$ tuning. Proceedings of The First IFAC Symposium on Fractional Differentiation and its Applications (FDA04), Bordeaux, France, 2004

111. C.N. Zhao, D.Y. Xue, Y.Q. Chen. A fractional order PID tuning algorithm for a class of fractional order plants. Proceedings of the International Conference on Mechatronics and Automation. Niagara, Canada, 2005, 216–221

112. A. Oustaloup, J. Sabatier, P. Lanusse. From fractional robustness to CRONE control. Fractional Calculus and Applied Analysis, 1999, 2(1):1–30

113. R.C. Dorf, R.H. Bishop. Modern Control Systems. Upper Saddle River: Pearson Education, 2005

114. Y. Tarte, Y.Q. Chen, W. Ren, et al. Fractional horsepower dynamometer — a general purpose hardware-in-the-loop real-time simulation platform for nonlinear control research and education. Proceedings of IEEE Conference on Decision and Control. San Diego, California, USA, 2006, 3912–3917

115. G. Franklin, J. Powell, A. Naeini. Feedback Control of Dynamic Systems. Addison-Wesley, 1986

116. Y.Q. Chen, C.H. Hu, K.L. Moore. Relay feedback tuning of robust PID controllers with iso-damping property. Proceedings of the 42nd IEEE Conference on Decision and Control. Hawaii, USA, 2003, 347–352

117. J. Pintér. Global Optimization in Action. The Netherlands: Kluwer Academic Publishers, 1996

118. W.S. Levine. The Control Handbook. CRC Press and IEEE Press, 1996

119. A. Oustaloup. From fractality to non integer derivation through recursivity, a property common to these two concepts: a fundamental idea from a new process control strategy. Proceedings of the 12th IMACS World Congress. Paris, France, 1998, 203–208

120. H.F. Raynaud, A. Zergaïnoh. State-space representation for fractional order controllers. Automatica, 2000, 36:1017–1021

121. S.B. Skaar, A.N. Michel, R.K. Miller. Stability of viscoelastic control systems. IEEE Transactions on Automatic Control, 1998, 33(4):348–357

122. C.A. Monje, A.J. Calderón, B.M. Vinagre, et al. The fractional order lead compensator. Proceedings of the IEEE International Conference on Computational Cybernetics. Vienna, Austria, 2004, 347–352

123. C. A. Monje, B. M. Vinagre, A. J. Calderón, et al. Auto-tuning of fractional lead-lag compensators. Proceedings of the 16th IFAC World Congress, Prague, Czech Republic, 2005

124. K.K. Tan, S. Huang, R. Ferdous. Robust self-tuning PID controller for nonlinear systems. Journal of Process Control, 2002, 12(7):753–761

125. C.C. Hang, K.J. Åström, Q.G. Wang. Relay feedback auto-tuning of process controllers — a tutorial review. Journal of Process Control, 2002, 12:143–162

126. D. Valério. Fractional Robust System Control. Ph.D. thesis, Instituto Superior Técnico, Universidade Técnica de Lisboa, 2005

127. I.D. Landau, D. Rey, A. Karimi, et al. A flexible transmission system as a benchmark for robust digital control. European Journal of Control, 1995, 1(2):77–96

128. D.F. Thomson. Optimal and Sub-Optimal Loop Shaping in Quantitative Feedback Theory. West Lafayette, IN, USA: School of Mechanical Engineering, Purdue University, 1990

129. A. Gera, I. Horowitz. Optimization of the loop transfer function. International Journal of Control, 1980, 31:389–398

130. C.M. Frannson, B. Lenmartson, T. Wik, *et al.* Global controller optimization using Horowitz bounds. Proceedings of the IFAC 15th Trienial World Congress, Barcelona, Spain, 2002

131. Y. Chait, Q. Chen, C.V. Hollot. Automatic loop-shaping of QFT controllers via linear programming. ASME Journal of Dynamic Systems, Measurements and Control, 1999, 121:351–357

132. W.H. Chen, D.J. Ballance, Y. Li. Automatic Loop-Shaping of QFT using Genetic Algorithms. Glasgow, UK: University of Glasgow, 1998

133. C. Raimúndez, A. Ba nos, A. Barreiro. QFT controller synthesis using evolutive strategies. Proceedings of the 5th International QFT Symposium on Quantitative Feedback Theory and Robust Frequency Domain Methods. Pamplona, Spain, 2001, 291–296

134. J. Cervera, A. Baños. Automatic loop shaping QFT using CRONE structures. Journal of Vibration and Control, 2008, 14:1513–1529

135. J. Cervera, A. Baños. Automatic loop shaping in QFT by using a complex fractional order terms controller. Proceedings of the 7th International Symposium on QFT and Robust Frequency, University of Kansas, USA, 2005

136. I. Horowitz. Quantitative Feedback Design Theory - QFT (Vol.1). Boulder, Colorado, USA: QFT Press, 1993

137. I. Horowitz. Optimum loop transfer function in single-loop minimum-phase feedback systems. International Journal of Control, 1973, 18:97–113

138. B.J. Lurie, P.J. Enright. Classical Feedback Control with MATLAB. New York, USA: Marcel Dekker, 2000

139. J. Cervera, A. Baños, C.A. Monje, *et al.* Tuning of fractional PID controllers by using QFT. Proceedings of the 32nd Annual Conference of the IEEE Industrial Electronics Society, IECON'06, Paris, France, 2006

140. C. Edwards, S.K. Spurgeon. Sliding Mode Control. Theory and Applications. Taylor and Francis, 1998

141. V. Utkin. Variable structure systems with sliding modes. IEEE Transactions on Automatic Control, 1977, 22:212–222

142. A.J. Calderón, B.M. Vinagre, V. Feliu. Fractional order control strategies for power electronic buck converters. Signal Processing, 2006

143. K.J. Astrom, B. Wittenmark. Adaptive Control. Second Edition. Reading: Addison-Wesley Publishing Company Inc., 1995

144. C.C. Hang, P.C. Parks. Comparative studies of model reference adaptive control systems. IEEE Transaction on Automatic Control, 1973, 18:419–428

145. J.C. Clegg. A nonlinear integrator for servomechanism. Trans AIEE, 1958, 77:41–42

146. J. Liu. Comparative Study of Differentiation and Integration Techniques for Feedback Control Systems. Ph.D. thesis, Cleveland State University, Cleveland, 2002

147. H. Hu, Y. Zheng, Y. Chait, *et al.* On the zero-input stability of control systems with Clegg integrators. Proceedings of the American Control Conference. Alburquerque, Nw Mexico, USA, 1997, 408–410

148. Y.Q. Chen, K.L. Moore. Discretization schemes for fractional-order differentiators and integrators. IEEE Transactions on Circuits and Systems-I: Fundamental Theory and Applications, 2002, 49(3):363–367

149. L. Zaccarian, D. Nesic, A.R. Teel. First order reset elements and the Clegg integrator revisited. Proceedings of the American Control Conference. Portland, Oregon, USA, 2005, 563–568

150. O. Beker, C.V. Hollot, Y. Chait, *et al.* Fundamental properties of reset control systems. Automatica, 2004, 40:905–915

151. A. Gelb, W.E. Vander. Multiple-Input Describing Functions and Nonlinear System Design. McGraw-Hill, 1968

152. D.P. Atherton. Nonlinear Control Engineering — Describing Function Analysis and Design. London: Van Nostrand Reinhold Co., 1975

153. K.H. Johansson. The quadruple-tank process: A multivariable laboratory process with an adjustable zero. IEEE Transactions on Control Systems Technology, 2000, 8(3):456–465

154. I. Petráš, I. Podlubny, P. O'Leary. Analogue Realization of Fractional Order Controllers. Fakulta BERG, TU Košice, 2002

155. G.W. Bohannan. Analog realization of a fractional controller, revisited. In: BM Vinagre, YQ Chen, eds, Tutorial Workshop 2: Fractional Calculus Applications in Automatic Control and Robotics, Las Vegas, USA, 2002

156. M. Ichise, Y. Nagayanagi, T. Kojima. An analog simulation of non-integer order transfer functions for analysis of electrode processes. Journal of Electroanalytical Chemistry, 1971, 33:253–265

157. K.B. Oldham. Semiintegral electroanalysis: Analog implementation. Analytical Chemistry, 1973, 45(1):39–47

158. M. Sugi, Y. Hirano, Y.F. Miura, et al. Simulation of fractal immittance by analog circuits: An approach to optimized circuits. IEICE Trans Fundamentals, 1999, E82-A(8):1627–1635

159. I. Podlubny, I. Petráš, B.M. Vinagre, et al. Analogue realizations of fractional-order controllers. Nonlinear Dynamics, 2002, 29(1-4):281–296

160. B.M. Vinagre, I. Podlubny, A. Hernández, et al. Some approximations of fractional order operators used in control theory and applications. Fractional Calculus and Applied Analysis, 2000, 3(3):231–248

161. D.Y. Xue, Y.Q. Chen. Sub-optimum $\mathcal{H}_2$ pseudo-rational approximations to fractional order linear time invariant systems. In: J. Sabatier, J. Machado, O. Agrawal, eds., Advances in Fractional Calculus: Theoretical Developments and Applications in Physics and Engineering. Springer Verlag, 2007, 61–75

162. D. Valério. Ninteger Toolbox. [online] http://www.mathworks.com/matlabcentral/fileexchange/8312-ninteger

163. D.Y. Xue, C.N. Zhao, Y.Q. Chen. A modified approximation method of fractional order system. Proceedings of IEEE Conference on Mechatronics and Automation. Luoyang, China, 2006, 1043–1048

164. B.M. Vinagre, I. Podlubny, A. Hernandez, et al. On realization of fractional-order controllers. Proc of the Conference Internationale Francophone d'Automatique, Lille, France, 2000

165. B.M. Vinagre, Y.Q. Chen, I. Petras. Two direct Tustin discretization methods for fractional-order differentiator/integrator. The Journal of Franklin Institute, 2003, 340(5):349–362

166. Y.Q. Chen, B.M. Vinagre. A new IIR-type digital fractional order differentiator. Signal Processing, 2003, 83(11):2359–2365

167. M. A. Al-Alaoui. Novel digital integrator and differentiator. Electronics Letters, 1993, 29(4):376–378

168. M. A. Al-Alaoui. A class of second-order integrators and low-pass differentiators. IEEE Trans on Circuit and Systems I: Fundamental Theory and Applications, 1995, 42(4):220–223

169. M. A. Al-Alaoui. Filling the gap between the bilinear and the backward difference transforms: an interactive design approach. Int J of Electrical Engineering Education, 1997, 34(4):331–337

170. C.C. Tseng, S.C. Pei, S.C. Hsia. Computation of fractional derivatives using Fourier transform and digital FIR differentiator. Signal Processing, 2000, 80:151–159

171. C.C. Tseng. Design of fractional order digital FIR differentiator. IEEE Signal Processing Letters, 2001, 8(3):77–79

172. Y.Q. Chen, B.M. Vinagre, I. Podlubny. A new discretization method for fractional order differentiators via continued fraction expansion. Proc. of The First Symposium

on Fractional Derivatives and Their Applications at The 19th Biennial Conference on Mechanical Vibration and Noise, the ASME International Design Engineering Technical Conferences & Computers and Information in Engineering Conference (ASME DETC2003). Chicago, Illinois, 2003, 1–8, DETC2003/VIB--48391

173. I. Petráš. Digital fractional order differentiator/integrator — FIR type. [online] http://www.mathworks.com/matlabcentral/fileexchange/3673

174. The MathWorks Inc. MATLAB Control System Toolbox, User's Guide, 2000

175. Y.Q. Chen. Contributed files to MATLAB Central. [online] http://www.mathworks.com/matlabcentral/fileexchange/authors/9097

176. The MathWorks, Inc. MATLAB Signal Processing Toolbox. User's Guide, 2000

177. D. Xue, D. P. Atherton. A suboptimal reduction algorithm for linear systems with a time delay. International Journal of Control, 1994, 60(2):181–196

178. K.J. Åström. Introduction to Stochastic Control Theory. London: Academic Press, 1970

179. W. H. Press, B. P. Flannery, S. A. Teukolsky. Numerical Recipes, the Art of Scientific Computing. Cambridge: Cambridge University Press, 1986

180. F.S. Wang, W.S. Juang, C.T. Chan. Optimal tuning of PID controllers for single and cascade control loops. Chemical Engineering Communications, 1995, 132:15–34

181. D.Y. Xue, Y.Q. Chen. Advanced Applied Mathematical Problem Solutions with MATLAB. Beijing: Tsinghua University Press, 2004. In Chinese

182. D.Y. Xue, Y.Q. Chen. Solving Applied Mathematical Problems with MATLAB. Boca Raton: CRC Press, 2008

183. D.Y. Xue, Y.Q. Chen, D.P. Atherton. Linear Feedback Control Analysis and Design with MATLAB. Philadelphia: SIAM Press, 2007

184. I. Podlubny. Mittag-Leffler function, [online] http://www.mathworks.com/matlabcentral/fileexchange/8738, 2005

185. A.K. Shukla, J.C. Prajapati. On a generalization of Mittag-Leffler function and its properties. Journal of Mathematical Analysis and Applications, 2007, 336(1):797–811

186. A.A. Kilbas, M. Saigob, R.K. Saxena. Generalized Mittag-Leffler function and generalized fractional calculus operators. Integral Transforms and Special Functions, 2004, 15(1):31–49

187. The MathWorks Inc. Simulink User's Guide, 2009

188. C.R. Houck, J.A. Joines, M.G. Kay. A Genetic Algorithm for Function Optimization: a MATLAB Implementation. Electronic version of the GAOT Manual, 1995

189. The MathWorks Inc. Genetic Algorithm and Direct Search Toolbox, 2009

190. V. Feliu, S. Feliu. A method of obtaining the time domain response of an equivalent circuit model. Journal of Electroanalytical Chemistry, 1997, 435(1–2):1–10

191. L. Ljung. System Identification. Theory for the User. Prentice Hall, 1987

192. R. Pintelon, P. Guillaume, Y. Rolain, et al. Parametric identification of transfer functions in the frequency domain: A survey. IEEE Transactions on Automatic Control, 1994, 39(11):2245–2259

193. Ch.M.A. Brett, A.M. Oliveira. Electrochemistry Principles. Methods and Applications. Oxford University Press, 1983

194. R. de Levie. Fractals and rough electrodes. Journal of Electroanalytical Chemistry, 1990, 281:1–21

195. R. de Levie, A. Vogt. On the electrochemical response of rough electrodes. Journal of Electroanalytical Chemistry, 1990, 281:23–28

196. A.A. Pilla. A transient impedance technique for the study of electrode kinetics. Journal of Electrochemical Society, 1970, 117(4):467–477

197. P. Agarwal, M.E. Orazem. Measurement models for electrochemical impedance spectroscopy. Journal of Electrochemical Society, 1992, 139(7):1917–1927

198. A. Sadkowski. Time domain responses of constan phase electrodes. Electrochimica Acta, 1993, 38(14):2051–2054

199. F.H. van Heuveln. Analysis of nonexponential transient response due to a constant-phase element. J Electrochem Soc, 1994, 141(12):3423–3428

200. A.A. Sagüés, S.C. Kraus, E.I. Moreno. The time-domain response of a corroding system with constant phase angle interfacial component: Application to steel in concrete. Corrosion Science, 1995, 37(7):1097–1113

201. H. Schiessel, R. Metzler, A. Blumen, et al. Generalized viscoelastic models: their fractional equations with solutions. J Physics A: Math Gen, 1995, 28:6567–6584

202. S.S. Rao. Mechanical Vibrations. Addison-Wesley, 1990

203. H.R. Pota, T.E. Alberts. Multivariable transfer functions for a slewing piezoelectric laminate beam. ASME Journal of Dynamic Systems, Measurements, and Control, 1995, 117

204. F. Wang, Y. Gao. Advanced Studies of Flexible Robotic Manipulators, Modeling, Design, Control and Applications. New Jersey, USA: World Scientific, 2003

205. A. Benosman, G. Le Vey. Control of flexible manipulators: A survey. Robotica, 2004, 22(5):533–545

206. M. Zinn, O. Khatib, B. Roth, et al. Playing it safe [Human-Friendly Robots]. IEEE Robotics and Automation Magazine, 2004, 11(2):12–21

207. S.K. Dwivedy, P. Eberhard. Dynamic analysis of flexible manipulators, a literature review. Mechanism and Machine Theory, 2006, 41(7):749–777

208. V. Feliu. Robots flexibles: Hacia una generación de robots con nuevas prestaciones. Revista Iberoamericana de Automática e Informática Industrial, 2006, 3(3):24–41

209. W.J. Book. Modeling, Design and Control of Flexible Manipulator Arms. Ph.D. thesis, Department of Mechanical Engineering, Massachusetts Institute of Technology, Cambridge MA, 1974

210. O. Maizza-Neto. Modal Analysis and Control of Flexible Manipulator Arms. Ph.D. thesis, Department of Mechanical Engineering, Massachusetts Institute of Technology, Cambridge MA, 1974

211. W.J. Book, O. Maizza-Neto, D.E. Whitney. Feedback control of two beam, two joint systems with distributed flexibility. Journal of Dynamic Systems, Measurement and Control, Transactions of the ASME, 1975, 97G(4):424–431

212. J. Denavit, R.S. Hartenberg. A kinematic notation for lower-pair mechanisms based on matrices. ASME Journal of Applied Mechanics, 1955, June:215–221

213. W.J. Book. Recursive Lagrangian dynamics of flexible manipulator arms. International Journal of Robotics Research, 1984, 3(3):87–101

214. M.J. Balas. Active control of flexible systems. Journal of Optimisation Theory and Applications, 1978, 25(3):415–436

215. M.J. Balas. Trends in large space structures control theory: Fondest hopes, wildest dreams. IEEE Transactions on Automatic Control, 1982, 27(3):522–535

216. R.H. Cannon, E. Schmitz. Initial experiments on the end-point control of a flexible robot. International Journal on Robotics Research, 1984, 3(3):62–75

217. A. De Luca, B. Siciliano. Trajectory control of a non-linear one-link flexible arm. International Journal of Control, 1989, 50(5):1699–1715

218. E. Bayo. A finite-element approach to control the end-point motion of a single-link flexible robot. Journal of Robotics Systems, 1987, 4(1):63–75

219. D. Wang, M. Vidyasagar. Transfer functions for a single flexible link. International Journal on Robotics Research, 1991, 10(5):540–549

220. W.J. Book. Controlled motion in an elastic world. Journal of Dynamic Systems, Measurement and Control, Transactions of the ASME, 1993, 115(2):252–261

221. J.H. Yang, F.L. Lian, L.C. Fu. Nonlinear adaptive control for flexible-link manipulators. IEEE Transactions on Robotics and Automation, 1997, 13(1):140–148

222. T.C. Yang, J.C.S. Yang, P. Kudva. Load adaptive control of a single-link flexible manipulator. IEEE Transactions on Systems, Man and Cybernetics, 1992, 22(1):85–91

223. J.J. Feliu, V. Feliu, C. Cerrada. Load adaptive control of single-link flexible arms based on a new modeling technique. IEEE Transactions on Robotics and Automation, 1999, 15(5):793–804

224. K. Takahashi, I. Yamada. Neural-network-based learning control of flexible mechanism with application to a single-link flexible arm. Journal of Dynamic Systems Measurement and Control, Transactions of the ASME, 1994, 116(4):792–795

225. M. Isogai, F. Arai, T. Fukuda. Modeling and vibration control with neural network for flexible multi-link structures. Proceedings of the IEEE International Conference on Robotics and Automation. San Diego, USA, 1999, 1096–1101

226. A.C. Huang, Y.C. Chen. Adaptive sliding control for single-link flexible-joint robot with mismatched uncertainties. IEEE Transactions on Control Systems Technology, 2004, 12(5):770–775

227. S.B. Choi, C.C. Cheong, H.C. Shin. Sliding mode control of vibration in a single-link flexible arm with parameter variations. Journal of Sound and Vibration, 1995, 179:737–748

228. V. Feliu, J. A. Somolinos, C. Cerrada, et al. A new control scheme of single-link flexible manipulators robust to payload changes. Journal of Intelligent and Robotic Systems, 1997, 20:349–373

229. S.S. Ge, T.H. Lee, G. Zhu. Energy-based robust controller design for multi-link flexible robots. Mechatronics, 1996, 6(7):779–798

230. V. Feliu, F. Ramos. Strain gauge based control of single-link very lightweight flexible robots to payload changes. Mechatronics, 2005, 15(5):547–571

231. I.D. Landau, J. Langer, D. Rey, et al. Robust control of a 360 degrees flexible arm using the combined pole placement/sensitivity function shaping method. IEEE Transactions on Control Systems Technology, 1996, 4(4):369–383

232. J. Daafouz, G. García, J. Bernussou. Robust control of a flexible robot arm using the quadratic D-stability approach. IEEE Transactions on Control Systems Technology, 1998, 6(4):524–533

233. H.R. Karimi, M.K. Yazdanpanah, R.V. Patel, et al. Modeling and control of linear two-time scale systems: Applied to single-link flexible manipulator. Journal of Intelligent and Robotics Systems, 2006, 45(3):235–265

234. L.Y. Pao, W.E. Singhose. Robust minimum time control of flexible structures. Automatica, 1998, 34(2):229–236

235. V. Feliu, K.S. Rattan, H.B. Brown Jr. Control of flexible arms with friction in the joints. IEEE Transactions on Robotics and Automation, 9(4):467–475

236. V. Feliu, B.M. Vinagre, C.A. Monje. Fractional control of a single-link flexible manipulator. Proceedings of the ASME International Design Engineering Technical Conference and Computer and Information in Engineering Conference, Long Beach, California, USA, 2005

237. C.A. Monje, A.J. Calderón, B.M. Vinagre, et al. On fractional PI$^\lambda$ controllers: Some tuning rules for robustness to plant uncertainties. Nonlinear Dynamics, 2004, 38(1-4):369–381

238. D. Valério, J. Sá Da Costa. Ziegler-Nichols type tuning rules for fractional PID controllers. Proceedings of the ASME International Design Engineering Technical Conference and Computer and Information in Engineering Conference, Long Beach, California, USA, 2005

239. C.O. Stockle. Environmental impact of irrigation. Proceedings of the IV International Congress of Agricultural Engineering, Chillan, Chile, 2001

240. P.O. Malaterre, D.C. Rogers, J. Schuurmans. Classification of canal control algorithms. Journal of Irrigation and Drainage Engineering, 1998, 124(1):3–10

241. X. Litrico, D. Georges. Robust continuous-time and discrete-time flow control of a dam-river system. (I) Modelling. Applied Mathematical Modelling, 1999, 23:809–827

242. X. Litrico, V. Fromion. Advanced control politics and optimal performance for irrigation canal. Proceedings of the European Control Conference. Cambridge, UK, 2003,

243. B.T. Wahlin, A.J. Clemmens. Automatic downstream water-level feedback control of branching canal networks: theory. Journal of Irrigation and Drainage Engineering, 2006, 132(3):198–207

244. P.O. Malaterre. Regulation of irrigation canals: characterization and classification. International Journal of Irrigation and Drainage Systems, 1995, 9(4):297–327

245. A.J. Clemmens. Canal automation. Resource Magazine, 2006, 9:7–8

246. F. Liu, F. Feyen, J. Berlamont. Downstream control of multireach canal systems. Journal of Irrigation and Drainage Engineering, 1995, 121(2):179–190

247. E. Weyer. System identification of an open water channel. Control Engineering Practice, 2001, 9(12):1289–1299

248. J. Schuurmans, A.J. Clemmens, S. Dijkstra, et al. Modeling of irrigation and drainage canals for controller design. Journal of Irrigation and Drainage Engineering, 1999, 125(6):338–344

249. X. Litrico, V. Fromion. $\mathcal{H}_\infty$ control of an irrigation canal pool with a mixed control politics. IEEE Transactions on Control Systems Technology, 2006, 14(1):99–111

250. R.R. Pérez, V. Feliu, L.S. Rodríguez. Robust system identification of an irrigation main canal. Advances in Water Resources, 2007, 130:1785–1796

251. X. Litrico. Nonlinear diffusive wave modeling and identification of open channels. Journal of Hydraulic Engineering, 2001, 127(4):313–320

252. J.L. Deltour, F. Sanfilippo. Introduction of Smith predictor into dynamic regulation. Journal of Irrigation and Drainage Engineering, 1998, 124(1):3–30

253. J.P. Baume, P.O. Malaterre, J. Sau. Tuning of PI controllers for an irrigation canal using optimization tools. Proceedings of the Workshop on Modernization of Irrigation Water Delivery Systems. Phoenix, Arizona, USA, 1999, 483–500

254. A.J. Clemmems, J. Schuurmans. Simple optimal downstream feedback canal controllers: theory. Journal of Irrigation and Drainage Engineering, 2004, 130(1):26–34

255. A.J. Clemmems, B.T. Wahlin. Simple optimal downstream feedback canal controllers: ASCE test case results. Journal of Irrigation and Drainage Engineering, 2004, 130(1):35–46

256. A. Montazar, P.J. Van Overloop, R. Brouver. Centralized controller for the Narmada main canal. Irrigation and Drainage, 2005, 54(1):79–89

257. K. Akouz, A. Benhammou, P.O. Malaterre, et al. Predictive control applied to ASCE canal 2. Proceedings of the IEEE International Conference on Systems, Man & Cybernetics. San Diego, USA, 1998, 3920–3924

258. X. Litrico, V. Fromion, J.P. Baume. Tuning of robust distant downstream PI controllers for an irrigation canal pool. II: Implementation issues. Journal of Irrigation and Drainage Engineering, 2006, 132(4):369–379

259. V. Feliu, R.R. Pérez, L.S. Rodríguez. Fractional robust control of main irrigation canals with variable dynamic parameters. Control Engineering Practice, 2007, 15(6):673–686

260. V. Feliu, R.R. Pérez, L.S. Rodríguez, et al. Robust fractional order PI controller implemented on a hydraulic canal. Journal of Hydraulic Engineering, 2009, 135(5):271–282

261. V. Feliu, R.R. Pérez, F.J.C. García, et al. Smith predictor based robust fractional order control: Application to water distribution in a main irrigation canal pool. Journal of Process Control, 2009, 19(3):506–519

262. V. Feliu, R.R. Pérez, F.J.C. García. Fractional order controller robust to time delay for water distribution in an irrigation main canal pool. Computers and Electronics in Agriculture, 2009, 69(2):185–197

263. V.T. Chow. Open-Channels Hydraulics. New York, USA: McGraw-Hill, 1988

264. X. Litrico, V. Fromion. Analytical approximation of open-channel flow for controller design. Applied Mathematical Modelling, 2004, 28:677–695

265. M.H. Chaudhry. Open-Channels Flow. Englewoods Clifs: Prentice-Hall, 1993

266. P.I. Kovalenko. Automation of Land Reclamation Systems. Moscow: Kolos, 1983

267. C.M. Burt, R.S. Mills, R.D. Khalsa, et al. Improved proportional integral (PI) logic for canal automation. Journal of Irrigation and Drainage Engineering, 1998, 124(1):53–57

268. V. Feliu, R. Rivas Pérez, L. Sánchez Rodríguez, *et al.* Robust fractional order PI controller for a main irrigation canal pool. Proceedings of the 17th International Federation of Automatic Control World Congress, Seoul, South Korea, 2008

269. X. Litrico, V. Fromion, J.P. Baume, *et al.* Experimental validation of a methodology to control irrigation canals based on Saint-Venant equations. Control Engineering Practice, 2005, 13(11):1341–1454

270. S.K. Ooi, M.P.M. Krutzen, E. Weyer. On physical and data driven modelling of irrigation channels. Control Engineering Practice, 2005, 13(4):461–471

271. G. Corriga, S. Sanna, G. Usai. Estimation of uncertainty in an open-channel network mathematical model. Applied Mathematical Modelling, 1989, 13:651–657

272. J.P. Baume, J. Sau, P.O. Malaterre. Modeling of irrigation channel dynamics for controller design. Proceedings of the IEEE International Conference on Systems, Man & Cybernetics (SMC98), San Diego, California, USA, 1998

273. R. Rivas Pérez, V. Feliu, F.J. Castillo García, *et al.* System identification for control of a main irrigation canal pool. Proceedings of the 17th International Federation of Automatic Control World Congress, Seoul, South Korea, 2008

274. V. Feliu, R. Rivas Pérez, F.J. Castillo García. Fractional robust control to delay changes in main irrigation canals. Proceedings of the 16th International Federation of Automatic Control World Congress, Prague, Czech Republic, 2005

275. B.T. Wahlin, A.J. Clemmens. Performance of historic downstream canal control algorithms on ASCE test canal 1. Journal of Irrigation and Drainage Engineering, 2002, 128(6):365–375

276. R. Rivas Pérez, J.R. Perán González, B. Pineda Reyes, *et al.* Distributed control under centralized intelligent supervision in the Gira de Melena Irrigation System. Ingeniería Hidráulica en México, 2003, 18(2):53–68

277. P.J. van Overloop. Model Predictive Control on Open Water Systems. The Netherlands: IOS Press Inc, 2006

278. M. J. Shand. Automatic downstream control systems for irrigation Canals. Ph.D. thesis, University of California, Berkeley, USA, 1971

279. Z.J. Palmor, Y. Halevi. On the design and properties of multivariable dead time compensators. Automatica, 1983, 19:255–264

280. D.L. Laughlin, D.E. Rivera, M. Morari. Smith predictor design for robust performance. International Journal of Control, 1987, 46:477–504

281. Z.J. Palmor. The Control Handbook. Time Delay Compensation: Smith Predictor and its Modifications. New York, USA: CRC Press and IEEE Press, 1996

282. F.J. Castillo García, R. Rivas Pérez, V. Feliu. Fractional I^α controller combined with a Smith predictor for effective water distribution in a main irrigation canal pool. Proceedings of the 17th International Federation of Automatic Control World Congress, Seoul, South Korea, 2008

283. R. Rivas Pérez. Automatic control of water distribution in irrigation systems. Ph.D. thesis, Institute of Hydraulic Engineering and Land Reclamation, Ukrainian Academy of Agricultural Sciences, Kiev, Ukraine, 1990

284. T. Hägglund. An industrial dead-time compensating PI controller. Control Engineering Practice, 1996, 4(6):749–756

285. K.J. Aström, B. Wittenmark. Computer Controlled Systems: Theory and Design. Englewoods Clifs, New Jersey, USA: Prentice-Hall, 1997

286. T.H. Lee, Q.G. Wang. Robust Smith-predictor controller for uncertain delay systems. AIChE Journal, 1996, 42(4):1033–1040

287. A.J. Calderón, C.A. Monje, B.M. Vinagre, *et al.* Implementación de controladores de orden fraccionario mediante autómatas programables. Proceedings of the XXV Jornadas de Automatica, Ciudad Real, Spain, 2004

288. T. Habetler, R. Harley. Power electronic converter and system control. Proceeding of the IEEE, 2001, 89(6):913–924

289. H. Sira-Ramirez. A geometric approach to pulse-width-modulated control design. Proceedings of the 26th IEEE Conference on Decision and Control. Los Angeles, CA, USA, 1987, 1771–1776

290. S.L. Jung, Y.Y. Tzou. Discrete sliding-mode control of a PWM inverter for sinusoidal output waveform synthesis with optimal sliding curve. IEEE Transactions on Power Electronics, 1996, 11(4):567–577

291. G. Spiazzi, P. Mattavelli, L. Rossetto. Sliding mode control of DC-DC converters. 4 Congresso Brasileiro de Electronica de Potencia, Belo Horizonte, Brasil, 1997, 59–68

292. R. Middlebrook, S. Cuk. A general unified approach to modelling switching-converter power stages. International Journal of Electronics, 1977, 42(6):521–550

293. J. Agrawal. Power Electronic Systems: Theory and Design. Prentice Hall, 2001

294. I. de la Nuez, V. Feliu. On the voltage pulse-width modulation control of L-C filters. IEEE Transaction on Circuits and Systems I: Fundamental Theory and Applications, 2000, 47(3):338–349

295. J. Alvarez-Ramírez, I. Cervantes, G. Espinosa-Pérez, et al. A stable design of PI for DC-DC converters with an RHS zero. IEEE Transaction on Circuits and Systems I: Fundamental Theory and Applications, 2001, 48(1):103–106

296. R. Martin, I. Aspiazu, I. de la Nuez. Sliding control of a buck converter with variable load. IASTED International Conference Control and Applications, Banf, Canada, 1999

297. M. Castilla, I. García de Vicuña, M. López. On the design of sliding mode control schemes for quantum resonant converters. IEEE Transactions on Power Electronics, 2000, 15(6):960–973

298. L. Dorčak, I. Petráš, I. Kostial. Modeling and analysis of fractional-order regulated systems in the state-space. Proceedings of the International Carpathian Control Conference. High Tatras, Podbanské, Slovak Republic, 2000, 185–188

299. A.J. Calderón. Fractional Control of Power Electronic Buck Converters. Ph.D. thesis, Industrial Engineering School, University of Extremadura, 2003

300. Y.Q. Chen, I. Petráš, B.M. Vinagre. A list of Laplace and inverse Laplace transforms related to fractional order calculus. [online] http://www.steveselectronics.com/petras/foc_laplace.pdf, 2007

Index

Other titles published in this series (continued):